LEŚNIEWSKI'S SYSTEMS

PROTOTHETIC

NIJHOFF INTERNATIONAL PHILOSOPHY SERIES

Volume 54

Leśniewski's Systems Protothetic

Edited by

JAN T.J. SRZEDNICKI – *General Editor*
University of Melbourne, Australia

and

ZBIGNIEW STACHNIAK – *Volume Editor*
York University, Canada

SPRINGER-SCIENCE+BUSINESS MEDIA, B.V.

Library of Congress Cataloging-in-Publication Data

ISBN 978-94-010-6419-4 ISBN 978-94-011-5736-0 (eBook)
DOI 10.1007/978-94-011-5736-0

Printed on acid-free paper

TABLE OF CONTENTS

EDITOR'S FOREWORD

This edition of papers concerning Leśniewski's logical system Protothetic completes the four volume project – *The Leśniewski Collection* – a collected edition of Leśniewski's papers and major contributions to Leśniewski's system of the foundations of mathematics. The three volumes published so far are:

(1) *Leśniewski's Systems. Ontology and Mereology*
Edited by J.T.J. Srzednicki, V.F. Rickey, and J. Czelakowski. Nijhoff International Philosophy Series, 13 (1984).
(2) *S. Leśniewski's Lecture Notes in Logic*
Edited by J.T.J. Srzednicki and Z. Stachniak. Nijhoff International Philosophy Series, 24 (1988).
(3) *Stanisław Leśniewski: Collected Works*
Edited by J.T.J. Srzednicki, S.J. Surma, and D. Barnett with an Annotated Bibliography by V.F. Rickey. Nijhoff International Philosophy Series, 44 (1992).

I.

In the period between the two world wars, Stanisław Leśniewski (1886–1939), one of the founders and a prominent member of the Warsaw School of Logic, created a system of the foundations of mathematics comprising three deductive theories: Protothetic, Ontology, and Mereology. The point of departure for the construction of this system was his study of logical paradoxes and, in this context, a distinction between the distributive and collective interpretations of a class. This distinction between the two interpretations was reflected in the development of two deductive theories, the theory of collective classes, which he eventually called Mereology, and the theory of distributive classes, called Ontology. Finally, in order to combine Mereology and Ontology into a logically rigorous system, he constructed Protothetic – the system of "first principles." Leśniewski's ambition was

> not to add one more calculus to the variety already invented, nor even to prove general metatheorems about alternative formal calculi, in the interests of "comparative logic"; it was

instead to perfect a universally valid classical system of logic and foundations of mathematics, in which he could rigorously formulate generalizations expressible only in the metalanguages of systems poorer in means of expression, [...] and on which he could rely as a true instrument of deduction and scientific investigation (Luschei, 1962, p. 24).*

This program was initiated by Leśniewski in 1914 with his studies on a general theory of sets (later to be named 'Mereology'). The first version of Mereology appeared in print in 1916 under the title *Podstawy Ogólnej Teoryi Mnogości. I.***

In 1919, Leśniewski joined the University of Warsaw as a professor of the philosophy of mathematics. He met a group of gifted mathematicians, Zygmunt Janiszewski, Stefan Mazurkiewicz, Wacław Sierpiński, whose research interests, like those of Leśniewski, were focused on the foundations of mathematics. In 1920 this group, joined by Jan Łukasiewicz, founded the mathematical journal *Fundamenta Mathematicae* with Mazurkiewicz and Sierpiński as editors, and Leśniewski and Łukasiewicz as members of the editorial board. The name, scope, and membership of the editorial board of the journal adequately reflected the research activities of the Warsaw schools of mathematics and logic during the first decade of the journal's existence.

The construction of Ontology in the period between 1919–1921, marked the next step in the formation of Leśniewski's system of the foundations of mathematics, although it was not until 1930 that Ontology appeared in print (cf. Leśniewski, 1930).

The construction of Protothetic began in 1922 and went quickly through numerous improvements and modifications, to be concluded in 1923. By then, Leśniewski's system of the foundations of mathematics was formally ready and, to quote Jordan, it was the

most thorough, original, and philosophically significant attempt to provide a logically secure foundation for the whole of mathematics (cf. Jordan, 1945).

Even such a critic of the importance of Leśniewski's contribution to modern logic as Grzegorczyk admitted that

Leśniewski's treatment of logic was in his times the most exact; it was simpler than *Principia* [*Mathematica*] and had it been published simultaneously with the second edition of the *Principia*, it would have played a considerable part in the development of logic (Grzegorczyk, 1955, p. 78).

The roots of Protothetic can already be found in Leśniewski's early writings between 1912 and 1914. The "deductions" in his 1916 work on the general theory of sets are based on his logical intuitions which eventually were captured in the axioms and directives of Protothetic and Ontology. In 1921, having developed his systems of Mereology and Ontology, Leśniewski was ready to

* For complete publication information for this and other works cited in Editor's Foreword, see the Protothetic Bibliography at the end of this volume.

** Cf. Stanisław Leśniewski, *Foundations of a General Theory of Sets. I.* Prace Polskiego Koła Naukowego w Moskwie, Sekcya Matematyczno–Przyrodnicza, No. 2, Moskwa, 1916 [in Polish].

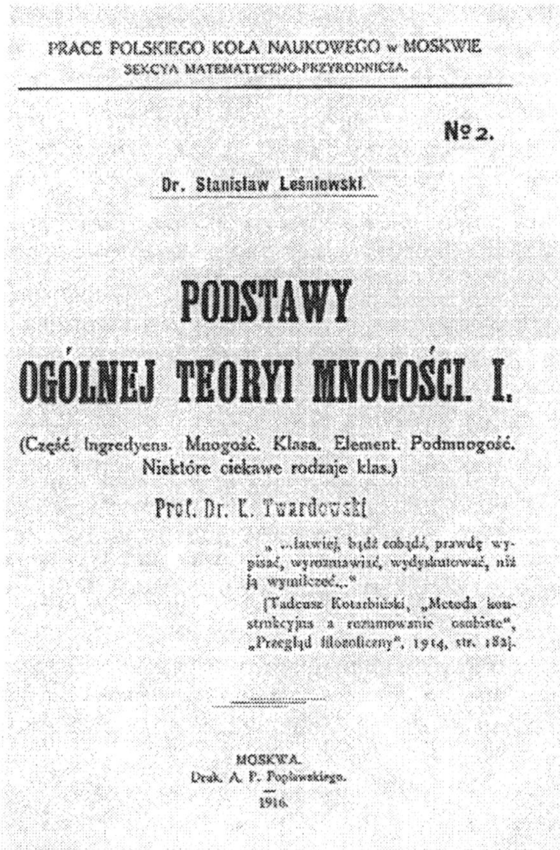

PRACE POLSKIEGO KOŁA NAUKOWEGO w MOSKWIE
SEKCYA MATEMATYCZNO-PRZYRODNICZA.

№ 2.

Dr. Stanisław Leśniewski.

PODSTAWY
OGÓLNEJ TEORYI MNOGOŚCI. I.

(Część. Ingredyens. Mnogość. Klasa. Element. Podmnogość. Niektóre ciekawe rodzaje klas.)

Prof. Dr. K. Twardowski

„...łatwiej, bądź cobądź, prawdę wypisać, wyrozumować, wydyskutować, niż ją wymilczeć..."
(Tadeusz Kotarbiński, „Metoda konstrukcyjna a rozumowanie osobiste", „Przegląd filozoficzny", 1914, str. 182).

MOSKWA.
Druk. A. P. Popławskiego.
1916.

The title page of *Podstawy Ogólnej Teoryi Mnogości. I.*

search for a general system of "first principles." Dissatisfied with the existing proposals (e.g., theories of deduction by J.G.P. Nicod or A.N. Whitehead and B. Russell), he embarked on constructing the calculus of propositions that would be based on a minimal set of primitive symbols, with rigorous directives of procedure, and the simplest possible axiom-system. Leśniewski carefully designed Protothetic to be general and universally valid, consistent, decidable, complete, expressively powerful, logically economic, pure, and elegant, in other words – perfect. In 1922, Alfred Tarski provided Leśniewski with just what was needed. He demonstrated (in his doctoral dissertation written under Leśniewski's supervision) that the "ordinary" theory of deduction can be based on the sign of equivalence as the sole primitive term. He proved that the logical product, and, hence, all the familiar functions of propositions, can be defined in terms of equivalence, the universal quantifier, and variable propositional functors.

Leśniewski began the development of Protothetic by constructing a weaker system $\mathfrak{S}$ (which is a complete two-valued equivalential system). In 1922 he began the task of formulating the directives for Protothetic from the perspective of his theory of semantic categories.

Testing and modifying various details of these directives lasted several years. But by 1922 the directives were precise enough to form the basis of a considerable amount of axiomatic research (Leśniewski, 1929; cf. Leśniewski, 1992, p. 422).

Still in 1922, Protothetic underwent several transformations. First, $\mathfrak{S}$ was strengthened to the system $\mathfrak{S}_1$, and later to $\mathfrak{S}_2$, which is

one of the many possible mutually equivalent systems of the theory that I name protothetic (Leśniewski, 1938a; cf. McCall, 1967, p. 122).

$\mathfrak{S}_2$, in turn, was succeeded by a more elegant system $\mathfrak{S}_3$. Finally, in 1923, Leśniewski proposed the directives for the system $\mathfrak{S}_5$, the most mature of the different dialects of Protothetic. In the years 1923–1927 the directives of $\mathfrak{S}_5$ were polished and supplemented with forty-nine "terminological explanations." Concerning the formalization of Protothetic, Leśniewski remarked that

Having no predilection for various 'mathematical games' that consist in writing out according to one or another conventional rule various more or less picturesque formulae which need not be meaningful, or even – as some of the 'mathematical gamers' might prefer – which should necessarily be meaningless, I would not have taken the trouble to systematize and to often check quite scrupulously the directives of my system, had I not imputed to its theses a certain specific and completely determined sense, in virtue of which its axioms, definitions, and final directives (as encoded for $\mathfrak{S}_5$), have for me an irresistible intuitive validity. I see no contradiction, therefore, in saying that I advocate a rather radical 'formalism' in the construction of my system even though I am an obdurate 'intuitionist' (Leśniewski, 1929; cf. Leśniewski, 1992, p. 487).

Satisfied with the directives and general properties of $\mathfrak{S}_5$, Leśniewski focused his attention on the problem of simplification of the single axiom of $\mathfrak{S}_5$. His collaboration with Mordchaj Wajsberg in the years 1923–1926, and later with Bolesław Sobociński in the years 1937–1938, resulted in a significant simplification of the axiom. In 1945, Sobociński found the single axiom for $\mathfrak{S}_5$ consisting of 54 symbols (when written in the authentic Leśniewski symbolism), which still remains the shortest known single axiom for this system.*

In the meantime, an implicational version of Protothetic appeared with the name $\mathfrak{S}_4$. Its axiomatization is the result of a fruitful collaboration between Leśniewski and Tarski. A number of "computative" or "algorithmic" dialects

* Leśniewski in (1929) and (1938a) discusses the details concerning the search for the shortest single axiom of Protothetic until 1937; Sobociński (1960) describes this search until the discovery of the shortest known single axiom in 1945.

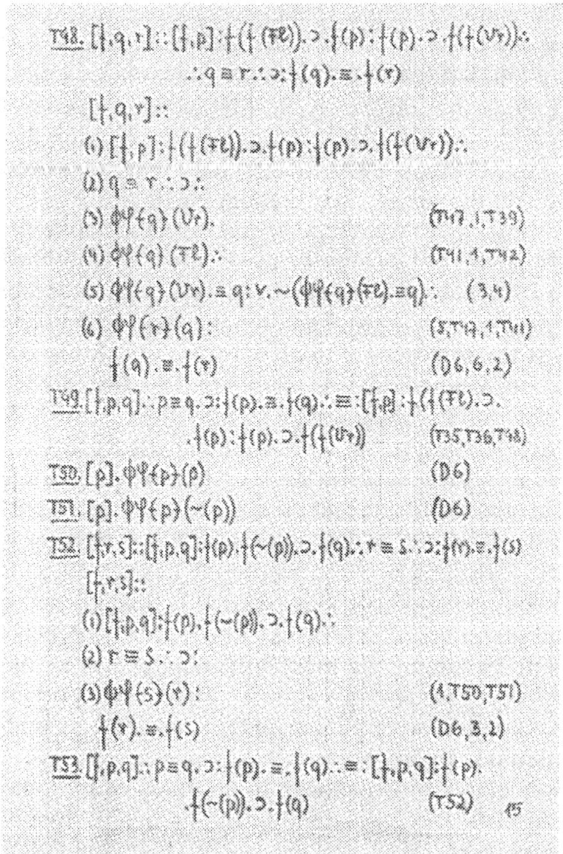

One of the pages of Choynowski's lecture notes taken during the course *From the Foundations of Protothetic* in 1932–1933.

of Protothetic were also developed between 1924 and 1935 (see Le Blanc, 1991).

From the very beginning, Protothetic attracted considerable attention from Leśniewski's colleagues and students. Leśniewski discussed the directives for Protothetic in detail in his lectures at the University of Warsaw, starting with his course on the *Foundations of Logistic* in the 1924/25 academic year. Lecture notes taken by Mieczysław Choynowski during the course *From the Foundations of Protothetic* in the academic year 1932–1933 have survived, giving a unique insight into the logical structure and successive developments of this system (cf. Leśniewski, 1988). In 1937, Łukasiewicz created a journal – *Collectanea Logica* – that was devoted exclusively to logic. (Sobociński was the managing editor.) The founders hoped that the role and the international stature of this new journal would be analogous to *Fundamenta Mathematicae*. Among the eleven papers of the first volume of the journal, eight, written by Leśniewski, Sobociński, and Słupecki, were devoted to Protothetic. Unfortunately, the first and only volume of *Collectanea Logica*, which was ready to appear, was destroyed during the first days of World War II. The documents which could have been used to reconstruct the volume perished during the Warsaw uprising in 1944.*

The postwar history of Protothetic begins with the heroic effort of former colleagues and students of Leśniewski, especially Hiż, Kotarbiński, Lejewski, Sobociński, and Słupecki, to continue Leśniewski's work, to reconstruct and salvage his logical contributions, and to make them widely known. Tadeusz Kotarbiński was the first person to attempt to publish the collected works of Leśniewski, a task which he was unable to complete. He knew that some of the notes taken during Leśniewski's lectures circulated in Warsaw during the war and were an important source of knowledge about Leśniewski's work. These notes represented the missing link between a few papers that Leśniewski managed to publish and the wealth of results he left unpublished. Kotarbiński collected some of the notes which survived the war and gave them to Słupecki with the task of reconstructing Leśniewski's logical systems. Słupecki's articles on Protothetic (see paper V in this volume) and Ontology, which appeared in volumes I and II of the newly created Polish journal *Studia Logica*, proved to be among the most influential papers in the development and further studies of Leśniewski's systems. The most important lecture notes collected by Kotarbiński were published by Kluwer Academic Publishers in 1988 (cf. Leśniewski, 1988).

Bolesław Sobociński's involvement in the study of Protothetic goes back to his prewar investigations concerning the axiomatic foundations of this system (cf. Sobociński, 1939; see also paper IV in this volume). After the war he continued his research on Protothetic and devoted his first postwar publications to this system (cf. the Protothetic Bibliography). His paper 'On the single axioms of protothetic' (paper VI in this volume) is regarded as one of the most important papers concerning Protothetic. Another work, 'An investigation of Protothetic' (paper IV in this volume), in addition to important technical contributions to Protothetic, gives a unique account of the ill-fated *Collectanea Logica*. At the same time, he lectured on logic attracting students and researchers to the logical ideas of Leśniewski. In 1960, Sobociński created the *Notre Dame Journal of Formal Logic* making it an important forum for promoting research investigations concerning Leśniewski's system of the foundations of mathematics.

* For more information on *Collectanea Logica* and its fate, see Sobociński (1949) (also paper IV in this volume).

II.

This volume is a collection of the major contributions to Protothetic. The title *Leśniewski's Systems: Protothetic* is not intended to suggest that all important papers devoted to Protothetic are included here, for many have been left out, either because they were similar in content to those included in the present volume or simply because they were too numerous to be included in a single volume. Also excluded are Leśniewski's own papers dealing with Protothetic, since they appear in Leśniewski (1992). The papers included in this volume are arranged by dates of publication. Those by Peter Simons on nominalism in Poland and by Frederick Rickey on the logical systems of Leśniewski, which open the volume, are the only exceptions to this order. Simons' paper discloses Leśniewski's philosophical commitments regarding his system of the foundations of mathematics, while Rickey's work surveys the most important technical contributions to Leśniewski's system. Both papers provide a philosophical and historical context for the technical contributions reported in the remaining papers of this volume.

Alfred Tarski's paper 'On the primitive term of logistic' is a new English translation of his doctoral thesis that was originally published in Polish in *Przegląd Filozoficzny* under the title 'O wyrazie pierwotnym logistyki' (cf. Tarski, 1923b). The well-known translation by J. H. Woodger, which is also entitled 'On the primitive term of logistic', in the collection Tarski (1956), is a compilation of three papers: Tarski (1923a, 1923b), and (1924). In the present version, we aimed at a literal translation of the Polish original rather than a smooth one based on Tarski's other publications. We tried to retain Tarski's individual style as well as the lay-out of the Polish publication.

Bolesław Sobociński's paper 'An investigation of Protothetic' was intended to appear under the title 'Z badań nad prototetyką' in Vol. 1 of *Collectanea Logica*. Its English translation, preceded by the 21-page long introduction which discusses the origins and fate of *Collectanea Logica* as well as the contents of its first volume, was eventually published in 1949 (cf. Sobociński, 1949). Another English translation from the Polish original by Z. Jordan appeared in 1967 in McCall's *Polish Logic* (see McCall, 1967). In this volume we include an edited version of Sobociński (1949), in which the part of the introduction that is devoted to *Collectanea Logica* joins Jordan's translation. To retain Sobociński's individual style only the necessary editorial alterations have been made.

The Protothetic Bibliography included in this volume lists only works in which Protothetic is explicitly discussed. Papers on subjects related to Protothetic, such as equivalential calculi, calculi with variable functors, theories of definitions, quantification, syntactic categories, etc., are excluded. Many of these exclusions can be found in Rickey's 'An Annotated Leśniewski Bibliography' included in Leśniewski (1992).*

ACKNOWLEDGMENTS

The editors of this volume are indebted to many individuals for their help during the editorial work. We wish to thank V. Frederick Rickey for many valuable suggestions concerning the contents of the volume. We thank Cynthia Bearlin, Ewa Stachniak, David Tremaine, and Michael Wharton for their generous help during the translation and edition of the material. Our gratitude is also due to Jarosław Malik for his many hours devoted to the preparation of the Protothetic Bibliography. Special thanks go to Maja S.M. de Keijzer of Kluwer Academic Publishers for her continuous assistance in the preparation of this volume, for her valuable support, and for her guidance through all the stages of the editorial process.

Finally, we would like to thank Jan Tarski and the following institutions and publishers for granting their reprint permissions for papers included in this volume: the Institute of Philosophy and Sociology of the Polish Academy of Sciences, The Notre Dame Journal of Formal Logic, Oxford University Press, and Rodopi. We have done our utmost to contact the copyright owners of all the material reprinted in this volume. In some cases, however, locating the copyright owner proved to be an impossible task after all these years. If you are a copyright owner whom we have failed to reach, then please contact Kluwer Academic Publishers.

ZBIGNIEW STACHNIAK

* An updated version of 'An Annotated Leśniewski Bibliography' is available from V.F. Rickey of the Department of Mathematics and Statistics at Bowling Green State University, Ohio, U.S.A.

Stanisław Leśniewski

PETER M. SIMONS

NOMINALISM IN POLAND *

Several prominent Polish philosophers were nominalists. This paper concentrates mainly on the views of Leśniewski and Kotarbiński, in the belief that their views are the most interesting in themselves, the most historically important, and present most clearly the difficulties and challenges that nominalism has to face. But first, some terminological clarifications.

1. THE SEVERAL SENSES OF THE TERM 'NOMINALISM'

The term 'nominalism' as used in twentieth-century philosophy has a number of distinct but connected senses. In the first and traditional sense, nominalism is the doctrine that everything that exists is particular, that there are no universals. The opposite to this is realism with respect to universals, whether of a strong (Platonistic) or a moderate (Thomistic) sort. Nominalism in this sense subsumes conceptualism, the doctrine that universals exist as concepts in the mind, which was the most common form of nominalism in antiquity and the Middle Ages. The name 'nominalism' is not inappropriate because it was held that concepts are mental names. A second and more restricted meaning of 'nominalism' takes universals to be words properly so called, as distinct from concepts. This kind of nominalism, sometimes called extreme nominalism, can be found in Roscelin and Hobbes. A third meaning extends the denial of universals to a denial of all abstract entities. This is the sense employed by Quine and Goodman in their essay "Steps Toward a Constructive Nominalism"[1] which begins with the trenchant statement "We do not believe in abstract entities". The fourth sense of the term 'nominalism' is due to Goodman, who wrote, some years after his joint essay with Quine, and in partial revision of it, "Nominalism for me consists specifically in the refusal to recognize classes".[2]

Let us summarize these four meanings:

Nu There are no real (mind-independent) universals.

Nu+ There are no real universals. Universals are in fact words.

* Bibliographical Note. Reprinted from *Polish Scientific Philosophy: The Lvov-Warsaw School*. F. Coniglione, R. Poli, and J. Woleński, editors. Poznań Studies in the Philosophy of the Sciences and the Humanities, Vol. 28, Editions Rodopi B.V., by permission of Rodopi.

Na There are no abstract entities.
Nc There are no classes.

If we want descriptive labels for three of these positions we can call *Nu* 'particularism', *Na* 'concretism' and *Nc* 'individualism'. The mnemonic letters used here and below for identifying theses serve not only as *aides mémoire*: if we read '*Nx*' as 'there are no *xs*', '*Ex*' as 'there exist *xs*', '*Axy*' as 'All *xs* are *ys*' and '*Sxy*' as '*xs* and *ys* are the same' (i.e. all *xs* are *ys* and all *ys* are *xs*), logical relations among the theses may be read off their names and those of the additional assumptions to follow.

Under the generally accepted additional assumptions
Aua Real universals are abstract entities
Aca Classes are abstract entities
this gives the following pattern of entailments among the nominalistic theses:

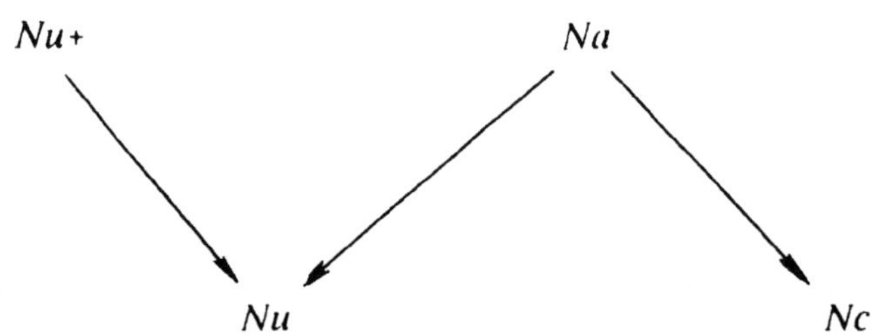

From an ontological point of view the thesis *Nu+* is not importantly different from *Nu*. Particularists (nominalists in the sense of *Nu*) are compelled to offer some explanation for the widespread (and in their opinion erroneous) view that there are real universals, and must attempt to explain how generality arises, either as a conceptual or as a linguistic phenomenon, or in some other terms. *Nu+* embodies one such explanation. Since it is a specification of *Nu*, we shall focus mainly on the latter and henceforth, unless explicitly stated otherwise, we shall understand 'universal' to mean 'real, mind-independent universal'. Under certain additional assumptions, the relationships among *Nu*, *Na* and *Nc* are simplified. Under the assumption

Acu All classes are universals

Nc entails *Nu*. Under

Auc All universals are classes

Nu entails *Nc*, so under the conjunction of *Acu* and *Auc* (*Scu*), *Nu* and *Nc* are equivalent. Under

Aac All abstract entities are classes

Na and Nc are equivalent, both entailing *Nu*; under

Aau All abstract entities are universals

Nu and Na are equivalent, both entailing *Nc*, so under any of the combinations of theses entailing

Sacu Abstract entities, classes, and universals are the same

all three nominalistic theses are equivalent.

So much for conceptual geography. It is important to distinguish the different senses of nominalism because not all people who have been called 'nominalists' agree in what they reject. Goodman and Quine originally pleaded for *Na*, but their views diverged afterwards, Quine embracing classes (affirming *Ec*, the contradictory of *Nc*) and effectively accepting *Auc*, discarding all but extensional universals, but regarding some classes (the singletons with concrete members) as concrete. Goodman later accepted universal and abstract entities, but continued to reject classes. However, it not infrequently happens that nominalists either embrace *Na* from the start, or are not especially concerned to distinguish the different theses. There are various possible combinations of acceptance and rejection, provided one does not subscribe to the simplifying assumption *Sacu*. For instance, one may accept universals and reject classes. It may be worth noting that even the generally accepted assumptions *Aua* and *Aca* have not always been accepted, at least according to the letter. The former was seemingly rejected by William of Champeaux, the latter by Leśniewski. We circumvent the problem by *stipulating* that 'universal' and 'class' be so understood that the supplementary theses *Aua* and *Aca* are analytically true.

2. LEŚNIEWSKI VERSUS TWARDOWSKI

Twardowski, the father of modern Polish philosophy, accepted what he called 'general objects'. In his *Zur Lehre vom Inhalt und Gegenstand der Vorstellungen*, Twardowski upholds, in the final chapter, the view that even general ideas have their own peculiar objects, general objects, which are distinct from the objects of singular ideas, individual objects, as Twardowski calls them.[3] Twardowski's arguments for general objects are not especially convincing: he appears to think it is fairly straightforward to show that common prejudices against them stem from misleading psychology and theory of language, it being possible to expose the prejudices provided our descriptive psychology of general ideas is properly executed. Positively, Twardowski puts forward several theses concerning general objects:

Tw1 There are general objects
Tw2 General objects are the objects of general ideas (presentations)
Tw3 General objects are not intuitable
Tw4 General objects are distinct from the objects of singular presentations (individual objects), which are intuitable
Tw5 The apprehension of general objects is invariably accompanied by singular auxiliary intuitions of individual objects of the kind which the general

object represents (which is mainly why general ideas have received less attention than they deserve)

Tw6 General objects do not exist (not in contradiction with *Tw1*!)

Tw7 The object of a general presentation is part of the objects of these singular presentations

Tw8 A general object is a unity consisting of constituents common to several presentations.

Historically speaking these theses are, as Twardowski readily admits, not especially novel. The main difference between this theory and Plato's theory of Ideas lies in thesis *Tw6*. Twardowski, before Meinong, and influencing the latter, held that every idea (presentation) has an object, but that not all objects exist. Twardowski's embryonic theory of general objects presages Meinong's more fully developed theory of incomplete objects, which was to emerge rather later. Obviously Twardowski, like Meinong, had in mind the controversy between Locke and Berkeley about abstract ideas. Twardowski transposed the psychological-semantic dispute (back) into an ontological key, and the result was to recall the nominalism/realism debate.

Leśniewski studied with Twardowski in Lwów and obtained his doctorate under him in 1912. At that time Meinong's work was also becoming known in Lwów through the offices of Jan Łukasiewicz, who had visited Meinong in Graz as a fellowship scholar and had heard Meinong lecture about incomplete objects.[4] Łukasiewicz was at this time inclined to take incomplete objects seriously, and suggested that they might provide a counterexample to the ontological principle of excluded middle.[5]

OEM For all objects A and all properties P: A has P or A has non-P where non-P is the property obtained from P by property-negation.

If the general triangle has neither equilaterality nor non-equilaterality, it would be a counterexample to *OEM*. Łukasiewicz was careful to distinguish *OEM* from the *logical* principle of excluded middle:

LEM For all p: either p or not-p.

Leśniewski too was concerned in his early work to distinguish these two, in the belief that *OEM* was true but *LEM* false. His reason for the latter belief is rather odd: he thought that both positive and negative existential statements are false.[6] Leśniewski argued against attempts to show *OEM* is false on the basis of there being general objects. He defines a general object as one which represents a number of other objects in having as its properties all and only those properties common to all the objects in question. If we suppose a number of objects $A_1, A_2, \ldots$, and a property P such that some of the objects have P and others do not, then the general object A representing $A_1, A_2, \ldots$ cannot have P, for some of the A_i do not have P, but neither can it have not-P, for some of the A_i have P and hence do not have not-P.[7] The only way for an object to represent other objects without contradiction is for it to represent only itself, in which case it is no general object.[8]

The questionable assumption in this reasoning has been exposed by Guido Küng.[9] Leśniewski assumes that there is no problem in taking a general object to share some of its material properties with the individuals it represents, an assumption which most advocates of realism with respect to universal would reject, mindful of the havoc wreaked by Plato's Third Man, which is predicated on this assumption. Twardowski's thesis *Tw8* is close to the position Leśniewski criticizes, and it is not implausible to suppose Twardowski believed it, taking the 'parts' in *Tw7* to be properties. *Tw8* does not, however, say exactly that the general object has all *and only* those properties all its individuals have, so Leśniewski does not refute Twardowski to the letter. It is indeed reasonable to expect a general object to have numerous properties its individuals do not have, such as being abstract, having more than one instance, not existing, being general, and so on. Leśniewski's argument succeeds in refuting *one* theory of general objects, but it is unlikely to be one seriously entertained by the advocate of general objects. If Leśniewski took himself to have refuted general objects once and for all, he was mistaken. Nevertheless, his "proof" raises wider questions for adherents of general and other abstract objects. What and whither are the characteristics of abstract objects? If abstract objects are indeterminate with respect to certain properties and not with respect to others, what principles govern this? Are we led to accept objects which do not exist, but are nevertheless objects? All of these are difficult issues for the proponent of abstract objects.

3. LEŚNIEWSKI VERSUS SET THEORY

If Leśniewski's critique of Twardowski lands him implicitly in the camp of particularism (*Nu*), his criticisms of set theory mark him clearly as an individualist (*Nc*). It is well known that Leśniewski was preoccupied with the paradoxes of mathematical logic, in particular Russell's Paradox, and attempted to expose what he considered to be the faulty assumptions in the reasoning leading to the paradox.[10] Leśniewski's analysis of the term 'class' led him to the conclusion that if the term 'class' refers to a single object, then every class is a member of itself, where 'member of' is taken to mean 'proper part of or identical with'. Russell's paradox cannot, that is, be constructed within that theory of classes according to which a class is the concrete whole of which its members are parts. Either a class exists, in which case it is a member of itself, or, if a class is not member of itself, there is no such class. This is the collective or mereological sense of 'class'. There is another, distributive sense of 'class' according to which to say that *A* is a member of the class of *b*s is to say no more or less than that *A* is a *b*. The latter form of expression is one with which Leśniewski deals in his Ontology, and it too makes no assumptions about the existence of abstract classes as rejected in *Nc*. Originally, under the influence of Mill, Leśniewski had regarded a simple sentence of the form '*A* is *b*' as true iff the object denoted by '*A*' has the property connoted by 'is *b*'.[11] But he soon rejected both properties and connotation and came to regard '*A* is a *b*' as true if and only if the object denoted by '*A*' is one of the one or more objects denoted by '*b*'. We need only accept the medieval doctrine that a name may signify many individuals.

Something akin to Leśniewski's distinction between the collective and distributive senses of 'class' had already been recognized by Frege,[12] except that Frege – to his cost – had taken classes as abstract objects quite seriously, which, in conjunction with other principles of his logic, led precisely to Russell's Antinomy. Leśniewski, who was willing enough to agree with Frege's criticisms of Schröder, and himself belaboured the point that there can be no null class, a class made up of or composed of nothing at all,[13] turns against Frege's idea that classes are the extensions of concepts, because – a typical Leśniewskian statement – he finds himself utterly unable to understand what might be meant by 'extension of a concept', nor is his understanding improved when Frege talks of the 'value-course of a function' instead. And even if Frege's conception made sense, the revision made to Frege's system in the light of Russell's Antinomy would abrogate the previous sense and return us to ignorance.

The way round Russell's problem which soon gained the approval of most mathematicians was that of Zermelo, who, starting probably from Cantor rather than Frege, restricted the principles of set-formation sufficiently to (presumably) exclude paradoxes like that of Russell. Zermelo's conception of set, however, likewise completely eludes Leśniewski's grasp.[14]

Finally, Whitehead and Russell attempt to use class symbolism in *Principia Mathematica* while not accepting classes into their ontology. Leśniewski's main complaint here is that Whitehead and Russell, by confusing use and mention, do not make it clear whether it is classes themselves or just class-symbols that are incomplete symbols, and he finds nothing clear enough to enable him to tell whether the classes Whitehead and Russell think do not exist are the same as his own classes, or mereological wholes. Leśniewski apparently does not penetrate far enough into the ontology of *PM* to see that class terminology is brought in on the back of an ontology of propositional functions, doubtless items which Leśniewski would likewise have failed to understand.

Leśniewski claims his own mereological view of sets is in agreement with that of Cantor.[15] It is true that some of Cantor's examples fit Leśniewski's understanding, but it is very doubtful whether Leśniewski would be happy with sets of objects such as numbers and order-types, which "arise" by abstraction, as in Cantor. One may also be forgiven for feeling that Leśniewski's analysis of Russell's Antinomy slides past the problem without addressing directly the issues which led to it in the first place. Whichever Leśniewskian sense of 'member' we take, whether the sense of 'is part of or identical with', or 'is one of', the result is left-reflexive and transitive: that is, taking '*E*' to cover either sense, we have

$$\text{Refl} \quad \forall x \forall y (x\ E\ y \rightarrow x\ E\ x)$$
$$\text{Trans} \quad \forall x \forall y \forall z ((x\ E\ y\ \&\ y\ E\ z) \rightarrow x\ E\ z).$$

The conception of set which gave rise to the paradox, as well as its successors as found in Zermelo and in Russell's type theory, are neither reflexive nor transitive. It is all very well for Leśniewski to complain that he does not understand what 'member' means unless it means what he thinks it means. Leśniewski's alternative to set theory is based on the (justified) assumption: "If you follow *me* in what you mean by 'member' and 'set', you won't get into logical trouble". That is true, but the proponents of inconsistent set theory were doing more than trying to stay out of trouble (which is why they got into it in the first place): they were trying to provide a logical foundation for mathematics, some of them with a view to showing at least part of mathematics to be just logic. Leśniewski's weaker system may, or, more likely, may not be adequate to that purpose; we have no explicit statement by him in this regard. Like Frege, whom he admired more than any other logician, Leśniewski is a very uncharitable interpreter of other philosophers and logicians, which makes it hard for him to see beyond their shortcomings to the often interesting and fruitful theories they were trying to articulate, and which it is often possible to express in ways which avoid the mistakes which irritated Leśniewski. Zermelo's set theory is perhaps the best example. Considered purely as a mathematical theory, without philosophical pretensions, it has been and continues to be extraordinarily fruitful, in a way which it must be admitted Leśniewski's logic has not, for all that a minority (myself included) would prefer that logic, including the logic of mathematics, be done more in the manner and conviction of Leśniewski. It was perhaps difficult at the time for logicians to develop any firm intuitions about the nature of sets. Over the decades since then, something of a consensus has emerged that Zermelo's cumulative conception of sets is not unintuitive, albeit that it does not have the transparency one might wish for and so would still probably not satisfy Leśniewski's requirement for intuitive content. Nevertheless, it is in his own positive contribution rather than in his criticisms of others that the chief value of Leśniewski's work lies, and to this we now turn.

4. NOMINALISM AND LEŚNIEWSKI'S LOGIC: SEMANTICS

As an upholder of both *Nu* and *Nc* one might well have expected Leśniewski to round his position out and accept *Na*, that there are no abstract entities at all. Surprisingly perhaps, he does not do so: despite his deprecatory remarks about the writings of many of those who do believe in various kinds of abstract entity, he does not commit himself to concretism. It appears that he was not completely satisfied beyond doubt that there are no abstract entities, citing the objects of dreams and after-images as examples.[16] Whether or not these are good candidates for the title 'abstract entity', they were enough to give him pause.

While Leśniewski was not prepared to commit himself, he was certainly not prepared to let logic make commitments for him, and he elevated the ontological neutrality of logic to the status of a regulative principle. Frege criticized and abandoned Kant's view that existence statements are synthetic, and in his logic purported to demonstrate the existence of infinitely many "logical objects", mainly the concept extensions Leśniewski (and he was not alone) found so puzzling. The manifest failure of Frege's logic was a salutary lesson that such extravagance could be dangerous. Theorem *24.52 of *Principia Mathematica*, which states that the class of all individuals has a member, likewise offends against the ontological neutrality of logic, and Russell later described it as embodying a "defect in logical purity".[17] No one seriously denies that at least one individual exists, but that this should not follow from logic alone seems a reasonable requirement.

Leśniewski's system of the foundations of mathematics comprised three theories: protothetic, Ontology, and mereology. While presumably not made with a view to completing Frege's or Russell's logicist programme, it was nevertheless intended to provide a suitable language for the formulation of mathematical theories. In his lectures in Warsaw, some of which have since been published,[18] Leśniewski formulates Peano arithmetic in the language of Ontology. It is Ontology which provides the framework for any theory which, unlike protothetic, deals with a particular subject matter. Leśniewski himself regarded logic as comprising protothetic and Ontology, mereology being already an extra-logical theory. In protothetic it is acceptable to quantify propositional variables and variables of propositional functors, to any finite degree, which makes protothetic equivalent in expressive potential to a theory of propositional types. Likewise in Ontology not only nominal variables are quantified, but also predicate variables and variables of any higher order. Indeed, apart from the universal quantifier and punctuation marks, all symbols of Ontology are categorematic and can be replaced by variables which can be quantified, which means that Ontology is equivalent in expressive potential to simple type theory.

If we accept Quine's views on quantification, to quantify variables is to turn them into variable names and to be ontologically committed to whatever entities the variables range over. To quantify predicate variables for instance is to "countenance" attributes (properties and relations). If we apply Quine's view to Leśniewski's logic, Leśniewski is ontologically committed to an infinite hierarchy of abstract entities. When Quine visited Warsaw in 1933, he tells us,

> With Leśniewski I would argue far into the night, trying to convince him that his system of logic did not avoid, as he supposed, the assuming of abstract objects.[19]

Quine's attempts to persuade Leśniewski were clearly unsuccessful. One of the chief difficulties facing those accustomed to standard predicate logic who try to understand Leśniewski's logic is the interpretation of the quantifiers, about which much has been written. The problem can be appreciated without first ascending to higher orders. There is a theorem of Ontology written

$$\exists a \cdot \sim ex(a)$$

which is sometimes read as 'there is something that does not exist' or 'some things do not exist'. Now although there is no strict incompatibility between accepting non-existents and denying the existence of abstract entities, so that it is possible to be a Meinongian nominalist, everything we know about Leśniewski tells us this reading cannot embody his view of what the formula means. A more accurate though still approximate rendering would be 'for some a, there exist no a's'. Clearly then 'for some' cannot mean 'there exist'.

Because of the way Leśniewski conceived his logical systems, he did not pose the question of their interpretation in the way a modern logician would. Leśniewski's systems arose as a regimented extension of natural language, and were never without an interpretation, having inherited their basic meaningfulness from natural language. Nevertheless, natural languages do not contain quantifiers binding variables, so *their* meaning is not directly inherited, and we are entitled to ask what they mean and whether they indeed embody ontological commitment to abstract entities. There is no direct answer forthcoming from Leśniewski's writings. Quine took away from Warsaw the impression that Leśniewski's quantification was substitutional.[20] Thus a sentence of the form $\exists X \cdot A$ is true iff some sentence obtained by substituting for 'X' in 'A' is true, where 'X' may belong to any admissible syntactic category. Quine's own view is that substitutional quantification involves indirect ontological commitment to expressions or to natural numbers, within which syntax may be arithmeticized,[21] I do not imagine Leśniewski would have agreed with this.

An alternative explanation of Leśniewskian quantification, due to Küng and Canty, is that Leśniewski is committed neither to objects in a domain of quantification nor expressions substitutable for variables, but to extensions of expressions.[22] This is no clearer than the term 'extension', to which, we noted, Leśniewski took exception. If extensions are taken, as they usually are, to be sets, this is quite unacceptable to Leśniewski, the upholder of *Nc*. There are other ways to construe extensions, e.g. as functions, but they all seem to end up hypostatizing extensions as some kind of abstract entity. Is it then, as Quine believed, impossible to interpret the higher-order quantifiers of Leśniewski's logic without explicitly or implicitly accepting that there are abstract entities of some sort or other?

Let us first jettison one piece of unnecessary ballast. Quine's view that in quantifying variables from categories other than that of names we somehow nominalize these variables may safely be rejected. Not only does it involve, as Prior puts it, "a quite eccentric account of namehood",[23] it also makes nonsense of the idea that variables can be congruously replaced by constants of non-nominal categories. The problem of having to assume abstract entities to account for non-nominal quantification remains to be solved even if we allow that predicate-variables, sentence-variables, etc. are not nominal, since the question is whether they can be accorded semantic values without being assigned abstract entities *as* their semantic values. The fact that Leśniewski's quantification of nominal variables is non-standard is what helped people to realize that the usual objectual interpretation does not suit his system, but even if Leśniewski had used perfectly normal nominal quantification the question of ontological commitment would have arisen for higher-order quantification. If a non-objectual interpretation can work there, it can work at the level of nominal quantification, whether standard or not, as well.

I believe that Leśniewski was quite right to resist Quine in their 1933 discussions, though I have no idea whether or not Leśniewski would have approved of the explanation I shall attempt.

Note first that officially Leśniewski has only one variable binder, the universal quantifier, and that he regards this as syncategorematic. I do not think he needed to accept this. We can distinguish many universal quantifiers, assigning each one its own category, depending on the categories of variables it binds. There will be typical or categorial ambiguity among these many quantifiers. Further, there are two aspects to each quantifier: its role as a variable binder, which it shares with other quantifiers of the same category (e.g. the dual, particular quantifier), and the aspect of meaning which makes it the *universal* and not some other quantifier of this category. It is clearer if we separate these roles. There is a precedent in the writings of Church and Ajdukiewicz:[24] take the variable-binding role to be performed by an abstraction operator, and let a universal *functor* perform the role distinguishing the universal meaning. Instead of

$$\forall XYZ\ldots(A)$$

we write

$$\Pi(\lambda XYZ\ldots(A)),$$

it being understood that the Π and the λ are of appropriate category. Here I would differ again from Church, Prior and others, who regard the abstractor λ as syncategorematic.[25] I think we can accept many abstractors of different categories, but we have to accept that these are not *functors* but *operators*, and that their effect is syntactically and semantically quite different from that of functors. By virtue of binding variables, they can reach arbitrarily deep inside their matrices and be nested to arbitrary degrees of complexity, so that their syntactic and semantic role is quite unlike that of a functor, which simply adds an extra layer of syntactic structure and spits out a semantic output value for each semantic input value. That being said, I do not think it is the variable binding role of quantifiers or other operators which leads to the issue of whether we need abstract entities in our semantic analysis. The semantic values possessed by complex expressions formed using abstractors are no different in kind from those of simple expressions of the same category. The difference in the semantics of the quantified sentence $\forall a \cdot F(a)$ and the related functor-predicate sentence $\Pi(F)$ is an inessential kink only: it is the difference between the latter and $\Pi(\lambda a \cdot F(a))$. Assuming (as I think we must) that Leśniewski subscribes to Frege's principle that the semantic value of a compound expression depends functionally on the semantic values of its component expressions, we have then to ask whether the semantic values of Π and F (for some fixed predicate F) require us to accept certain abstract entities. If we replaced the general nominal variable 'a' by the singular nominal variable 'x' we know that one would answer in the negative: all we need assume for the truth or otherwise of this sentence are individuals, of which the sentence matrix $F(x)$ is true or false. Because of his way of understanding quantifiers, Leśniewski does not have this way available to him. Nevertheless, a semantic account of names excludes their namehood if it leaves out of account that they do, or at least may, name individuals, not, after all, a controversial assumption. Suppose, again truly, that some individuals are named. Note further that for Leśniewski names are not confined to singular denoting names: there are empty names like 'Pegasus', which name no individual, and plural names (usually, but necessarily, complex), like 'Benelux' or 'Russell and Whitehead', which name more than one individual. There is also a name, defined by Leśniewski, usually written either 'V' or '*ob*', which names *every* individual. So, it turns out, every individual has a name, though the overwhelming majority of individuals do not have their own singular or proper names. But for any collection of individuals, there *could* be a name naming just them and nothing else. Given a fixed domain of individuals then, we automatically have a fixed range of ways in which names could name, where two ways of naming count for Leśniewskian purposes as the same when the same individuals are named. The possible semantic values for Leśniewskian names are therefore not individuals but possible ways of naming (zero or more) individuals. In putting it this way, to avoid commitment to abstract entities one seems forced to accept modal discourse. I think this modal component in the metalanguage is probably unavoidable, but accepting modal facts does not run against *Na*.

Now consider a monadic predicate F. Its (presumed fixed) semantic value is not an entity, but a way of yielding truth or falsehood when input a certain way of naming. Assuming neither ways of naming nor being true or false are themselves entities, just shorthand for telling us what certain expressions do, how they work, the semantic value of a functor like F is not an entity either. The only entities required are the individuals (if any) which delimit the possible semantic values of names. (If there are no individuals, there is only one way a name can name, namely emptily!) Among the possible ways a predicate can mean is the following: it yields a true sentence no matter how the name means that completes it. Label this way a predicate can mean 'U'. Now consider Π. It yields truth when taking a predicate whose value is U, otherwise it yields falsehood. That is why it is called the *universal* quantifier (functor). But this way for such a functor to mean is just one of the many which are appropriate to its category: that of its dual, for instance, is another. So we may also quantify variables of the category of Π, introduce higher-order universal quantifier functors, and so on. At each stage the range of possible ways expressions may mean is fixed by there being a domain of individual nameables, the ways other functors may mean being determined by these and by the fact that sentences are allowed only the two values true and false. Of course the whole language is extensional, though its metalanguage is ostensibly and probably inevitably modal.[26]

There is another and more familiar way to tell the story. One starts with a set (possibly empty) of individuals, and two abstract objects called the True and the False. Names are assigned subsets of the domain, predicates are assigned functions from such subsets to the truth-values, higher-order functors are assigned higher-order functions, and so on. From a combinatorial point of view the two stories are interchangeable, but from an ontological point of view they are worlds apart. The more familiar one employs the machinery of sets and/or functions. It is mathematically handy and ontologically extravagant. The first version takes pains to distinguish names from non-names and entities from possible roles. Of course in *talking* about it, even now, I am compelled by the English language to use nominal expressions like 'possible roles'. I do not think we need take seriously the idea that there are abstract entities named by such abstract nouns. If Olivier can play Hamlet and can play Othello, (but not Ophelia or Desdemona) we may say there are at least two Shakespearian roles he can play and two he cannot, but I cannot take seriously the idea that in so counting (which is a form of quantifying of course) we are ontologically committed to abstracta called Shakespearian roles. The nominalizations are much too far from grass-roots talk about what Olivier can or cannot do for them to be taken ontologically seriously.

5. NOMINALISM AND LEŚNIEWSKI'S LOGIC: SYNTAX

There is another respect in which Leśniewski's logic is nominalistic which is less controversial and can be dealt with more briefly. It would be of no avail to attempt to avoid commitment to abstracta in logic and in the semantics of logic if the language of logic were itself composed of abstracta. Leśniewski's views on the constituents of language are anchored to *terra firma*. Expressions, their components and the wholes they constitute are one and all concrete entities: marks on paper, blackboards, etc., or certain noises, perhaps also covert neuro-physiological events, though I know of no place where Leśniewski considers the latter. Tokens, not types, are the stuff that language is made of.

Leśniewski is thus led to formulate his metalogic in a way which expressly takes this point of view into account. The terminological explanations in which he describes his logical syntax and presents the directives of his logical system[27] refer only to actual marks and otherwise use general terms covering expressions. Leśniewski does not, as is common metalogical practice, assume there are infinitely many expressions of every category available. A system of logic for him is no less concrete than any other chunk of language, and his job as a metalogician is to produce objects which are the beginnings of logical systems and issue instructions ("directives") on how to go about extending or adding to systems like the one in question. A logical system is not something abstract: it is a concrete collection of signs which grows as its master adds new signs to it according to the directives. In practice, this has no restrictive effect on the logician at all. Whereas a Platonist logician thinks of himself as creating tokens corresponding to types of expression which already exist as abstract entities, the role of the tokens being the didactic one of putting us in cognitive contact with the pre-existing types which form the logical system proper, Leśniewski takes the more down-to-earth view of taking the logician to create the logical system itself. This approach to the syntax of logical systems is known as 'inscriptional syntax' and has been sporadically developed in Leśniewski's wake.[28] That Platonistic syntax has been overwhelmingly preferred by logicians and mathematicians is due probably more to its greater simplicity and ease of pedagogic transmission than to principled objections to the inscriptional approach. Platonists will not in any case scruple to accept expression types alongside other abstracta. It may be that an adequate metalogical account of inscriptional syntax requires a modal idiom, since one thinks in terms of how a system *can* be extended (even if it actually is not). Again, while I think there are problems involved in working this out, it does not of itself commit the Leśniewskian to abstract entities, albeit that it must be admitted that most Leśniewskians are at least as opposed to modal talk as they are to abstracta.

6. KOTARBIŃSKI

Kotarbiński's ontological views were more extreme than those of Leśniewski. He believed that everything is an object with physical characteristics located in time and space. Since time, space, and properties are among the items Kotarbiński denies exist, we have to rephrase this to avoid apparent reference to them, and say his view is that

Everything is an object which is somewhere, somewhen, and physically somehow or other.

This view went under several names at different times: reism, because everything is a thing (*res*), concretism, because everything is concrete, nothing is abstract, and pansomatism, because everything is a body. To contrast his view with that of the later Brentano, who was also apparently a reist, but believed in the existence of souls, *res cogitantes*, Kotarbiński sometimes called his view 'materialistic reism', and to contrast it with Leśniewski's view that there are (or might be) images which are not spatio-temporal, he also called it 'radical realism'. We shall stick to 'reism'. Reism implies all the forms of nominalism considered in Section 1. Kotarbiński himself confined 'nominalism' to *Nu*, seconding Leśniewski's argument against general objects:

Concretism ... joined the current of nominalism, if by nominalism we mean the view that universals do not exist. An essential point in this borrowed proof (sc. that of Leśniewski against general objects) was the assumption that each object of a given set of objects has a specific property of its own, and hence a universal can by definition have neither that property nor its negation, which runs contrary to the law of excluded middle, and indirectly contrary to the law of contradiction.[29]

Reism goes beyond *Nu*:

Not only do properties not exist, but neither do relations, states of things, or events, and the illusion of their existence has its source in the existence of certain nouns, which suggest the erroneous idea of the existence of such objects, in addition to things.[30]

Kotarbiński goes further than Leśniewski is prepared to go, denying the existence of images:

People commonly think that when recall something or when they dream of something, then images come into being, which they call inner images ... Now reism dared to storm the stronghold of that doctrine, without availing itself of Leśniewski's assistance, since that assistance was in this matter refused ... immanent images do not exist at all, since they exist neither in a person's head nor outside his head ... concretism attempts to handle immanent images in the same way as it handles properties, relations, states of things, events and universals: whenever reference is apparently made to them, it is in fact things that are meant.[31]

Kotarbiński found he had no need to construct a logic adequate to express his viewpoint, since Leśniewski, having taken care that his logic carried no ontological commitment, had already delivered him the instrument he needed:

Reism, having found the fine invention represented by Leśniewski's *Ontology*, had no need to make its own instruments in the field of formal logic, but obtained them ready-made from a firm that enjoyed an excellent reputation.[32]

Finally, Kotarbiński was prepared to deny that there are classes, at least abstract classes which cannot be interpreted in either of the senses discerned by Leśniewski. Thus embracing *Nu* and *Nc*, and affirming that everything is concrete and spatio-temporal, Kotarbiński accepted *Na*.

In what it denies, Kotarbiński's reism goes well beyond the demands of *Na*, the denial of abstract objects, since events and states may certainly be regarded as concrete individuals in the manner of Davidson. Kotarbiński called states of affairs (things) 'static facts' and events 'kinetic facts',[33] and since it is common to consider facts abstract this may explain his view. However, events should be distinguished from facts if the latter are taken to be abstract. It also stretches the term 'body', perhaps beyond its elastic limit, if like Kotarbiński we call flames 'bodies',[34] or be prepared nowadays to accept some of the objects quantum theory tells us exist, but which are very unlike what otherwise pass for bodies. Even the most recherché objects of quantum theory nevertheless have physical characteristics and are in space and time, even if their location is not such a straightforward affair as that of a lump of rock.

Kotarbiński's views developed in stages, the most critical stage being when his views on the status of sentences denying that there are objects of this or that category came under fire from Ajdukiewicz.[35] Ostensibly, reism is an ontological thesis about what there is, but Kotarbiński always combined it with a semantic thesis to the effect that nouns apparently referring to objects he denies exist are not genuine nouns but pseudo-nouns or onomatoids, and that the sentences in which they occur can in principle be replaced by synonymous sentences in which apparent reference to such entities disappears. Here is a sample of such paraphrastic reductions:[36]

Roundness is an attribute of spheres
Spheres are round

The relationship of seniority holds between John and Peter
John is older than Peter

The state of affairs that London lies on the Thames obtains
London lies on the Thames

John made a journey from London to New York
John travelled from London to New York

The class of mammals is included in the class of vertebrates
Every mammal is a vertebrate

The motion of this vehicle is rapid
This vehicle is moving rapidly

The greenness of this leaf is light
This leaf is light green

The number of M's is even
M's are pairwise numerous.

Kotarbiński is bound to admit that the elimination of onomatoids from mathematical theories like set theory is a problem which he has not solved,[37] and he is likewise forced to accept that reism takes on the status of a perpetual programme, that of taking up the challenge to eliminate all terms except those denoting concrete individuals.[38]

The final example above indicates some of the difficulty. A paraphrase into ordinary language will often fail because there are no expressions of ordinary language into which the paraphrase can go, it being precisely the way of ordinary language to use nominalizations to express complex ideas. It is more promising to look to a language like that of Leśniewski's Ontology as a source of suitable sentences. The following will do for 'the number of M's is even':

$$\exists R(\forall ab(Rab \rightarrow (a \varepsilon M \ \& \ b \varepsilon M \ \& \ Rba \ \& \ a \neq b \ \& \ \forall c(Rac \rightarrow b = c)) \\ \& \ \forall a(a \varepsilon M \rightarrow \exists b(Rab)))$$

Note the need to quantify a relational variable: evenness is not first-order definable. Thus the success of Kotarbiński's attempt to eliminate onomatoids from even simple arithmetic turns on whether higher-order quantification in Leśniewski's Ontology (or some equally powerful system) can be interpreted without invoking abstract objects. Kotarbiński was aware of the problem of ontological commitment associated with quantification binding variables of higher order, but his remarks on the question are sketchy and unhelpful.[39]

It is doubtful whether Kotarbiński's reism can provide an adequate account of simple predication. Of a true simple predication like 'This leaf is green' the question must be raised – assuming we are not idealists – *in virtue of what* the sentence is true, or, to put it another way, what in the world *makes it true*? Kotarbiński's answer is drawn from his understanding of Leśniewski's Ontology: 'This leaf is green' is true because the individual named by 'this leaf' is one of the individuals named by 'green' (or perhaps 'green object'). This is unsatisfactory, because it offers us no account as to *why* this leaf is named by 'green'. Presumably it is not a matter of whim or convention. Again, if Kotarbiński replies that it is because the leaf really is green, he has not answered the query. What makes the leaf green, better, in what does the leaf's being green consist? This is the request for a *fundamentum in re* for the leaf's being green.[40] Colours being in fact not just a matter of objects but also of our visual and cortical make-up, the answer is likely to be extremely complicated, much more complicated than the sentence. Nevertheless, even for predications to which the answer is straightforward, we shall need to invoke entities which, while particular, are not concrete things like the leaf, but individual accidents.[41] A nominalistic account of predication will be forced to go beyond reism.

There is another problem facing reism which falls in the same area. Presumably part of what makes it correct to say just of things that are spherical that they are spherical, whereas things shaped otherwise are not, is that spherical things are all similar to one another. But just to say this is too little, since everything is similar to everything else. We have to be told in what *respect* spheres are similar to one another, namely, in shape.[42] If we wish to avoid reference to abstract shapes, or to universal sphericity, we had better accept individual accidents, "thin" individuals, inhering in concrete things, to provide the basis for similarities in different respects. A red sphere, a green sphere, a red cube and a green cube can then be sorted into similarity classes appropriate to the predicates of shape and colour, by invoking individual moments of redness, sphericity etc., or other moments accounting for a sphere's being spherical, a red thing's being red, etc. Again, reism's ontology is simply too spartan to provide the resources for a nominalistic account with any hope of success.

Another problem faced by reism, which for reasons of space I shall not follow up, is how to give an account of propositional attitudes. This has a bearing on the nominalism issue, in that any solution favouring an ontology of abstract propositions entails rejecting *Na*.

7. TARSKI

It may come as something of a surprise to see Tarski's name included in a discussion of Polish nominalism. A glance at Tarski's bibliography reveals that he worked in many areas of mathematics where it would be naturally assumed no nominalist would care or dare to tread, such as set theory, large cardinal theory, infinitary logic. Tarski was of course taught by Leśniewski and Kotarbiński, but he went his own way in logic and one might expect him to do so in philosophy as well. No one who knows Tarski's views only from what he published would discover that he was a nominalist. So let us quote from the recollections of Patrick Suppes, who knew Tarski:

> As is evident to anyone who has read any significant part of Tarski's published writings, he was extraordinarily cautious and careful in giving any direct philosophical interpretation of his work. In contrast, he was in conversation willing to express a much wider range of philosophical opinions – I know this from my own experience and also from reports of colleagues. On the basis of his extensive publications in set theory, it might be expected that he would be a Platonist as far as the foundations of mathematics are concerned – to my knowledge he never committed himself in print on this issue. But certainly in conversation he often expressed skepticism of Platonism and would set forth views that would be congenial to formalism in the philosophy of mathematics and nominalism in general philosophy.[43]

More succinctly, according to Henry Hiż, Tarski was a reist.[44] Whether the latter is true I cannot tell, but it is made less probable neither by the testimony of Suppes,[45] nor by the fact that Tarski dedicated the collection of English translations of his early essays to Kotarbiński, nor by his effort to make Kotarbiński's work better known.[46] At all events, we cannot expect to find a systematic discussion of nominalism in Tarski. Nevertheless, there are hints in his early works, before he turned almost completely to mathematics, that he attempted to avoid positions irrevocably incompatible with nominalism. The big truth paper for instance, where Leśniewski's influence is evident in several places, shows how to define not 'truth' but 'is a true sentence of language *L*' for suitable *L*. Where Tarski makes metalogical assumptions which are uncongenial from the point of view of inscriptional syntax, e.g. in supposing that expressions are classes of tokens rather than individual tokens,[47] or that there are infinitely many expressions,[48] he is almost apologetic and suggests the problem may be got round with more effort. Tarski also indicates at one point that one reason for being dissatisfied with the idea of admitting semantic concepts as primitive is that "it would then be difficult to bring ... into harmony with the postulates of the unity of science and of physicalism".[49] Physicalism is not the same as nominalism, but it is often regarded as likewise incompatible with Platonism.

So while we can find no argued defence of nominalism in Tarski, the fact that he was apparently a nominalist is of significance for two reasons. The first is historical: it suggests that the influence of Tarski's teachers, especially Kotarbiński, was more long-lived than the published works would allow us to infer. The second has to do with the psychology of mathematics. Many of the great mathematicians and logicians have been Platonists, and their firm belief in the existence and independence of the objects they were investigating perhaps helped to give them the stamina and confidence they needed to press on and achieve their results. Frege, Cantor and Gödel may serve as recent examples. Among the Poles, we find Łukasiewicz is a Platonist.[50] Great mathematicians and logicians who are nominalists are rarer, and Tarski serves as an excellent modern example to put alongside Leibniz.

8. CONCLUSION

The influence of Leśniewski and Kotarbiński goes beyond Tarski of course. Another of their students, Czesław Lejewski, is a staunch defender of a reism which is committed to unverifiable theses which even Kotarbiński was reluctant to accept.[51] Lejewski has also, more than any other follower of Leśniewski, gone out of his way to show how to accommodate within Ontology talk which is ostensibly about abstract entities like classes,[52] and to furnish language within which a Leśniewskian nominalist can carry on a dialogue with his Platonist opponent.[53] No doubt there are other Polish nominalists whose thinking was influenced by Leśniewski and Kotarbiński. A more exhaustive cataloguing of the extent of Polish nominalism is something I shall not undertake here, because what is important is not a head-count but the issue of principle: does Polish nominalism contribute essentially to answering the question whether nominalism is correct? My answer is plainly that it does, and the main problem which needs to be overcome is the one at the heart of the discussions between Quine and Leśniewski: can a powerful higher-order logical language avoid commitment to abstract entities? Thanks to the work of Leśniewski, Kotarbiński, their associates and students, we are closer to an answer than before.

NOTES

[1] Goodman and Quine (1947), p. 105 (reprint p. 173).

[2] Goodman (1972), p. 156. Cf. p. 157: "Nominalism as I conceive it ... does not involve excluding abstract entities". Goodman rejects classes not because they are abstract but because one can generate more than one entity from the same basic individuals, e.g. $\{a, b\}$, $\{\{a\}, \{b\}\}$, $\{\{a, b\}\}$ are three entities all generated from a and b (cf. ibid., 171). It is this appearance of new entities from old, as if by magic, to which Goodman objects.

[3] Twardowski (1894), ch. 15, pp. 102–111; (1977), pp. 97–105.

[4] Łukasiewicz (1987a), p. 112. Łukasiewicz reported on Meinong's work in Lwów in a lecture whose summary was published as Łukasiewicz (1909).

[5] Cf. Łukasiewicz (1910a, 1987b).

[6] That negative existentials are all false is asserted in Leśniewski (1911); that positive existentials are also false is asserted in Leśniewski (1913a). Cf. Küng (1967), p. 103.

[7] Leśniewski (1913a), pp. 319–320.

[8] For a reconstruction cf. Luschei (1962), pp. 308–310.

[9] Küng (1967), p. 104.

[10] Cf. Leśniewski (1914), pp. 182–189; (1983), pp. 14–17.

[11] Cf. Leśniewski (1913b), p. 497; (1968), p. 26. Cf. Kotarbiński (1966), p. 430.

[12] The distinction is made in Frege's 1895 review of Schröder's *Vorlesungen zur Algebra der Logik*, cf. Frege (1984), p. 210ff.

[13] Leśniewski (1927), pp. 192-196; (1983), pp. 18–20.

[14] Ibid., pp. 201–202; (1983), p. 22.

[15] Ibid., p. 190; (1983), p. 17.

[16] Personal communication from Czesław Lejewski. It is confirmed by what Kotarbiński writes about Leśniewski's refusal to help him argue against images (see below).

[17] Russell (1919), p. 203.

[18] Leśniewski (1988). Cf. esp. ch. 4, "Primitive Terms of Arithmetic", pp. 129–152.

[19] Quine (1985), p. 104.

[20] Quine (1969), p. 63.

[21] Ibid., p. 107.

[22] Cf. Küng and Canty (1970), Küng (1977).

[23] Prior (1971), p. 165.

[24] Church (1932), Ajdukiewicz (1935).

[25] Church (1940), p. 57; Prior (1971), p. 46.

[26] Cf. Küng (1967), p. 182: "The formal structures represented by an extensional system must have an intensional background". Accepting modal discourse does not of itself commit one to possibilia, but there is a special problem for the concretist: finding an acceptable account of the meaning of modal operators which does not involve substituting abstracta (such as universals) for possibilia. Concretists tend not to be possibilists, but there is no strict incompatibility between the two positions. Like most concretists, I should like to think one could account for the meaning of modal discourse without possibilia. Cf. Simons (1985), pp. 214–215.

[27] Cf. Leśniewski (1929), pp. 59–77; (1930), pp. 115–128.

[28] Cf. Luschei (1962), pp. 167–288; Rickey (1972–1973).

[29] Kotarbiński (1966), p. 430. The first concurrence with Leśniewski's argument is in Kotarbiński (1920). It should be noted that Kotarbiński's understanding of 'concretism' is narrower than that proposed in Section 1 above, since it is possible to accept concreta that are not individual things in Kotarbiński's sense.

[30] Kotarbiński (1966), p. 430.

[31] Ibid., p. 432.

[32] Ibid.

[33] Ibid, p. 430.

[34] Ibid., p. 434.

[35] Cf. Lejewski (1979).
[36] The paraphrases all come from Kotarbiński (1958a); (1966), pp. 429–437.
[37] Kotarbiński (1966), pp. 436-437.
[38] Ibid., p. 435; "In its mature form, concretism absolutely insists on its programme only."
[39] Cf. Kotarbiński (1965), pp. 236–238. Kotarbiński merely begs the question, saying that since existence in the primary (philosophical) sense means being able to act on something else, it is clear that the particular quantifier does not express existence.
[40] Cf. Küng (1967), pp. 173, 186.
[41] Cf. Mulligan, Simons and Smith (1984), where accidents are called 'moments'.
[42] Goodman calls this "the difficulty of imperfect community": Goodman (1977), p. 119.
[43] Suppes (1988), p. 81.
[44] Personal communication.
[45] Suppes (1988), p. 80 mentions that Tarski often referred to the influence of Kotarbiński in conversation. According to Jan Tarski (personal communication), Tarski always had a photograph of Kotarbiński in his study.
[46] Tarski assisted in the translation of Kotarbiński (1955), and it was at Tarski's insistence that this translation was included in his own *Collected Papers*.
[47] Tarski (1956), p. 156 n.
[48] Ibid., p. 174 n.
[49] Ibid., p. 406.
[50] Cf. Łukasiewicz (1970), pp. 241, 249.
[51] Cf. Lejewski (1976). Lejewski told me that when Kotarbiński received a copy of this he replied that he was not prepared to go as far.
[52] Lejewski (1985).
[53] Lejewski (1974).

REFERENCES

Ajdukiewicz, K. (1935). 'Die syntaktische Konnexität', *Studia Philosophica* 1, 1–27. English translation Ajdukiewicz (1967).

Ajdukiewicz, K. (1967). 'Syntactic Connection', translated by H. Weber, in McCall (ed.) (1967), 207–231.

Bocheński, J.M., Church, A. and Goodman, N. (1956). *The Problem of Universals. A Symposium*, University of Notre Dame Press, Notre Dame, Indiana.

Church, A. (1932). 'A Set of Postulates for the Foundations of Logic', *Annals of Mathematics* 2nd ser. 33, 346–366.

Church, A. (1940). 'A Formulation of the Simple Theory of Types', *Journal of Symbolic Logic* 5, 56–68.

Frege, G. (1984). *Collected Papers on Mathematics, Logic, and Philosophy*, Blackwell, Oxford.

Goodman, N. (1956). 'A World of Individuals'. In Bocheński, Church and Goodman (1956), 13–31. Reprinted in Goodman (1972), 155–172. Page references are to the reprint.

Goodman, N. (1972). *Problems and Projects*, Bobbs-Merrill, Indianapolis.

Goodman, N. (1977). *The Structure of Appearance*, 3rd ed., Reidel, Dordrecht.

Goodman, N. and Quine, W.V. (1947). 'Steps Toward a Constructive Nominalism', *Journal of Symbolic Logic* 12, 105–122. Reprinted in Goodman (1972), 173–198.

Kotarbiński, T. (1920). 'Sprawa istnienia przedmiotów idealnych (The Question of the Existence of Ideal Objects)', *Przegląd Filozoficzny* XXIII, 149–170. Reprinted in Kotarbiński (1958b), Vol. 2, 7–39.

Kotarbiński, T. (1929). *Elementy teorii poznania, logiki formalnej i metodologii nauk*, Ossolineum, Lwów. English translation in Kotarbiński (1966). Reprinted 1990.

Kotarbiński, T. (1935). 'Zasadnicze myśli pansomatyzmu', *Przegląd Filozoficzny* XXXVIII, 283–294. English translation in Kotarbiński (1955).

Kotarbiński, T. (1955). 'The Fundamental Ideas of Pansomatism', translated by A. Tarski and D. Rynin. *Mind* LXIV, 488–500.

Kotarbiński, T. (1958a). 'Fazy rozwojowe konkretyzmu (The Developmental Stages of Concretism)', *Studia Filozoficzne* 4, 3–13. English translation in Kotarbiński (1966), 429–438.

Kotarbiński, T. (1958b). *Wybór pism (Selected Writings)*, 2 vols., PWN, Warsaw.

Kotarbiński, T. (1965). *Leçons sur l'histoire de la logique*, translated by A. Posner, PWN, Warsaw.

Kotarbiński, T. (1966). *Gnosiology. The Scientific Approach to the Theory of Knowledge*, translated by O. Wojtasiewicz, Pergamon, Oxford.

Küng, G. (1967). *Ontology and the Logistic Analysis of Language*, Reidel, Dordrecht.

Küng, G. (1977). 'The Meaning of the Quantifiers in the Logic of Leśniewski', *Studia Logica* XXXVI, 309–322.

Küng, G. and Canty, J.T. (1970). 'Substitutional Quantification and Leśniewskian Quantifiers', *Theoria* XXXVI, 165–182.

Lejewski, C. (1974). 'A System of Logic for Bicategorial Ontology', *Journal of Philosophical Logic* 3, 265–283.

Lejewski, C. (1976). 'Outline of an Ontology', *Bulletin of the John Rylands University Library of Manchester* 59, 127–147.

Lejewski, C. (1979). 'On the Dramatic Stage in the Development of Kotarbiński's Pansomatism', in Weingartner and Morscher (eds.), 197–214.

Lejewski, C. (1985). 'Accommodating the Informal Notion of Class within the Framework of Leśniewski's Ontology', *Dialectica* 39, 217–241.

Leśniewski, S. (1911). 'Przyczynek do analizy zdań egzystencjalnych', *Przegląd Filozoficzny* XIV, 329–345. English translation 'A Contribution to the Analysis of Existential Propositions' in Leśniewski (1992), 1–19.

Leśniewski, S. (1912). 'Próba dowodu ontologicznej zasady sprzeczności', *Przegląd Filozoficzny* XV, 206–226. English translation 'An Attempt at a Proof of the Ontological Principle of Contradiction' in Leśniewski (1992), 20–46.

Leśniewski, S. (1913a). 'Krytyka logicznej zasady wyłączonego środka', *Przegląd Filozoficzny* XVI, 315–322. English translation 'The Critique of the Logical Principle of the Excluded Middle' in Leśniewski (1992), 47–85.

Leśniewski, S. (1913b). 'Czy prawda jest tylko wieczna czy też i wieczna i odwieczna?', *Nowe Tory* 18, 493–528. English translations: [1] Leśniewski (1968), and [2] 'Is All Truth Only Eternal or Is It Also True without a Beginning' in Leśniewski (1992), 86–114.

Leśniewski, S. (1914). 'Czy klasa klas, nie podporządkowanych sobie, jest podporządkowana sobie?', *Przegląd Filozoficzny* XVII, 63–75. English translation 'Is the Class of Classes Not Subordinated to Themselves, Subordinated to Itself?' in Leśniewski (1992), 115–128.

Leśniewski, S. (1927). 'O podstawach matematyki I', *Przegląd Filozoficzny* XXX, 164–206. English translations: [1] in Leśniewski (1983), 7–24, and [2] 'On the Foundations of Mathematics' in Leśniewski (1992), 174–226.

Leśniewski, S. (1929). 'Grundzüge eines neuen Systems der Grundlagen der Mathematik', *Fundamenta Mathematicae* 14, 1–81. English translation 'Fundamentals of a New System of the Foundations of Mathematics' in Leśniewski (1992), 410–605.

Leśniewski, S. (1930). 'Über die Grundlagen der Ontologie', *Comptes rendus des Séances de la Société des Sciences et des Lettres de Varsovie, Classe III* 23, 111–132. English translation 'On the Foundations of Ontology' in Leśniewski (1992), 606–628.

Leśniewski, S. (1968). 'Is Truth Eternal or Is It Eternal and Since Eternity?', *The Polish Review* 8, 23–43.

Leśniewski, S. (1983). 'On the Foundations of Mathematics', translated by V.F. Sinisi, *Topoi* 2, 7–52.

Leśniewski, S. (1988). *S. Leśniewski's Lecture Notes in Logic*, edited by J.T.J. Srzednicki and Z. Stachniak, Kluwer, Dordrecht.

Leśniewski, S. (1992). *Stanisław Leśniewski: Collected Works*, edited by S.J. Surma, J.T.J. Srzednicki and D.I. Barnett, with an annotated bibliography by V.F. Rickey, Kluwer, Dordrecht.

Łukasiewicz, J. (1909). 'O poglądach filozoficznych Meinonga (On Meinong's Philosophical Views)', *Przegląd Filozoficzny* XII, 559.

Łukasiewicz, J. (1910a). 'O zasadzie wyłączonego środka', *Przegląd Filozoficzny* XIII, 372–373. English translation in Łukasiewicz (1987b).

Łukasiewicz, J. (1910b). *O zasadzie sprzeczności u Arystotelesa. Studium krytyczne* (*On the Principle of Contradiction in Aristotle. A Critical Study*), Polska Akademia Umiejętności, Kraków. Reprint in Łukasiewicz (1987a).

Łukasiewicz, J. (1970). *Selected Works*, edited by L. Borkowski, North-Holland/PWN, Amsterdam–Warsaw.

Łukasiewicz, J. (1987a). *O zasadzie sprzeczności u Arystotelesa*, edited by J. Woleński, PWN, Warsaw.

Łukasiewicz, J. (1987b). 'On the Principle of the Excluded Middle', translated by J. Woleński and P.M. Simons, *History and Philosophy of Logic* 8, 69.

Luschei, E.C. (1962). *The Logical Systems of Leśniewski*, North-Holland, Amsterdam.

McCall, S. (ed.) (1967). *Polish Logic 1920–1939*, Clarendon Press, Oxford.

Mulligan, K., Simons, P.M. and Smith, B. (1984). 'Truth-Makers', *Philosophy and Phenomenological Research* 44, 278–321.

Prior, A.N. (1971). *Objects of Thought*, edited by P.T. Geach and A.J.P. Kenny, Clarendon Press, Oxford.

Quine, W.V. (1969). *Ontological Relativity and Other Essays*, Columbia University Press, New York.

Quine, W.V. (1985). *The Time of My Life. An Autobiography*, The MIT Press, Cambridge, MA.

Rickey, V.F. (1972–1973). 'Axiomatic Inscriptional Syntax', *Notre Dame Journal of Formal Logic* XIII (1972), 1–33, XIV (1973), 1–52.

Russell, B.A.W. (1919). *Introduction to Mathematical Philosophy*, George Allen & Unwin, London.

Simons, P.M. (1985). 'A Semantics for Ontology', *Dialectica* 39, 193–216.

Suppes, P. (1988). 'Philosophical Implications of Tarski's Work', *Journal of Symbolic Logic* 53, 80–91.

Tarski, A. (1956). *Logic, Semantics, Metamathematics. Papers from 1923 to 1938*, translated by J.H. Woodger, Clarendon Press, Oxford; 2nd ed., edited by J. Corcoran, Hackett, Indianapolis (1983).

Twardowski, K. (1894). *Zur Lehre vom Inhalt und Gegenstand der Vorstellungen*, Hölder, Vienna. Reprinted: Philosophia, Munich (1982). English translation in Twardowski (1977).

Twardowski, K. (1977). *On the Content and Object of Presentations*, translated by R. Grossmann, Nijhoff, The Hague.

Weingartner, P. and Morscher, E. (eds.) (1979). *Ontologie und Logik* (*Ontology and Logic*), Duncker & Humblot, Berlin.

V. FREDERICK RICKEY*

A SURVEY OF LEŚNIEWSKI'S LOGIC**

Dedicated to my teacher
Professor Bolesław Sobociński
on his seventieth birthday

In the period between the two world wars the eminent Polish logician Stanisław Leśniewski (1886–1939) created a hierarchy of logical systems which are unparalleled for their generality, precision, and intuitive foundations. By way of a careful and insightful analysis of the Russell Antinomy he came to distinguish between two notions of class, the distributive class and the collective class. Investigation of these concepts led to the creation of his systems of Ontology and Mereology, respectively. Then, in order to secure the foundations of these systems, he created the most fundamental system, Protothetic.

We intend to survey the most important technical contributions to these three disciplines. The restriction to technical results is necessary to restrict this paper to manageable proportions. Moreover, the non-technical aspects of Leśniewski's systems are more accessible. There is no doubt that the philosophical aspect of Leśniewski's work is extremely important and interesting. This, together with historical matters, is enthusiastically discussed in Luschei's book (1962) on *The Logical Systems of Leśniewski*. Applications to philosophy and linguistics are also omitted. Hopefully, all of these matters will receive full consideration at this conference.‡

As with any survey, this paper is biased by my own background and particular interests; of course, it is influenced by those areas which I know best. I studied with Sobociński from 1962 to 1968, and so am well acquainted with his work and with his views concerning the Leśniewskian systems; hopefully, I represent his views accurately here. In this paper the work of

* This work was partially supported by the Faculty Research Committee of Bowling Green State University.

** Bibliographical Note: This paper originally appeared in *Studia Logica* XXXVI (1977), 407–426, by permission of the Institute of Philosophy and Sociology of the Polish Academy of Sciences.

‡ [Ed. Note: The author refers to *XXIInd Conference on the History of Logic*, July 5–9, 1976, Kraków, Poland.]

Sobociński and his students will receive most of my attention. The work of the young Polish logicians – especially Iwanuś and Kubiński – will not receive the attention it deserves, for I do not know their work as well as I would like to. In the spirit of an international meeting it is perhaps fitting if I tell you what is happening in my country and you tell me what is happening in yours.

The plan of this paper will be to consider the three systems separately. But before that is done, one point about the origin of the systems must be discussed. In 1911 Leśniewski learned of the Russell Antinomy through Łukasiewicz's book (1910) on the Principle of Contradiction in Aristotle. This had a profound effect on Leśniewski. To understand this let me give the definition of Antinomy which Leśniewski adopted from Leonard Nelson: An Antinomy is a contradiction which arises from axioms, which we believe to be correct, by the use of deductive rules, which we believe to be valid. The problem is not just the contradiction. The real problem is the psychological one caused by the realization that something is wrong with one's foundational beliefs and also that part of one's intellectual heritage must be repudiated. Russell and Zermelo were only concerned about eliminating the contradiction; they were not concerned with what caused the contradiction. This is where Leśniewski made an important and lasting contribution. He felt that the Antinomy must be solved, i.e., one must analyze carefully the axioms and deductive rules which were involved in the derivation of the contradiction and decide what mistake is inherent in them. Then, on the basis of this experience, new axioms and rules must be chosen. This was a long and painful process for Leśniewski.

Leśniewski made three different analyses of the Antinomies before he found an entirely satisfactory one. In the process he made the distinction, which is usually credited to Ramsey, between logical and semantical Antinomies, but did not publish it. Also in the early twenties, he was the first to make the distinction between language and metalanguage. I would like to discuss these matters carefully here, but this would take us too far afield. Leśniewski's three analyses of the Russell Antinomy are discussed in Luschei (1962), Sinisi (1976) and Sobociński (1949b), respectively. The paper of Sobociński contains the definitive analysis and is worthy of careful consideration.

I believe that there is one thing lacking in the Nelson definition of an Antinomy. That is that it fails to account for the intellectual culture of the time when it was discovered. The logical climate today is very different from the early decades of this century. Our logical sophistication has modulated the effect of the Russell Antinomy. Today our students are not upset by it. Thus, today the Russell Antinomy is not a true Antinomy in the sense of Nelson. It is a mere paradox, i.e., something which is other than it seems. Consequently, I advocate amending the definition of Antinomy to account for the historical setting of its discovery and to realize that, hopefully, what was once an Antinomy will not always remain so.

I. PROTOTHETIC

Protothetic was formulated by Leśniewski in 1923. In the previous two years earlier versions of the system underwent a complicated series of changes before they evolved into the system which Leśniewski called $\mathfrak{S}_5$, and which we call, simply, Protothetic. For the history and some of the details of these changes see Leśniewski (1929c) and (1938a) and Sobociński (1960).

Leśniewski's aim was to construct an extension of the propositional calculus which is as strong as possible. In fact, Protothetic is often characterized as a generalization of the propositional calculus wherein there are quantifiers binding variables of all semantic categories.

The theory of semantic categories was originated by Leśniewski in 1922, but the idea stems from his work in 1912–1914. While the theory of semantic categories is similar to, but not identical with, the simple theory of types, it is intuitively much closer to the "parts of speech" of traditional grammar. Leśniewski found his semantic categories so intuitively appealing that he would feel himself obligated to accept them even if the Antinomies had not been discovered. As is customary today we shall use the notion of Ajdukiewicz (1935) to identify the different semantical categories. In Protothetic the category S of sentences is basic and the higher categories or types are generated using the following rule: If $\alpha, \beta_1, \ldots, \beta_n$ are types, then $\alpha/\beta_1 \ldots \beta_n$ is a type. It is the type that takes n arguments of types $\beta_1, \ldots, \beta_n$ respectively and gives a value of type α. In Protothetic quantification is allowed over each of the types which is generated from S according to this rule, once the types have been introduced via definition. This includes quantification of the types of the many link functors, i.e., over those types where α is different from S. As a word of warning, I should add that very often in the literature the semantical categories are not correctly characterized.

Protothetic can be characterized as a system containing

1. the propositional calculus
2. the law of bivalency for propositions, i.e., the thesis

$$[f] : f(0).f(1). \supset .[p].f(p),$$

plus enough rules to make the system strongly complete. These rules are the rules of detachment, substitution, definition, distribution of quantifiers, and extensionality. Protothetic is strong enough to contain the usual laws of quantifiers and bivalency for each semantical category.

Probably the most important lesson to be learned from Leśniewski is his utmost regard for precision. Nowhere is this better illustrated than in his formulation of the rule of procedure of Protothetic (1929c) and Ontology

(1930). There is no other system of logic for which the rules have been so carefully and precisely formulated. The rules of procedure for Protothetic have been simplified considerably by myself in Rickey (1973). Reportedly, Hiż worked on this problem earlier, but nothing was published. It would be very nice to have a simple presentation of the rule of procedure for Protothetic, for this is one of the most difficult things about the system. Machover (1966) hints that his notion of contextual determinacy can provide one. I tried to take up his idea, but am not satisfied with the results. For someone interested in the terminological explanations which lead up to the rule of Protothetic, I suggest that they first study them for the propositional calculus as in Leśniewski (1931) and Rickey (1975).

Among the rules for Protothetic I consider the rule of definition to be the most important. This rule plays a crucial and creative role in the development of the system. Unfortunately, the strength of this rule is not well understood, let alone carefully studied. I consider this a very promising area for research.

Leśniewski (1929c) was the first to give a correct presentation of the rule of substitution in logics with more than one type of variable. This is not generally known and often, e.g. Church (1956), statements to the contrary are made.

The first paper on Protothetic is Tarski's doctoral thesis under Leśniewski; incidentally, Tarski was the only person who received a Ph.D. degree from Leśniewski. The most important result of Tarski (1923) is the definition of conjunction in terms of equivalence. He gave two definitions:

$$[pq] :: p . q . \equiv : . [f] : . p \equiv : [r] . p \equiv f(r) . \equiv . [r] . q \equiv f(r)$$

and

$$[pq] : . p . q . \equiv : [f] : p \equiv . f(p) \equiv f(q) .$$

As is evident, these definitions are given in the propositional calculus extended with quantifiers and variable functors. The latter is also based on what Tarski called the law of substitution or what we call extensionality for the category of propositions, i.e.,

$$[pqf] : p \equiv q . \supset . f(p) \equiv f(g) .$$

The latter definition is called an extensional definition of conjunction, because without extensionality the usual laws of conjunction cannot be derived.

Many simplifications of these definitions have been given by Sobociński (1949a), but he has been unable to use any of them to simplify the axiom of Protothetic. The most intuitive, but not the shortest, of these is

$$[pq] : . p . q . \equiv : [f] : f(pq) . \equiv . f(11) ,$$

where here, and elsewhere, 1 is an abbreviation of $[u] . u : \equiv : [u] . u$. We could give a definition of 1, but we prefer to avoid unnecessary definitions for it makes the deductions easier.

The importance of these definitions is that they made Protothetic possible; i.e., it is possible to base a propositional logic with quantifiers on equivalence as sole primitive. This was the goal of Leśniewski, for he objected to stating definitions with a special definition symbol, and then not counting it among the primitive terms of the theory.

Perhaps the best introduction to Protothetic is given by Słupecki (1953). In reading this paper one should keep in mind that it is based on the late Mr. Jerzy Billich's notes of Leśniewski's lecture course entitled "On Certain Problems of Protothetic", which was given at Warsaw University in 1932–1933.* I do not know whether this was an elementary or advanced course, but I suspect the former. This suspicion is supported by reading Leśniewski (1938a) where it is mentioned that at that same time considerable work was being done on simplifying the axioms of Protothetic.

Słupecki considers four systems, none of which is equivalent to the full system of Protothetic $\mathfrak{S}_5$. The main deficiency of these systems is that they do not contain all of the semantical categories of Protothetic. Słupecki claims that the system E of Elementary Protothetics contains only the types S and S/S^n. Thus he omits the many link functors altogether (although on page 66 he uses the type $(S/S)/S$, and on page 70 the type $(S/S)/(S/SS)S$**; there are other examples also). If Słupecki's treatment of the semantical categories is considered to be in error and if the rules are strengthened then he has the full system of Protothetic. These systems do, however, have certain pedagogical advantages in that they are less complex and hence, presumably, more understandable. My view is that such systems are worthy of further study. If more light is shed on such subsystems it will help us to understand the full system better. However, in such work it is very important to make it clear exactly what system is being considered. It is especially necessary to carefully specify what deductive rules are being used. This is not done in Słupecki's papers nor in many other papers.

The most valuable feature of this paper is that it gives many examples of deductions in Protothetic. This is the only paper that provides a variety of elementary examples. (The papers of Leśniewski (1938b) and Sobociński (1960) provide more complicated examples.) To mention just one example, a proof is given that the following theses of extensionality are equivalent:

$$[pq] : .p \equiv q. \equiv: [f] : f(p). \equiv .f(q)$$

$$[pqrs] : .p \equiv q.r \equiv s. \equiv: [f] : f(pr). \equiv .f(qs).$$

Neither of these is admitted according to the rule of extensionality, for they are provable from the axiom of Protothetic. It would be interesting to know all of the dependencies which exist between the different theses of extensionality. To put it the other way around, I know of no categories for which the

* [Ed. Note: These lecture notes were published in 1988 by Kluwer Academic Publishers (see Leśniewski (1988) in the Protothetic Bibliography, pp. 3–28).]

** [Ed. Note: Cf. paper V in this volume, pp. 106 and 110.]

extensionality theses have been proven to be non-equivalent. If this question were answered, it might be possible to simplify the rule of extensionality.

System S is based on the rule of verification. Some instances of this are:

$$[f] : f(0).f(1). \equiv .[p].f(p)$$

$$[F] : F\langle Vr\rangle.F\langle Fl\rangle.F\langle As\rangle.F\langle N\rangle. \equiv .[f].F\langle f\rangle.$$

Some ideas about how this rule can be formulated are given by Słupecki in Chapter 10.

Leśniewski states that in order to formulate this rule "I needed a complicated apparatus of numerous terminological explanations" (1938a, p. 122). Leśniewski abandoned the rule of verification for the rule of extensionality because the rule of extensionality is easier to formulate. There could be some value in reconstructing the formulation of the rule of verification, for to do so one would have to specify in advance how each constant function in each semantical category could be defined. This could provide an interesting normal form theorem.

This valuable and interesting paper of Słupecki also contains (p. 75)* the germ of what Sobociński (1960) calls Metatheorem L. Słupecki says that a system of axioms for (elementary) Protothetic should contain the propositional calculus and the law of bivalency. The only item missing is that the four laws of logical conjunction should also be provable. It would be interesting to know if the result as stated is a complete characterization of the system E of elementary Protothetics.

As was mentioned above, considerable effort has gone into simplifying the axioms of Protothetic. The shortest known single axiom for Protothetic is the following:

$$[pq] :: p \equiv q. \equiv: .[f] : .f(pf(p[u].u)). \equiv: [r] : f(qr). \equiv .q \equiv p.$$

This was discovered by Sobociński in 1945 and published in Sobociński (1953) and then with full details in Sobociński (1960). Observe that this axiom contains (1) symmetry for equivalence and (2) the law of extensionality in the form

$$[f] : f(pf(p[u].u)). \equiv .[r].f(qr).$$

A variant of this, namely,

$$[f] : f(f(0)). \equiv .[r].f(r)$$

was shown by Łukasiewicz (1951) to imply the full propositional calculus.

In the spring of 1965 Sobociński taught a course entitled "Advanced Propositional Calculus." This course dealt mainly with Protothetic. The interesting

* [Ed. Note: Cf. paper V in this volume, p. 115.]

and difficult deductions from Sobociński's 1945 axiom were presented. These deductions are published in Sobociński (1960), which is, in my opinion, the most important paper concerning Protothetic. Unfortunately, only three parts of this paper have been published, but the contents of the other two were presented in this course. With Sobociński's permission, I will outline these results here.

If one accepts (1) (a fragment of) the equivalential calculus, (2) extensional reasoning for the category of propositions as a rule, i.e., if $\Phi(1)$ and $\Phi(0)$ are true, then $[p].\Phi(p)$ is true, and (3) the four laws of conjunction, i.e., $0.0 \equiv 0$, $0.1 \equiv 0$, $1.0 \equiv 0$, $1.1 \equiv 1$, together with rules of detachment, substitution, quantifiers, definition, but not extensionality, then one can derive the following:

(1) The full bi-valued propositional calculus. If one starts with the definition

$$[pq] : p.q \equiv p. \equiv .p \supset q$$

then it is not too difficult to derive the Tarski–Bernays axioms for implication, the law of bivalency, the rule of detachment for implication, and the usual laws involving negation.

(2) Extensionality as a law:

$$[pqf] : p \equiv q. \supset .f(p) \equiv f(q).$$

(3) The usual laws of quantifiers. These are summed up in the following thesis:

$$[fp] : .[x] : p \supset f(x) :\equiv: p \supset [x].f(x),$$

where x belongs to any category whatsoever, including those not belonging to Protothetic. Thus, in any system based on Protothetic the usual quantifier rules hold. In particular the ω-order functional calculus can be built up within Protothetic without any additional axioms. All that need be added is a new type I of individuals and a new constant f of type S/I.

(4) A metarule which permits the alphabetic variation of bound variables. Of course we are speaking of interior quantifiers; the rule of quantification permits the relettering of the variables in the main general quantifier.

As I understand it, the above are the principle results which are to be included in the completion of Sobociński's paper on single axioms of Protothetic. In his course on Protothetic a number of other interesting topics were discussed and some of them are now considered.

A proof of the strong completeness of Protothetic was discussed. By strong completeness we mean that if α is any sentence expressed in the vocabulary that has been introduced into a given development of Protothetic, and if α has no free variables, then either α or $\sim \alpha$ is provable.

The completeness theorem sketched in Słupecki (1953) is deficient in that the higher semantical categories are not considered. It would be nice to have a complete proof published.

Leśniewski developed computable Protothetic to provide an easy completeness proof for Protothetic. Such systems are very intuitive and easy to work with. Unfortunately, Leśniewski (1938a) contains the only published discussion of this interesting topic.* Sobociński sketched a system of computable Protothetic based on equivalence and the false proposition Λ. The single axiom is $\Lambda \equiv \Lambda$ and there are nine rules: four to describe the truth table for equivalence, three involving definitions, and two involving generalization. In some 24 steps the law of extensionality for propositions can be derived. I believe that further research in this area would be most interesting.

A great deal of time was spent discussing systems of many valued Protothetics. For finite valued logics one writes rules describing each position in the matrix. Then the development proceeds as in computable Protothetic. Unfortunately, this work is much too complicated to discuss here. I will only note that Scharle (1971) has shown that many valued Protothetics are also strongly complete.

In this short chapter the most important results have been mentioned, but there are other areas related to Protothetic which have not been mentioned even though they are interesting. I think especially of work on the equivalential calculus and work on the propositional calculus with variable functors.

There are many promising areas of research, but one of the things that would be nicest to have is a presentation of Protothetic which is accessible to beginners. I am not thinking of just an introductory paper, although this too would be useful, but of a system of Protothetic which is simple to understand and easy in development. As to research areas, subsystems of Protothetic seem most inviting to me: What if the rule of extensionality is deleted? What if only propositional variables are allowed? What is the structure of Słupecki's Elementary Protothetic? I invite you to continue.

II. ONTOLOGY

Leśniewski's Ontology is the most general theory of connections between names, and for this reason is sometimes called a calculus of names. Structurally, but not historically, it is based on Protothetic. Ontology is obtained from Protothetic by adding a new primitive semantical category N, that of names, a new primitive symbol ε of type S/NN, a single new axiom, and two new rules, the rules of Ontological definitions and Ontological extensionality.

In 1921 Łukasiewicz prompted Leśniewski to create Ontology when Łukasiewicz expressed his dissatisfaction with the way the cupola "is" is

* [Ed. Note: Further discussion concerning the systems of computable Protothetic can be found in Le Blanc (1991) in the Protothetic Bibliography.]

used by Peano and in set theory. Łukasiewicz asked Leśniewski how he used the word "is." Leśniewski replied that he used the word in the sense of ordinary (Polish) language. Then, in one of Łukasiewicz's few careless uses of language, he makes one of the most radically nominalistic statements ever made: "Some time later Leśniewski told me that he had found the required axiom sitting on a bench in the Warsaw-Saxon Park" (Łukasiewicz, 1953, p. 78).

Now that the issue of nominalism has come up, let me remark that Leśniewski was a nominalist and that he interpreted his theories that way.* But that is no reason that we must interpret, say, Ontology nominalistically. There is absolutely nothing wrong with thinking of the names of Ontology as naming some kind of abstract entities. This does not affect the theory in the least. There is one point, however, where it seems to me that Leśniewski's nominalistic viewpoint is useful, and perhaps even crucial. That is in the formulation of the rules of the system. Since the systems contain definitions as theses of the system and since definitions are ordered by the order that they are introduced into the system, it seems to me that the inscriptional view of syntax is the proper one. I have argued this point in Rickey (1972). If someone has a way of formulating a rule of definition without using inscriptional syntax I would be most interested in seeing it.

The axiom that Leśniewski found for Ontology is the following intuitive one:

$$\begin{aligned}[Aa] : .A \varepsilon a. \equiv: \ & [\exists B]. B \varepsilon A : \\ & [BC] : B \varepsilon A. C \varepsilon A. \supset . B \varepsilon C : \\ & [B] : B \varepsilon A. \supset . B \varepsilon a.\end{aligned}$$

This says that A is a if and only if A is unempty, A is unique, and anything which is A is a. Here the variables represent names, be they empty, singular, or general, i.e., whether they denote zero, one, or many individuals. The use of the word "name" in Ontology is nicely explained by use of Lejewski's ontological diagrams (1958).

The epsilon of Ontology has precisely the same meaning as the Polish "jest" or Latin "est," but it does not agree with the ambiguous English "is." This causes confusion for English speaking people. To avoid such problems one must use the English "is" in the technical sense specified in Ontology.

The simplifications of this axiom are not nearly so complicated as those for Protothetic. By 1929 the axiom had been simplified to

$$[Aa] : A \varepsilon a. \equiv . [\exists B]. A \varepsilon B. B \varepsilon a.$$

This work is due to Leśniewski, Sobociński and Tarski. It is nicely set out in Sobociński (1934). It should be remarked that the rule of extensionality

* [Ed. Note: Cf. paper I in this volume.]

must be used to prove these two axioms equivalent. This axiom still stands today as the shortest for Ontology. It probably is the shortest, but no one has attempted a proof of this as Łukasiewicz did for the implicational and equivalential calculi. Ontology also has the property that many of the terms definable in it can act as primitive terms. For some of these separate axioms have been given by Lejewski (1958).

Leśniewski published very little on Ontology. In Chapter XI of *O podstawach matematyki* (1927) the intuitive foundations of Ontology are discussed and in *Über die Grundlagen der Ontologie* (1930) the deductive rules are formalized. But nowhere are any deductions published. Leśniewski's two papers on the axiomatization of group theory (1929a, 1929b) do contain nice examples of the way that Leśniewski presented his proofs. The natural deduction technique that Leśniewski used seems to me to be one of the nicest in existence. Someone should adapt it to first order logic so that it can be more widely appreciated. Leśniewski did, however, lecture on Ontology a number of times (see Słupecki, 1955, pp. 7–8, and Leśniewski, 1927, Ch. XI, pp. 159–160).

For years the best informal outline of Ontology was in Kotarbiński (1929), yet this was very informal.

The first really useful paper on Ontology is Słupecki's paper on "S. Leśniewski's Calculus of Names" which was published in 1955. This paper is based on notes of a lecture course given by Leśniewski in 1929–1930 entitled "An Outline of Elementary Ontology." It is not at all clear how closely Słupecki follows the notes.*

In this paper a system of "Elementary Ontology" is presented. It is based on the long axiom of Ontology, but the deductive rules are not very carefully described. Consequently, I am not sure if I am interpreting him correctly. Only nominal variables occur in Elementary Ontology and the only constants which occur are those from the propositional calculus and those of types N/N^k and S/N^k (although a footnote on page 11 indicates otherwise). Because of this restriction the rule of definition must be severely limited. He also puts on the restriction that the definiens must occur in primitive notation. This seems like a harmless restriction, and may indeed be alright here, but in general this is a true restriction. As an example, if this restriction to primitive notation is made, then Sobociński's deductions (1960) from his axiom of Protothetic fail at D2, where the definiens is not, and necessarily so, in primitive notation.

This limitation of the semantic categories necessitates that the rule of extensionality be excluded entirely from Elementary Ontology. This raises an interesting point. Sobociński, in his lectures, calls Elementary Ontology that part of Ontology which can be derived without use of the rule of extensionality. I suspect the notion is garbled in Słupecki's paper; indeed his restriction on the types is quite artificial. In Chapter III on non-elementary Ontology

* [Ed. Note: These lecture notes were published in 1988 by Kluwer Academic Publishers (see Leśniewski (1988) in the Protothetic Bibliography, pp. 29–58).]

Słupecki does use the rule of extensionality, but then only for types S/N^k and N/N^k. This is an error on his part, for he says that Ontology has no functor forming functors (p. 9), whereas it certainly does and indeed he uses the type $(S/N)/N$ (p. 51). This confusion raises the question of the role that the rule of extensionality plays in Ontology. It would be interesting to know the exact differences between the two systems of "Elementary" Ontology.

These faults notwithstanding, Słupecki's paper is a very important one, for it contains a good deal of information about Ontology that is not otherwise available, among which is the following:

1. Derivation of the most basic consequences of the axiom regarding existence and uniqueness.
2. The development of syllogistic within Ontology.
3. The distinction between equality and identity.
4. The algebra of names, i.e., theorems involving intersection, union and nominal negation.
5. Derivation, without the rule of extensionality, of the law of extensionality for individuals, i.e.,

   $$[AB\Phi] : A = B. \supset .\Phi\{A\}. \equiv .\Phi\{B\}.$$

   Here "$=$" denotes identity; i.e., A and B are both individual names and both names denote the same thing.
6. A proof that Ontology contains creative definitions.
7. The consistency proof of Ontology of Kruszewski (1925). In this proof ε is interpreted as conjunction. Then the axiom of Ontology becomes a theorem of Protothetic. He does not verify that the rules of Ontology become derived rules of Protothetic.

On Słupecki's paper is based the important work of Iwanuś "On Leśniewski's Elementary Ontology" (1973). He views Ontology as a first order theory, i.e., as based on the predicate calculus without equality. In this theory if one starts with the long axiom of Ontology then there are only two creative definitions, those of union and nominal negation. This work raises several interesting questions: How does this presentation of Ontology based on first order logic compare with Słupecki's system of Elementary Ontology? Is this result true in Elementary Ontology? Then there is the most interesting question of all: What definitions are creative in Ontology? I have conjectured that there are infinitely many. While speaking of creative definitions I should mention an unpublished distinction of Sobociński. He distinguishes between weak creativity and strong creativity. A definition is weakly creative if its creativity can be avoided by a reformulation of the rule of substitution. Other creative definitions are strongly creative. For example, to obtain $p \equiv q$ from $f(p)$ one cannot substitute, as in *Principia Mathematicae*, "$- \equiv q$," for Leśniewski's systems do not contain incomplete symbols. One must introduce the many link definition $[pq] : p \equiv q. \equiv . \phi\{q\}(p)$ and then substitute $\phi\{q\}$ for f (observe that they are both of type S/S) in $f(p)$ to obtain $\phi\{q\}(p)$.

Then by use of the definition one obtains $p \equiv q$. This is an example of a weakly creative definition. Clearly, Ontology contains infinitely many such. The question is whether Ontology contains infinitely many strongly creative definitions.

Lejewski's paper "On Leśniewski's Ontology" (1958) contains a very nice introduction to Ontology. He discusses some of Leśniewski's views about his systems. In particular he mentions that Leśniewski intended his systems to be interpreted systems and that he did not believe that formalization of the systems destroyed this interpretation. Consequently, Leśniewski could describe himself as being simultaneously a confirmed intuitionist and a radical formalist. Unfortunately, from his published works, only the latter shows through. Lejewski's paper contains a very interesting "Ontological Table" which is designed to convey to the reader the intuitions underlying Ontology and the precise meanings in which the terms of Ontology are used. By means of this table Lejewski carefully distinguishes between the various functors of inclusion, exclusion, identity, and existence which Ontology contains. He also gives axioms for Ontology based on a variety of primitive terms.

More recently, there have been two books which provide good introductions to Ontology. They are Küng (1967) and Henry (1972). But a comprehensive introduction to Ontology does not yet exist.

In his thesis Canty (1969a, 1969b) shows that Gödel's incompleteness theorem is applicable to Ontology, if Ontology is strengthened by an axiom stating that there are infinitely many individuals. Canty needs this axiom to prove the analogues of the Peano postulates in Ontology. But there is a stronger reason that Canty does not mention but that Sobociński once remarked to me: If there are precisely n individuals then Ontology is strongly complete. If however we add to Ontology an axiom saying that there are finitely many individuals, then the system is not strongly complete, for we cannot prove that there are two individuals, i.e., $[\exists AB].A \neq B$, nor can we prove its negation. These facts certainly must be known, but I do not remember ever seeing them in the literature.

In his thesis Davis (1975) has done some very interesting work on the axiom of choice in Ontology. He has shown that in Ontology it can he expressed in either of the following equivalent forms:

$$[\exists f] : .[aA] : A \mathrel{\varepsilon} a . \supset .f(a) \mathrel{\varepsilon} a$$

$$[\exists f] : .[a\theta] : \theta\{a\}. \supset .\theta\{f\langle\theta\rangle\}.$$

The second of these is the more interesting, for the type τ of a is (if we ignore the parentheses) ambiguous, and so represents the axiom of choice for an arbitrary type. We shall refer to it as AC_τ. It is really surprising that the axiom of choice can assume such a simple form. The first of these forms also generalizes to higher types if the ε is taken to be a higher epsilon.

This work has been followed up in the dissertation of Kowalski (1977). He shows that the axiom of choice, Zorn's lemma, and the well ordering principle are equivalent in Ontology. But most importantly he formulates, in the style of Leśniewski, a rule which permits the addition of each of the theses AC_τ, τ any type, to Ontology. It should be remarked that if τ is a purely Protothetical type, i.e., if its Ajdukiewicz symbol involves only S, then AC_τ is provable in Protothetic.

An obvious interesting line of research is open and is being actively pursued by several people: What dependencies and independencies exist between the various AC_τ? On the dependencies side it seems that choice propagates downward among the types. As yet no work has been done on adapting forcing techniques to Ontology.

I have attended several courses of Sobociński which dealt mainly with Ontology. These courses contained a great wealth of unpublished material. They begin with the basic deductions which are done in a much nicer fashion and in much greater detail than in the literature. It is shown that syllogistic and Boolean algebra form subsystems of Ontology. Relations, functions, and equinumerosity are discussed in careful detail. An elementary course would terminate in the proof of the Schröder–Bernstein theorem, as a corollary of the theorem of the mean: $[abcde] : a \subset b.b \subset c.d \subset e.a \infty d.c \infty e. \supset .[\exists n].d \subset n.n \subset e.b \infty n$, where $a \infty d$ means that a and d are equinumerous.

More advanced courses contain information about higher epsilons, various definitions of finite and infinite and their equivalence, ordered and well ordered sets, comparison of Ontology and set theories, and many other topics. I regret to say that more details will have to await a monograph on the subject.

In concluding this chapter let me say that, in my opinion, Ontology is the most interesting of Leśniewski's systems. I maintain that it is a very intuitive system, but then I was raised on it, rather than on the first order language of *ZFC*. The most promising area of research is metalogical investigations of Ontology. Since the system contains many distinctive features – the semantical categories, definitions, and extensionality – such work will be difficult, but rewarding.

III. MEREOLOGY

Historically, Mereology is the first of Leśniewski's systems, having been developed in the period from 1914 to 1916. Structurally it is the last, for it is based on both Protothetic and Ontology. Etymologically, Mereology is the study of parts of things and so most naturally it is built on the single primitive term *pr* of type N/N as in $A \,\varepsilon\, pr(B)$, which is read "A is part of B." We now give Leśniewski's original (1916) axiomatization of Mereology (except that we formalize it and drop the redundant clause $[\exists B].B \,\varepsilon\, a$ from the definition of class):

M1. $[ABC] : A \,\varepsilon\, pr(B).\ B \,\varepsilon\, pr(C). \supset .\ A \,\varepsilon\, pr(C)$
M2. $[AB] : A \,\varepsilon\, pr(B). \supset .\sim (B \,\varepsilon\, pr(A))$

M3. $[AB] : A \varepsilon pr(B). \supset . B \varepsilon B$
D1. $[AB] : .A \varepsilon el(B). \equiv: A \varepsilon A : A \varepsilon pr(B). \vee .A = B$
D2. $[Ab] : .A \varepsilon Kl(b). \equiv: A \varepsilon A : [C] : C \varepsilon b. \supset . C \varepsilon el(A) :$
$[D] : D \varepsilon el(A). \supset .[\exists EF].E \varepsilon b.F \varepsilon el(D).F \varepsilon el(E)$
M4. $[ABa] : A \varepsilon Kl(a). B \varepsilon Kl(a). \supset . A = B$
M5. $[Ab] : A \varepsilon b. \supset .[\exists B]. B \varepsilon Kl(b).$

These axioms say that *pr* is transitive (M1), antisymmetric (M2), and that only individuals have parts (M3). Then, and this is very distasteful, there are definitions of element and collective or mereological class. The final axioms say that classes are unique (M4) and that we can form the class of things denoted by any unempty name.

The unhappiness about having definitions among the axioms is not just esthetic; it could be that these definitions are creative (in this case they are not). For example, in the case of the propositional calculus, see my paper (Rickey, 1975). This matter as well as the other esthetic concerns of Leśniewski regarding axiom systems are discussed in an interesting and readable – but apparently little read – paper of Sobociński (1955). Some of these ideas seem to be related to Goodman's notion of simplicity of primitive terms (1951). It is rather amazing that these esthetic concerns, e.g., minimizing the number of semantical categories that occur in an axiom system, or insisting on organic axioms, have in fact led to very simple axiom systems. I have often wondered if these esthetic criteria could be turned into theorems. But this is probably hopeless.

The literature for Mereology is richer than for the other systems. This is probably because Leśniewski published more on this system than on his other two systems. His publications on Mereology (1916, 1927) are both long works with considerable detail. They were both published in his pre-symbolic period, but his language is so precise that one can immediately translate his proofs to formal ones.

The best introduction to Mereology is Sobociński's paper "Studies in Leśniewski's Mereology" (1954). In this paper he defines the most commonly used terms of Mereology and discusses their intuitive meaning. He also indicates which of them can serve as primitive terms of Mereology. This theme is taken up in greater detail in Welsh (1971), who is a student of Clay. Sobociński also discusses other axiomatizations of Mereology. Grzegorczyk discovered the first single axiom of Mereology in 1946. By taking advantage of this result which Sobociński learned from Hiż, he was able to find the following axiom which is based on *el* as its sole primitive:

$$[AB] :.: A \varepsilon el(B). \equiv :: B \varepsilon B :: [fa] :: [c] : .C \varepsilon f(a). \equiv: [D] : D \varepsilon a. \supset . D \varepsilon el(c) : [D] : D \varepsilon el(c). \supset .[\exists EF].E \varepsilon a. F \varepsilon el(E). F \varepsilon el(D) : .B \varepsilon el(B). B \varepsilon a : . \supset . A \varepsilon el(f(a)).$$

Many axiomatizations of Mereology have been given (see Lejewski, 1954, 1955, 1963 and 1967).

In the first axiom system for Mereology $Kl(a)$ is essentially the least upper bound of the a's. Part of the definition of Kl, viz., $[D] : D \varepsilon a . \supset . D \varepsilon el(A)$ says that $Kl(a)$ or A is an upper bound for the a's. To say it is least requires

$$[D] : D \varepsilon a . \supset . D \varepsilon el(B) : \supset . A \varepsilon el(B).$$

These can be put together into

$$[Aa] :: A \varepsilon Kl(a) . \equiv : . A \varepsilon A : . [B] : . [D] : D \varepsilon a . \supset . D \varepsilon el(B) : \equiv . A \varepsilon el(B)$$

or more compactly

M6. $[Aa] :: A \varepsilon Kl(a) . \equiv : . A \varepsilon A : . [B] : a \subset el(B) . \equiv . A \varepsilon el(B).$

Tarski showed this for strong inclusion ($\sqsubset$) and then Lejewski strengthened it to weak inclusion ($\subset$) in 1960, when he was at Notre Dame. Note that this suffices as a new definition of Kl in terms of el. However, Clay (1966) has shown that if M6, the short definition of Kl, replaces D2 in the axiom system, then a weaker system is obtained. Investigations of the connections between Mereology and the theory of least upper bounds are made in Kubiński (1971).

Mereology was used by Tarski (1929) to form a foundation for the geometry of solids. He takes the notion of a solid sphere as primitive. Then, with some rather ingenious definitions, he is able to define concentric spheres. Finally, a point is considered to be a distributive class of concentric spheres. The axiomatization of Mereology given by Tarski is deficient in that M3 is not provable from his axioms. Thus his remark in footnote 2, p. 25 is incorrect. The definitions given by Tarski have been simplified by Jaśkowski (1948) and even more by Clay (unpublished). Moreover, this idea has been extensively developed by Sullivan (1971–1973b).

Another paper on the Mereological foundations of geometry is that of Glibowski (1969). At the Notre Dame Logic Seminar in 1969 I discussed this paper and indicated several ways that it could be improved. I provided a new primitive term for the system which replaced the two that he had and also showed how the axiom of continuity could be stated in Mereology (Glibowski stated that it couldn't be). But the suggestion that I made that turned out to be useful was that I suggested Atomistic Mereology be used as a base for Geometry. To obtain Atomistic Mereology one first defines atom:

$$[A] : . A \varepsilon atm . \equiv : [B] : B \varepsilon el(A) . \supset . B = A : A \varepsilon A$$

and then adds an axiom which states that every object is a class of atoms:

$$[A] : A \varepsilon A . \supset . [\exists a] . a \subset atm . A \varepsilon Kl(a).$$

Finally, I defined the term

$$[AB] : A \varepsilon atm(B). \equiv .A \varepsilon atm.A \varepsilon el(B).$$

Then I showed that this term could be used as a primitive for Atomistic Mereology by proving the theorems:

$$[Aa] : .A \varepsilon Kl(a). \equiv: A \varepsilon A : [B] : B \varepsilon atm(A). \equiv .[\exists c].B \varepsilon atm(c).c \varepsilon a$$

$$[A] : A \varepsilon atm. \equiv .A \varepsilon atm(A)$$

$$[AB] : .A \varepsilon el(B). \equiv: A \varepsilon A : [C] : C \varepsilon atm(A). \supset .C \varepsilon atm(B).$$

This definition of class is exactly analogous to my definition of Klass in my system of inscriptional syntax (Rickey, 1972). It was only at this time that I realized that Atomistic Mereology is a subsystem of my system of inscriptional syntax. The effect of this was that it prompted Sobociński's work on the axiomatization of Atomistic Mereology (1971a, 1971b, 1971c). This then inspired the work on Atomistic and Atomless Mereology by Lejewski (1973) and Clay (1975).

A subject which is always said to be "akin" to Mereology is the Calculus of Individuals of Leonard and Goodman (1940). The exact relationship between these two theories should be carefully explored, although it is fairly clear that the Calculus of Individuals is a subsystem of Mereology.

Up to this point I have concentrated on the best papers regarding Leśniewski's systems, but now I wish to consider one bad paper. This is the paper of Grzegorczyk (1955) on "The Systems of Leśniewski in Relation to Contemporary Logical Research." On the formal side there are a number of errors: his treatment of the semantical categories is almost entirely incorrect. He characterizes Ontology as the theory of atomic and complete Boolean algebra; this totally disregards the higher types of Ontology. His axiomatization of Mereology is incorrect. Not only does he completely ignore the Ontological base but also he give it the weakest possible logical base by making it an elementary theory. He says that the models of Mereology are the models of complete Boolean algebra with zero deleted. Clay (1974a) has carefully discussed this statement and shown where it is incorrect. Luschei (1962) has also written a condemnation of this paper. But the real problem is not the incorrect results it contains, but the views it purports. To mention but one point he says that "Leśniewski's systems are almost entirely unsuited for metalogical investigations owing to the fact that they are presented not as completed systems but as systems developing in the course of time." I would argue that this property of the system provides a rather interesting object of study. It is true that first order model theory has been most extensively developed; one reason is that it is the easiest to do. In the *Model Theory* book of Chang and Keisler they list model theory for second order logic as an open problem worthy of study. Surely then, model theory for Ontology is worthy of study.

I would say that the model theoretic aspects of Mereology are the most promising line of research. The consistency of the system with respect to the real number system has been demonstrated by Clay (1968).* Lejewski (1969) gave a consistency proof based on Protothetic. Clay has also given a number of models for Mereology in his papers, e.g., (1971, 1974b). Further work in this area should produce valuable and interesting results.

REFERENCES

Ajdukiewicz, Kazimierz (1935). 'Die Syntaktische Konnexität', *Studia Philosophica* 1, 1–27.

Canty, John Thomas (1969a). 'The Numerical Epsilon', *Notre Dame Journal of Formal Logic* X, 47–63.

Canty, John Thomas (1969b). 'Leśniewski's Terminological Explanations as Recursive Concepts', *Notre Dame Journal of Formal Logic* X, 337–369.

Chang, C.C. and Keisler, H.J. (1973). *Model Theory*, North-Holland Publ. Co.

Church, Alonzo (1956). *Introduction to Mathematical Logic*, Princeton University Press.

Clay, Robert E. (1966). 'On the Definition of Mereological Class', *Notre Dame Journal of Formal Logic* VII, 359–360.

Clay, Robert E. (1968). 'The Consistency of Leśniewski's Mereology Relative to the Real Number System', *Journal of Symbolic Logic* 33, 251–257.

Clay, Robert E. (1971). 'A Model for Leśniewski's Mereology in Functions', *Notre Dame Journal of Formal Logic* XII, 467–478.

Clay, Robert E. (1974a). 'Relation of Leśniewski's Mereology to Boolean Algebra', *Journal of Symbolic Logic* 39, 638–648.

Clay, Robert E. (1974b). 'Some Mereological Models', *Notre Dame Journal of Formal Logic* XV, 141–146.

Clay, Robert E. (1975). 'Single Axioms for Atomistic and Atomless Mereology', *Notre Dame Journal of Formal Logic* XVI, 345–351.

Davis, Charles C., Jr. (1975). 'An Investigation Concerning the Hilbert–Sierpiński Logical Form of the Axiom of Choice', *Notre Dame Journal of Formal Logic* XVI, 145–184.

Glibowski, Edmund (1969). 'The Application of Mereology to Grounding of Elementary Geometry', *Studia Logica* XXIV, 109–129.

Goodman, Nelson (1951). *The Structure of Appearance*, Cambridge, MA.

Grzegorczyk, Andrzej (1955). 'The Systems of Leśniewski in Relation to Contemporary Logical Research', *Studia Logica* III, 77–95.

Henry, Desmond Paul (1972). *Medieval Logic and Metaphysics: A Modern Introduction*, Hutchinson University Library.

Iwanuś, Bogusław (1973). 'On Leśniewski's Elementary Ontology', *Studia Logica* XXXI, 73–125.

Jaśkowski, Stanisław (1948). 'Une modification des définitions fondamentales de la géométrie des corps de M.A. Tarski', *Annales de la Société Polonaise de Mathématique* 21, 298–301.

Kotarbiński, Tadeusz (1929). *Elementy teorji poznania, logiki formalnej i metodologji nauk* (*Elements of Epistomology, Formal Logic, and Methodology*), Ossolineum, Lwów.

Kowalski, James G. (1977). 'Leśniewski's Ontology Extended with the Axiom of Choice', *Notre Dame Journal of Formal Logic* XVIII, 1–78.

Kruszewski, Z. (1925). 'Ontologja bez aksjomatów', *Przegląd Filozoficzny* XXVIII, 136.

Kubiński, Tadeusz (1971). 'A Report on Investigations Concerning Mereology', *Acta Universitatis Wratislaviensis* No. 139, Prace Filozoficzne VIII, 47–68.

* [Ed. Note: Leśniewski's proof of this fact can be found in *S. Leśniewski's Lecture Notes in Logic*, pp. 109–125 (see Leśniewski (1988) in the Protothetic Bibliography).]

Küng, Guido (1967). *Ontology and the Logistic Analysis of Language*, The Humanities Press.
Lejewski, Czesław (1954). 'A Contribution to Leśniewski's Mereology', *Polskie Towarzystwo Naukowe na Obczyźnie. Rocznik* 5, 43–50.
Lejewski, Czesław (1955). 'A New Axiom of Mereology', *Polskie Towarzystwo Naukowe na Obczyźnie. Rocznik* 6, 65–70.
Lejewski, Czesław (1958). 'On Leśniewski's Ontology', *Ratio* (Oxford) 1, 150–176.
Lejewski, Czesław (1963). 'A Note on a Problem Concerning the Axiomatic Foundations of Mereology', *Notre Dame Journal of Formal Logic* IV, 135–139.
Lejewski, Czesław (1967). 'A Single Axiom for the Mereological Notion of Proper Part', *Notre Dame Journal of Formal Logic* VIII, 279–285.
Lejewski, Czesław (1969). 'Consistency of Leśniewski's Mereology', *Journal of Symbolic Logic* 34, 321–328.
Lejewski, Czesław (1973). 'A Contribution to the Study of Extended Mereologies', *Notre Dame Journal of Formal Logic* XIV, 55–67.
Leonard, Henry S. and Goodman, Nelson (1940). 'The Calculus of Individuals and Its Uses', *Journal of Symbolic Logic* 5, 45–55.
Leśniewski, Stanisław (1916). *Podstawy ogólnej teoryi mnogości. I.* (*Foundations of a General Theory of Sets*), Prace Polskiego Koła Naukowego w Moskwie, Moscow.
Leśniewski, Stanisław (1927). 'O podstawach matematyki (On the Foundations of Mathematics)', *Przegląd Filozoficzny* XXX (1927), 164–206; XXXI (1928), 261–291; XXXII (1929), 60–101; XXXIII (1930), 77–105 and XXXIV (1931), 142–170.
Leśniewski, Stanisław (1929a). 'Über Funktionen, deren Felder Gruppen mit Rücksicht auf diese Funktionen sind', *Fundamenta Mathematicae* XIII, 319–332.
Leśniewski, Stanisław (1929b). 'Über Funktionen, deren Felder Abelsche Gruppen in bezug auf diese Funktionen sind', *Fundamenta Mathematicae* XIV, 242–251.
Leśniewski, Stanisław (1929c). 'Grundzüge eines neuen Systems der Grundlagen der Mathematik', *Fundamenta Mathematicae* XIV, 1–81.
Leśniewski, Stanisław (1930). 'Über die Grundlagen der Ontologie', *Comptes rendus des Séances de la Société des Sciences et des Lettres de Varsovie, Classe III* 23, 111–132.
Leśniewski, Stanisław (1931). 'Über Definitionen in der sogenannten Theorie der Deduktion', *Comptes rendus des Séances de la Société des Sciences et des Lettres de Varsovie, Classe III* 24, 289–309.
Leśniewski, Stanisław (1938a). 'Einleitende Bemerkungen zur Fortsetzung meiner Mitteilung u. d. T. "Grundzüge eines neuen Systems der Grundlagen der Mathematik"', *Collectanea Logica* I (Offprint 1938), 1–60.
Leśniewski, Stanisław (1938b). 'Grundzüge eines neuen Systems der Grundlagen der Mathematik, §12', *Collectanea Logica* I (Offprint 1938), 61–144.
Łukasiewicz, Jan (1910). 'Über den Satz des Widerspruchs bei Aristoteles', *Bulletin International de l'Académie des Sciences de Cracovie, Classe de philologie, Classe d'histoire et de philosophie*, 15–38.
Łukasiewicz, Jan (1951). 'On Variable Functors of Propositional Arguments', *Proceedings of the Royal Irish Academy* 54, sect. A, 25–35.
Łukasiewicz, Jan (1953). 'Symposium: The Principle of Individuation I', *Aristotelian Society*, Supp. Vol. 27, 69–82.
Luschei, Eugene C. (1962). *The Logical Systems of Leśniewski*, North-Holland Publ. Co.
Machover, M. (1966). 'Contextual Determinacy in Leśniewski's Grammar', *Studia Logica* XIX, 47–57.
Rickey, V. Frederick (1972). 'Axiomatic Inscriptional Syntax, Part I: General Syntax', *Notre Dame Journal of Formal Logic* XIII, 1–33.
Rickey, V. Frederick (1973). 'Axiomatic Inscriptional Syntax, Part II: The Syntax of Protothetic', *Notre Dame Journal of Formal Logic* XIV, 1–52. [Paper VII in this volume – Ed.]

Rickey, V. Frederick (1975). 'Creative Definitions in Propositional Calculi', *Notre Dame Journal of Formal Logic* XVI, 273–294.

Scharle, Thomas W. (1971). 'Completeness of Many-Valued Protothetic', *Journal of Symbolic Logic* 36, 363–364.

Sinisi, Vito F. (1976). 'Leśniewski's Analysis of Russell's Antimony', *Notre Dame Journal of Formal Logic* XVII, 19–34.

Słupecki, Jerzy (1953). 'St. Leśniewski's Protothetics', *Studia Logica* I, 44–112. [Paper V in this volume – Ed.]

Słupecki, Jerzy (1955). 'S. Leśniewski's Calculus of Names', *Studia Logica* III, 7–73.

Sobociński, Bolesław (1934). 'O kolejnych uproszczeniach aksjomatyki "ontologji" prof. St. Leśniewskiego', *Fragmenty Filozoficzne* I, 143–160.

Sobociński, Bolesław (1949a). *An Investigation of Protothetic*, Cahiers de l'Institut d'Études Polonaises en Belgique, no. 5, Bruxelles. [Paper IV in this volume – Ed.]

Sobociński, Bolesław (1949b). 'L'analyse de l'antinomie Russellienne par Leśniewski', *Methodos* 1 (1949), no. 1, 94–107; no. 2, 220–228; no. 3, 308–316; 2 (1950), 237–257.

Sobociński, Bolesław (1953). 'Z badań nad aksjomatyką prototetyki Stanisława Leśniewskiego', *Polskie Towarzystwo Naukowe na Obczyźnie. Rocznik* 4, 18–20.

Sobociński, Bolesław (1954). 'Studies in Leśniewski's Mereology', *Polskie Towarzystwo Naukowe na Obczyźnie. Rocznik* 5, 34–48.

Sobociński, Bolesław (1955). 'On Well Constructed Axiom Systems', *Polskie Towarzystwo Naukowe na Obczyźnie. Rocznik* 6, 54–65.

Sobociński, Bolesław (1960). 'On the Single Axioms of Protothetic I, II, III', *Notre Dame Journal of Formal Logic* I (1960), 52–73; II (1961), 110–126 and 129–148. [Paper VI in this volume – Ed.]

Sobociński, Bolesław (1971a). 'Lattice-Theoretical and Mereological Forms of Hauber's Law', *Notre Dame Journal of Formal Logic* XII, 81–85.

Sobociński, Bolesław (1971b). 'Atomistic Mereology I, II', *Notre Dame Journal of Formal Logic* XII, 89–103 and 203–213.

Sobociński, Bolesław (1971c). 'A Note on an Axiom-System of Atomistic Mereology', *Notre Dame Journal of Formal Logic* XII, 249–251.

Sullivan, Theodore F. (1971). 'Affine Geometry Having a Solid as Primitive', *Notre Dame Journal of Formal Logic* XII, 1–61.

Sullivan, Theodore F. (1972). 'The Name Solid as Primitive in Projective Geometry', *Notre Dame Journal of Formal Logic* XIII, 95–97.

Sullivan, Theodore F. (1973a). 'The Geometry of Solids in Hilbert Spaces', *Notre Dame Journal of Formal Logic* XIV, 575–580.

Sullivan, Theodore F. (1973b). 'Tarski's Definition of Point in Banach Spaces', *Journal of Geometry* 3, 179–189.

Tarski, Alfred (1923). 'O wyrazie pierwotnym logistyki (On the Primitive Term of Logistic)', *Przegląd Filozoficzny* XXVI, 68–89. [Paper III in this volume – Ed.]

Tarski, Alfred (1929). 'Les fondements de la géométrie des corps', Księga Pamiątkowa Pierwszego Polskiego Zjazdu Matematycznego, supplement to *Annales de la Société Polonaise de Mathématique*, Kraków, 29–33.

Welsh, Paul J., Jr. (1971). *Primitivity in Mereology*, Ph.D. dissertation, University of Notre Dame.

ALFRED TAJTELBAUM-TARSKI

ON THE PRIMITIVE TERM OF LOGISTIC *

Doctoral Dissertation[1]

INTRODUCTION

The considerations carried out in the present work belong to the area of logistic; the sentences on which I based these considerations are generally accepted among researchers working in this field of knowledge. I do not, however, carry out my considerations on the basis of any specific system of logistic; in particular, I do not make my reasonings dependent on the best known theory of logical types by Russell. Although it is not possible, as I see it, to develop a consistent system of logistic without this or that theory of types, among all the theories of types which could be constructed[2] there unquestionably exist those according to which my arguments, in their present general form, are faultless. One such theory was developed by S. Leśniewski during his lectures on the foundations of arithmetic at the University of Warsaw (in 1920–1921).[3]

The main objective of the present work is to settle the following problem: *is it possible to construct a system of logistic with the sign of equivalence as the sole primitive term (in addition, of course, to the quantifiers*[4]*)?*

This problem seems interesting to me for the following reasons. It is known that one can construct a system of logistic by means of only one primitive term by following the footsteps of Russell[5] and adopting the symbol of implication as such a term; by making use of the idea of Sheffer,[6] who adopts the sign of rejection as the sole primitive term, the sign introduced by him specifically for this purpose; or, finally, by using the unpublished results of Łukasiewicz.[7]

---

* Bibliographical Note. This paper appeared in print in Polish under the title 'O wyrazie pierwotnym logistyki', *Przegląd Filozoficzny* XXVI (1923), 68–89, by permission of Jan Tarski. Translated by Z. Stachniak.

However, for the goal just stated to be indeed attainable, i.e., to construct the whole system in terms of only one primitive term, definitions should be formulated using (apart from variables) only this primitive term, terms previously defined, and the term which a given definition introduces to the system; we must not, however, use any constant term which does not belong to any of these three categories.[8] From this standpoint, in my opinion, the sign of equivalence has a significant advantage over the terms used by the authors cited above; this sign, when adopted as primitive, permits strict compliance with the rule in question while, at the same time, it grants definitions a natural and convenient form of equivalences.

The theorem, which I will prove in Section 1 of this article, namely:

$$[p, q] :: p.q. \equiv .\because [f] \because p \equiv: [r].p \equiv f(r). \equiv .[r].q \equiv f(r),$$ [9]

gives an *affirmative answer* to the question raised above. In fact, it can serve as a definition of the sign of logical product in terms of the sign of equivalence and the universal quantifier. As soon as we are allowed to use the sign of logical product, defining other terms of logistic does not present any difficulties, as it is evident, for instance, from the following sentences:

$$Vr \equiv .[p].p \equiv p,$$

$$Fl \equiv .[p].p,$$ [10]

$$[p] :\sim (p) \equiv .p \equiv Fl,$$

$$[p, q] \because p \supset q. \equiv: p \equiv .p.q,$$

$$[p, q] : p \vee q. \equiv . \sim (p) \supset q.$$

It will be seen from the considerations carried out in Section 2, that my results could be considerably simplified on the basis of such a system of logistic which would contain the following sentence among its postulates or theorems:

$$[p, q, f] : p \equiv q.f(p). \supset f(q).$$

This sentence, however, can be neither proved nor disproved on the basis of any of the systems of logistic known from the literature. This fact motivated me to further studies of the sentence in question; in particular, it motivated me to search for other sentences equivalent to it. I will present the results of these studies in Sections 3–6 of this work.

1. FUNDAMENTAL THEOREM

I begin by introducing a few definitions, **Def. 1–3**, which I will be using in this and the following sections. Then, I will prove a number of lemmas, **Th. 1–10**, and finally the fundamental theorem of this work, **Th. 11**. The proofs I give are, naturally, incomplete; they can be regarded as commentaries explaining the course of reasoning. The structure of these commentaries, partially modeled after the structure of proofs in Whitehead and Russell's *Principia Mathematica*, does not, I think, require more detailed explanations.[11]

Def. 1. $[p] : vr(p) \equiv .p \equiv p$

Def. 2. $[p] : as(p) \equiv p$

Def. 3. $[p] : fl(p) \equiv .p \equiv\sim (p)$

For reasons of symmetry, we could also introduce the following definition:

Def. 2′. $[p].ng(p) \equiv\sim (p)$,

which would, however, be useless, since the term 'ng' would have the same meaning as the negation symbol '$\sim$' already occurring in logic.

Th. 1. $[p].vr(p)$ (**Def. 1**)

Th. 2. $[p] : [q].p \equiv vr(q). \supset p$

Com.

$[p].\because.Hp. \supset:$

(1) $p \equiv .[q].vr(q) :$

Ts (1, **Th. 1**)

Th. 3. $[p, q] : p \supset .p \equiv vr(q)$

Com.

$[p] : Hp. \supset .$

(1) $vr(q).$ (**Th. 1**)

Ts (Hp, 1)

Th. 4. $[p] : [q].p \equiv vr(q). \equiv p$ (**Th. 2, Th. 3**)

Th. 5. $[p, q] :: [f] .\therefore. p \equiv: [r].p \equiv f(r). \equiv .[r].q \equiv f(r) .\therefore \supset q$

Com.

$[p, q] :: Hp .\therefore \supset .\therefore$

(1) $p \equiv: [r].p \equiv vr(r). \equiv .[r].q \equiv vr(r) .\therefore$

(2) $[r].p \equiv vr(r). \equiv p :$ **(Th. 4)**

(3) $[r].q \equiv vr(r). \equiv q :$ **(Th. 4)**

(4) $p \equiv .p \equiv q :$ (1, 2, 3)

(5) $p \equiv p. \equiv q :$ (4)[12]

(6) $p \equiv p.$

Ts (5, 6)

Th. 6. $[p]. \sim ([q].p \equiv as(q))$

Com.

$[p].$

(1) $\sim (p \equiv\sim (p)).$

(2) $\sim ([q].p \equiv q).$ (1)

$\sim ([q].p \equiv as(q))$ (1, **Def. 2**)

Th. 7. $[p, q] : [r].p \equiv as(r). \equiv .[r].q \equiv as(r)$

Com.

$[p, q] :$

(1) $\sim ([r].p \equiv as(r)).$ **(Th. 6)**

(2) $\sim ([r].q \equiv as(r)) :$ **(Th. 6)**

$[r].p \equiv as(r). \equiv .[r].q \equiv as(r)$ (1, 2)

Th. 8. $[p, q] :: [f] .\because p \equiv : [r] . p \equiv f(r) . \equiv . [r] . q \equiv f(r) .\therefore \supset p$

Com.

$[p, q] :: Hp .\therefore \supset .\therefore$

(1) $p \equiv : [r] . p \equiv as(r) . \equiv . [r] . q \equiv as(r) .\therefore$

Ts **(1, Th. 7)**

Th. 9. $[p, q] :: [f] .\because p \equiv : [r] . p \equiv f(r) . \equiv . [r] . q \equiv f(r) .\therefore \supset . p . q$
(Th. 8, Th. 5)

Th. 10. $[p, q, f] :: p . q . \supset .\therefore p \equiv : [r] . p \equiv f(r) . \equiv . [r] . q \equiv f(r)$

Com.

$[p, q, f] :: Hp . \supset .\therefore$

(1) $p .$

(2) $q .$ (Hp)

(3) $p \equiv q .\therefore$ (1,2)

(4) $[r] : p \equiv r . \equiv . q \equiv r .\therefore$ (3)

(5) $[r] : p \equiv f(r) . \equiv . q \equiv f(r) .\therefore$ (4)

(6) $[r] . p \equiv f(r) . \equiv . [r] . q \equiv f(r) .\therefore$ (5)

Ts (1,6)

Th. 11. $[p, q] :: p . q \equiv .\therefore [f] .\because p \equiv : [r] . p \equiv f(r) . \equiv . [r] . q \equiv f(r)$
(Th. 10, Th. 9)

2. TRUTH-FUNCTIONS AND THE LAW OF SUBSTITUTION

In their work mentioned above, Russell and Whitehead call a function f a *truth-function*[13] if its argument is a sentence and if it satisfies the condition:

(a) $[p, q] : p \equiv q . f(p) . \supset f(q) .$

I will call the sentence:

(A) $[p, q, f] : p \equiv q . f(p) . \supset f(q)$,

which expresses the fact that every function f has the property mentioned above, the *law of substitution.*

In definition 4, I am introducing the symbol '$\vartheta\rho$'; by virtue of this definition, the expression '$\vartheta\rho\{f\}$' has the same meaning as 'f is a truth-function'. Definition 5 makes it possible to replace the law of substitution by a single term 'Sb'. Among the theorems proved in this section, **Th. 17** plays an important role; it demonstrates that in the system of logistic, which contains the law of substitution among its postulates or theorems, there exists another sentence:

$$[p, q] .\therefore p . q . \equiv : [f] : p \equiv . f(p) \equiv f(q),$$

simpler than **Th. 11**, which can also serve as the definition of the sign of logical multiplication in terms of the sign of equivalence and the universal quantifier.

From **Th. 20** it follows that to define the sign of logical sum in the system of logistics that contains the law of substitution, an analogous sentence can be used:

$$[p, q] .\therefore p \vee q . \equiv : [\exists f] : p \equiv . f(p) \equiv f(q) .^{*}$$

The remaining theorems of this section serve as lemmas.

Def. 4. $[f] .\therefore \vartheta\rho\{f\} \equiv : [p, q] : p \equiv q . f(p) . \supset f(q)$

Def. 5. $Sb \equiv . [f] . \vartheta\rho\{f\}$

Th. 12. $[f] .\therefore \vartheta\rho\{f\} \equiv : [p, q] : p \equiv q . \supset . f(p) \equiv f(q)$ **(Def. 4)**

Th. 13. $Sb \equiv : [p, q, f] : p \equiv q . f(p) . \supset f(q)$ **(Def. 5, Def. 4)**

Th. 14. $Sb \equiv : [p, q, f] : p \equiv q . \supset . f(p) \equiv f(q)$ **(Def. 5, Th. 12)**

Th. 15. $Sb \equiv .\therefore [p, q, f] : p . q . \supset . f(p) \equiv f(q) .\therefore$
$[p, q, f] : \sim (p) . \sim (q) . \supset . f(p) \equiv f(q)$ **(Th. 14)**

The theorems just stated demonstrate that the law of substitution is equivalent to the logical product of two sentences, the first of which could be called the *law of substitution for true sentences*, while the second one the *law of substitution for false sentences*. I cannot solve the problem, arising at this point, of

* [Ed. Note: The quantifier '$[b, d]$' originally occurring in this definition has been replaced by '$[p, q]$'.]

whether either of these two sentences alone is equivalent to the general law of substitution.

Th. 16. $[p, q, f] : p.q. \supset .f(p) \equiv f(q) :\supset .\therefore [p, q] .\because p.q. \equiv:$
$[f] : p \equiv .f(p) \equiv f(q)$

Com.

$Hp :\supset ::$

(1) $[p, q] :: p.q. \supset .\therefore$

(a) $p.q :$

(b) $[f].f(p) \equiv f(q).\therefore$ (Hp, a)

(c) $[f] : p \equiv .f(p) \equiv f(q) ::$ (a, b)

(2) $[p, q] :: [f] : p \equiv .f(p) \equiv f(q) :\supset .\therefore$

(d) $[f] : p \equiv .f(p) \equiv f(q).\therefore$

(e) $p \equiv .vr(p) \equiv vr(q) :$ (d)

(f) $vr(p) \equiv vr(q).$ (**Th. 1**)

(g) $p :$ (e, f)

(h) $p \equiv .as(p) \equiv as(q) :$ (d)

(i) $p \equiv .p \equiv q :$ (h, **Def. 2**)

(j) $p \equiv p. \equiv q :$ (i)

(k) $q.$ (j)

(l) $p.q ::$ (g, k)

Ts (1-c, 2-l)

Th. 17. $Sb \supset .\therefore [p, q] .\because p.q. \equiv: [f] : p \equiv .f(p) \equiv f(q)$ (**Th. 15, Th. 16**)

Th. 18. $[p, q, f] :\sim (p). \sim (q). \supset .f(p) \equiv f(q) :\supset .\therefore [p, q] .\therefore \sim (p).$
$\sim (q). \equiv: [f] :\sim (p) \equiv .f(p) \equiv f(q)$ (**Th. 1, Def. 2**)

I am omitting the proof of **Th. 18** which is fully analogous to that of **Th. 16**.

Th. 19. $[p, q, f] :\sim (p). \sim (q). \supset .f(p) \equiv f(q) :\supset .\therefore [p, q] .\therefore$
$p \vee q. \equiv: [\exists f] : p \equiv .f(p) \equiv f(q)$ (**Th. 18**)

Th. 20. $Sb \supset .\because [p, q] .\because p \vee q . \equiv : [\exists f] : p \equiv .f(p) \equiv f(q)$ **(Th. 15, Th. 19)**

Theorems 16 and 19 can easily be inverted. Hence, it can be directly concluded that the sentence:

$$[p, q] .\because p.q. \equiv : [f] : p \equiv .f(p) \equiv f(q)$$

not only follows from the law of substitution for true sentences but is equivalent to it. Similarly, the sentence:

$$[p, q] .\because p \vee q. \equiv : [\exists f] : p \equiv .f(p) \equiv f(q)$$

is equivalent to the law of substitution for false sentences, while the logical product of the two above mentioned sentences is equivalent to the general law of substitution.

3. INDEPENDENCE OF THE LAW OF SUBSTITUTION OF THE AXIOMS OF LOGISTIC; THE LAW OF THE NUMBER OF FUNCTIONS

Regarding the question of whether every function whose argument is a sentence is a truth-function, or, in other words, whether the law of substitution is a true sentence, Russell and Whitehead give a negative answer using in their justification arguments of merely intuitive nature.[14] Their answer does not seem to me quite convincing, especially since Leśniewski constructed a general method which makes it possible to eliminate from a language all functions which are not truth-functional.[15]

On the other hand, there seems to be no doubt that the law of substitution can be neither proved nor refuted on the basis of any of the systems of logistic known from the literature. Moreover, one can even "prove" the independence of this sentence of the known systems of axioms of logistic, e.g. of the system of axioms of Russell and Whitehead[16] – "prove" in the way one typically proves in mathematics the independence of a certain sentence of other sentences, i.e., by means of a suitably selected interpretation. However, I will not present my "proof" here, as it is based on hitherto unpublished results of Łukasiewicz's investigations concerning the three-valued logic, and its demonstration would present serious technical problems.

From what has been said above, it is clear that anyone who regards the law of substitution as a true sentence and wants to introduce it into the system of logistic, must admit as an axiom either the law itself or any other sentence which, together with other axioms of the system, can be used to prove the law of substitution. For exactly this reason, I find theorems (which I will state in the sequel of the present work) which establish equivalence between the law of substitution and certain other sentences, interesting. I think that such theorems can be helpful in the construction of the system of logistic, strengthened in the way described above, to the same extent as, for example, theorems concerning the equivalent forms of Euclid's postulate are helpful in research on the foundations of geometry.*

The theorems which I have just mentioned are sometimes direct conclusions of logically stronger theorems which establish the necessary and sufficient conditions for a function f to be a truth-function. For instance, as it follows from **Th. 40** of this section, the following condition is of this kind:

$$\text{(b)}\ [p].f(p) \equiv vr(p). \vee .[p].f(p). \equiv as(p). \vee .[p].f(p) \equiv \sim (p). \vee . [p].f(p) \equiv fl(p).$$

An immediate conclusion from **Th. 40** is **Th. 42**, from which it follows that the law of substitution is equivalent to the sentence:

$$\text{(B)}\ [f] : [p].f(p) \equiv vr(p). \vee .[p].f(p) \equiv as(p). \vee .[p].f(p) \equiv \sim (p). \vee . [p].f(p) \equiv fl(p).$$

Due to its intuitive content, this sentence can be called the *law of the number of functions*, for it states that every function whose argument is a sentence is equivalent, for every value of the argument, to one of the following four functions: $vr, as, \sim, fl$.

Def. 6. $[f, g].\because = \{f, g\} \equiv .[p].f(p) \equiv g(p)$

Th. 21. Vr

Th. 22. $\sim (Fl)$

Theorems 21 and 22 immediately follow from the definitions of the terms 'Vr' and 'Fl' given in the Introduction.

Th. 23. $[p] : p \equiv .p \equiv Vr$ **(Th. 21)**

Th. 24. $[p] :\sim (p) \equiv .p \equiv Fl$ **(Th. 22)**

Th. 25. $[p] : p \equiv Vr. \vee .p \equiv Fl$

Com.

$[p]$:

(1) $p \vee \sim (p)$:

* [Ed. Note: Tarski presumably refers to the famous Euclid's fifth postulate (of parallels).]

(2) $p \supset .p \equiv Vr :$ (**Th. 23**)

(3) $\sim (p) \supset .p \equiv Fl :$ (**Th. 24**)

$p \equiv Vr. \vee .p \equiv Fl$ (1, 2, 3)

Th. 26. $[p, f] : \vartheta\rho\{f\}.f(Vr).f(Fl). \supset f(p)$

Com.

$[p, f].\therefore Hp. \supset:$

(1) $p \equiv Vr. \vee .p \equiv Fl :$ (**Th. 25**)

(2) $p \equiv Vr. \supset f(p) :$ (**Def. 4**, Hp)

(3) $p \equiv Fl. \supset f(p) :$ (**Def. 4**, Hp)

Ts (1, 2, 3)

Th. 27. $[f] : \vartheta\rho\{f\}.f(Vr).f(Fl). \supset =\{f, vr\}$

Com.

$[f].\therefore Hp. \supset:$

(1) $[p].f(p) :$ (**Th. 26**)

(2) $[p].f(p) \equiv vr(p) :$ (1, **Th. 1**)

Ts (**Def. 6**, 2)

Th. 28. $[f] : \vartheta\rho\{f\}.f(Vr). \sim (f(Fl)). \supset =\{f, as\}$

Com.

$[f] :: Hp. \supset .\therefore$

(1) $[p] : p \supset .$

(a) $p \equiv Vr.$ (**Th. 21**)

(b) $f(p).\therefore$ (**Def. 4**, Hp, a)

(2) $[p] :\sim (p) \supset .$

(c) $p \equiv Fl.$ **(Th. 22)**

(d) $\sim (f(p)).\therefore$ (**Def. 4**, Hp, c)

(3) $[p].p \equiv f(p) :$ (1-b, 2-d)

(4) $[p].f(p) \equiv as(p) :$ (**Def. 2**, 3)

Ts (**Def. 6**, 4)

Th. 29. $[f] : \vartheta\rho\{f\}. \sim (f(Vr)).f(Fl). \supset = \{f, \sim\}$
(Th. 21, Def. 4, Th. 22, Def. 6)

The proof of this theorem, analogous to the proof of the previous theorem (**Th. 28**), does not present any difficulties.

Th. 30. $[p]. \sim (fl(p))$ **(Def. 3)***

Th. 31. $[p, f] : \vartheta\rho\{f\}. \sim (f(Vr)). \sim (f(Fl)). \supset \sim (f(p))$
(Th. 25, Def. 4)

The proof is fully analogous to that of **Th.** 26.

Th. 32. $[f] : \vartheta\rho\{f\}. \sim (f(Vr)). \sim (f(Fl)). \supset = \{f, fl\}$
(Th. 31, Th. 30, Def. 6)

The proof is fully analogous to that of **Th. 26.**

Th. 33. $[f] : \vartheta\rho\{f\} \supset . = \{f, vr\} \vee = \{f, as\} \vee = \{f, \sim\} \vee = \{f, fl\}$

Com.

$[f].\therefore Hp \supset:$

(1) $f(Vr).f(Fl). \vee .f(Vr). \sim (f(Fl)). \vee .$
$\sim (f(Vr)).f(Fl). \vee . \sim (f(Vr)). \sim (f(Fl)) :$

Ts (Hp, **Th. 27, Th. 28, Th. 29, Th. 32**, 1)

---

* [Ed. Note: The quantifier '[f]' originally occurring in theorem 30 has been replaced by '[p]'.]

Th. 34. $\vartheta\rho\{vr\}$

Com.

(1) $[p,q]: p \equiv q.vr(p). \supset vr(q).\because$ **(Th. 1)**

$\vartheta\rho\{vr\}$ (**Def. 4**, 1)

Th. 35. $\vartheta\rho\{as\}$

Com.

(1) $[p,q]: p \equiv q.as(p). \supset .$

(a) $p \equiv q.p.$ **(Def. 2)**

(b) $q.$ (a)

(c) $as(q).\because$ (**Def. 2**, b)

$\vartheta\rho\{as\}$ (**Def. 4**, 1-c)

Th. 36. $\vartheta\rho\{\sim\}$

Com.

(1) $[p,q]: p \equiv q. \sim(p). \supset\sim(q).\because$

$\vartheta\rho\{\sim\}$ (**Def. 4**, 1)

Th. 37. $\vartheta\rho\{fl\}$

Com.

(1) $[p,q]: p \equiv q.fl(p). \supset fl(q).\because$ **(Th. 30)**

$\vartheta\rho\{fl\}$ (**Def. 4**, 1)

Th. 38. $[f,g]: =\{g,f\}.\vartheta\rho\{f\}. \supset \vartheta\rho\{g\}$

Com.

$[f,g] :: Hp. \supset \therefore$

(1) $[r].g(r) \equiv f(r).\therefore$ (**Def. 6**)

(2) $[p, q] : p \equiv q.g(p). \supset .$

(a) $p \equiv q.f(p).$ (1)

(b) $f(q).$ (**Def. 4**, Hp, a)

(c) $g(q).\therefore$ (1, b)

Ts (**Def. 4**, 2-c)

Th. 39. $[f] : =\{f, vr\} \vee =\{f, as\} \vee =\{f, \sim\} \vee =\{f, fl\}. \supset \vartheta\rho\{f\}$ (**Th. 38, Th. 34, Th. 35, Th. 36, Th. 37**)

Th. 40. $[f] : \vartheta\rho\{f\} \equiv . =\{f, vr\} \vee =\{f, as\} \vee =\{f, \sim\} \vee =\{f, fl\}$ (**Th. 33, Th. 39**)

Th. 41. $Sb \equiv .[f]. =\{f, vr\} \vee =\{f, as\} \vee =\{f, \sim\} \vee =\{f, fl\}$ (**Th. 40, Def. 5**)

Th. 42. $Sb \equiv: [f] : [p].f(p) \equiv vr(p). \vee .[p].f(p) \equiv as(p). \vee .[p].f(p) \equiv\sim (p). \vee .[p].f(p) \equiv fl(p)$ (**Th. 41, Def. 6**)

4. THE LAW OF DEVELOPMENT

The main theorems of this section are **Th. 49** and **Th. 50**. In the first of these theorems, I give a new necessary and sufficient condition for a given function to be a truth-function, that is:

(c) $[p]\therefore f(p) \equiv: f(Vr).p. \vee .f(Fl). \sim (p).$

Th. 50, which directly follows from **Th. 49**, expresses the fact that the law of substitution is equivalent to the sentence:

(C) $[p, f]\therefore f(p) \equiv: f(Vr).p. \vee .f(Fl). \sim (p).$

In this last sentence, we can easily recognize the *law of development* well known from the algebra of logic.[17]

Th. 43. $[p, f]\therefore \vartheta\rho\{f\}.f(p). \supset: f(Vr).p. \vee .f(Fl). \sim (p)$

Com.

$[p, f] \therefore Hp. \supset:$

(1) $p \equiv Vr. \vee .p \equiv Fl:$ (**Th. 25**)

(2) $p \equiv Vr. \supset .f(Vr).p:$ (**Def. 4**, Hp, **Th. 23**)

(3) $p \equiv Fl. \supset .f(Fl). \sim (p):$ (**Def. 4**, Hp, **Th. 24**)

Ts (1, 2, 3)

Th. 44. $[p, f] : \vartheta\rho\{f\}.f(Vr).p. \supset f(p)$ (**Def. 4, Th. 23**)

Th. 45. $[p, f] : \vartheta\rho\{f\}.f(Fl). \sim (p). \supset f(p)$ (**Def. 4, Th. 24**)

Th. 46. $[p, f] \therefore \vartheta\rho\{f\} : f(Vr).p. \vee .f(Fl). \sim (p) :\supset f(p)$ (**Th. 44, Th. 45**)

Th. 47. $[f] :: \vartheta\rho\{f\} \supset \therefore [p] \therefore f(p) \equiv: f(Vr).p. \vee .f(Fl). \sim (p)$
(**Th. 43, Th. 46**)

Th. 48. $[f] :: [p] \therefore f(p) \equiv: f(Vr).p. \vee .f(Fl). \sim (p) \therefore \supset \vartheta\rho\{f\}$

Com.

$[f] :: Hp. \therefore \supset ::$

(1) $[p, q] \therefore p \equiv q.f(p). \supset:$

(a) $p \equiv q.f(p):$

(b) $f(Vr).p. \vee .f(Fl). \sim (p):$ (Hp, a)

(c) $f(Vr).q. \vee .f(Fl). \sim (q):$ (b, a)

(d) $f(q) ::$ (Hp, c)

Ts (**Def. 4**, 1-d)

Th. 49. $[f] :: \vartheta\rho\{f\} \equiv \therefore [p] \therefore f(p) \equiv: f(Vr).p. \vee .f(Fl). \sim (p)$
(**Th. 47, Th. 48**)

Th. 50. $Sb \equiv \therefore [p, f] \therefore f(p) \equiv: f(Vr).p. \vee .f(Fl). \sim (p)$ (**Th. 49,*** **Def. 5**)

* [Ed. Note: **Th. 39** originally appearing in the text has been replaced by **Th. 49**.]

5. THE FIRST THEOREM ON THE BOUNDS OF A FUNCTION

Apart from the law of development, truth-functions satisfy many other laws concerning functions known from the algebra of logic. In particular, I will prove theorems 53, 55, and 56, according to which every truth-function satisfies the following conditions:

(d) $[p].f(p). \equiv .f(Vr).f(Fl)$,

(e) $[\exists p].f(p). \equiv .f(Vr) \vee f(Fl)$,

or, in equivalent formulations:

$$[p] : f(Vr).f(Fl). \supset f(p),$$

$$[p] : f(p) \supset .f(Vr) \vee f(Fl),$$

so, also the condition:

(f) $[p] : f(Vr).f(Fl). \supset f(p) \supset .f(Vr) \vee f(Fl)$.

The above theorems, however, express only necessary conditions for a given function to be a truth-function; it is impossible to prove that these conditions are sufficient. One can even "prove" the independence of these sentences, which express the fact that the above conditions are sufficient, of the axioms of logistic; I will not give any "proofs" here for the reasons already explained in Section 3.

I will demonstrate, however, that if every function (whose argument is a sentence) satisfies condition (d) or (e), hence also (f), then every function is a truth-function, and so the law of substitution holds true. This will enable me to prove in theorems 59, 61, and 62, that the law of substitution is equivalent to any of the following sentences:

(D) $[f] : [p].f(p). \equiv .f(Vr).f(Fl)$,

(E) $[f] : [\exists p].f(p). \equiv .f(Vr) \vee f(Fl)$,

(F) $[p, f] : f(Vr).f(Fl). \supset f(p) \supset .f(Vr) \vee f(Fl)$.*

I will call sentence (F), known from the algebra of logic,[18] the *first theorem on the bounds of a function* – "*first*" because of some theorem from the next section. By way of analogy, the sentence (D) could be called the *first theorem*

* [Ed. Note: The quantifier '$[p]$' originally occurring in formula (F) has been replaced by '$[p, f]$'.]

on the lower bound of a function while sentence (E) – the *first theorem on the upper bound of a function.*[19]

Th. 51. $[f]\therefore[p].f(p).\equiv.f(Vr).f(Fl):\equiv:[p]:f(Vr).f(Fl).\supset f(p)$

Com.

$[f]\therefore$

(1) $[p].f(p).\equiv.f(Vr).f(Fl):\equiv:[p].f(p).\supset.f(Vr).f(Fl):$
$f(Vr).f(Fl).\supset.[p].f(p)\therefore$

(2) $[p].f(p).\supset.f(Vr).f(Fl)\therefore$

(3) $[p].f(p).\equiv.f(Vr).f(Fl):\equiv:f(Vr).f(Fl).\supset.[p].f(p)\therefore$ (1, 2)

$[p].f(p).\equiv.f(Vr).f(Fl):\equiv:[p]:f(Vr).f(Fl).\supset f(p)$ (3)

Th. 52. $[f]\therefore[\exists p].f(p).\equiv.f(Vr)\vee f(Fl):\equiv:[p]:f(p)\supset.f(Vr)\vee f(Fl)$

The proof is similar to that of the previous theorem.

Th. 53. $[f]\therefore\vartheta\rho\{f\}\supset:[p].f(p).\equiv.f(Vr).f(Fl)$ **(Th. 26, Th. 51)**

Th. 54. $[p,f]:\vartheta\rho\{f\}.f(p).\supset.f(Vr)\vee f(Fl)$ **(Th. 31)**

Th. 55. $[f]\therefore\vartheta\rho\{f\}\supset:[\exists p].f(p).\equiv.f(Vr)\vee f(Fl)$ **(Th. 54, Th. 52)**

Th. 56. $[f]\therefore\vartheta\rho\{f\}\supset:[p]:f(Vr).f(Fl).\supset f(p)\supset.f(Vr)\vee f(Fl)$
(Th. 26, Th. 54)

Th. 57. $[p,q,g]::[f]:[r].f(r).\equiv.f(Vr).f(Fl)\therefore g(Vr,Vr).$
$g(Vr,Fl).g(Fl,Vr).g(Fl,Fl)\therefore\supset g(p,q)$

Com.

$[p,q,g]::Hp\therefore\supset:$

(1) $[r].g(Vr,r).\equiv.g(Vr,Vr).g(Vr,Fl):$

(2) $[r].g(Vr,r):$ (1, Hp)

(3) $g(Vr, q):$ (2)

(4) $[r].g(Fl, r). \equiv .g(Fl, Vr).g(Fl, Fl):$ (Hp)

(5) $[r].g(Fl, r):$ (4, Hp)

(6) $g(Fl, q):$ (5)

(7) $[r].g(r, q). \equiv .g(Vr, q).g(Fl, q):$ (Hp)

(8) $[r].g(r, q):$ (7, 3, 6)

Ts (8)

Th. 58. $[f]:[p].f(p). \equiv .f(Vr).f(Fl). \supset .[f].\vartheta\rho\{f\}$

Com.

$Hp :\supset ::$

(1) $[f].\therefore$

(a) $Vr \equiv Vr.f(Vr). \supset f(Vr):$

(b) $\sim (Vr \equiv Fl):$ (**Th. 21, Th. 22**)

(c) $Vr \equiv Fl.f(Vr). \supset f(Fl):$ (b)

(d) $Fl \equiv Vr.f(Fl). \supset f(Vr):$ (b)

(e) $Fl \equiv Fl.f(Fl). \supset f(Fl).\therefore$

(f) $[p, q]: p \equiv q.f(p). \supset f(q) ::$ (**Th. 57**, a, c, d, e)

Ts (**Def. 4**, 1-f)

Th. 59. $Sb \equiv: [f]:[p].f(p). \equiv .f(Vr).f(Fl)$ (**Th. 53, Th. 58, Def. 5**)

Th. 60. $[f]:[p].f(p). \equiv .f(Vr).f(Fl) :\equiv: [f]:[\exists p].f(p).$
$\equiv .f(Vr) \vee f(Fl)$

Com.

(1) $[f]:[p].f(p). \equiv .f(Vr).f(Fl) :\supset \therefore$

(a) $[f]:[p]. \sim (f(p)). \equiv . \sim (f(Vr)). \sim (f(Fl)).\therefore$

(b) $[f] : [\exists p].f(p). \equiv .f(Vr) \vee f(Fl) ::$ (a)

(2) $[f] : [\exists p].f(p). \equiv .f(Vr) \vee f(Fl) :\supset .\therefore$

(c) $[f] : [\exists p]. \sim (f(p)). \equiv . \sim (f(Vr)) \vee \sim (f(Fl)).\therefore$

(d) $[f] : [p].f(p). \equiv .f(Vr).f(Fl) ::$ (c)

$[f] : [p].f(p). \equiv .f(Vr).f(Fl) :\equiv: [f] : [\exists p].f(p). \equiv .$
$f(Vr) \vee f(Fl)$ (1-b, 2-d)

Th. 61. $Sb \equiv: [f] : [\exists p].f(p). \equiv .f(Vr) \vee f(Fl)$ **(Th. 59, Th. 60)**

Th. 62. $Sb \equiv: [p, f] : f(Vr).f(Fl). \supset f(p) \supset .f(Vr) \vee f(Fl)$
(Th. 59, Th. 61, Th. 51, Th. 52)

In addition to the properties of functions which have been investigated in this section, one can investigate more general properties:

(g) $[q] : [p].f(p). \equiv .f(q).f(\sim (q))$,

(h) $[q] : [\exists p].f(p). \equiv .f(q) \vee f(\sim (q))$,

(i) $[p, q] : f(q).f(\sim (q)). \supset f(p) \supset .f(q) \vee f(\sim (q))$.

By applying proof methods similar to those used by us before, we can easily arrive at a number of theorems analogous to theorems 53, 55, 56, 59, 61, and 62, proved in this section. Conditions (g), (h), and (i) constitute, therefore, the necessary conditions for a given function to be a truth-function, and the corresponding sentences (G), (H), and (I), which attribute the above stated conditions to every function, are equivalent to the law of substitution.

6. THE SECOND THEOREM ON THE BOUNDS OF A FUNCTION

With the help of **Th. 47**, I shall now prove (theorems 65, 66, and 67) that every truth-function f possesses the following properties:

(k) $[p].f(p). \equiv f(f(Fl))$,

(l) $[\exists p].f(p). \equiv f(f(Vr))$,

or, in equivalent formulations:

$[p].f(f(Fl)) \supset f(p)$,

$$[p].f(p) \supset f(f(Vr)),$$

and, hence, the property:

(m) $[p].f(f(Fl)) \supset f(p) \supset f(f(Vr))$.

The above conditions, like conditions (d), (e), and (f) of Section 5, are only necessary for a function f to be a truth-function. Furthermore, in contrast to considerations of the preceding section, I cannot even prove that the law of substitution is equivalent to any of the following sentences, which express the fact that every function satisfies condition (k) or (l):

(K) $[f] : [p].f(p). \equiv f(f(Fl))$,

(L) $[f] : [\exists p].f(p). \equiv f(f(Vr))$.

However, we shall see from theorems 75 and 76 that the law of substitution is equivalent to the logical product of sentences (K) and (L), and, hence, to the sentence:

(M) $[p, f].f(f(Fl)) \supset f(p) \supset f(f(Vr))$.

Sentence (M) is known in the algebra of logic[20]; I shall call it the *second theorem on the bounds of a function*. By analogy, sentences (K) and (L) will be given the names: the *second theorem on the lower bound of a function* and the *second theorem on the upper bound of a function*.

Th. 63. $[f].\cdot.[p].f(p). \equiv f(f(Fl)) :\equiv .[p].f(f(Fl)) \supset f(p)$

Th. 64. $[f].\cdot.[\exists p].f(p). \equiv f(f(Vr)) :\equiv .[p].f(p) \supset f(f(Vr))$

I omit the proofs of these theorems as they are analogous to the proof of **Th. 51**.

Th. 65. $[f].\cdot.\vartheta\rho\{f\} \supset: [p].f(p). \equiv f(f(Fl))$

Com.

$[f]::Hp \supset::$

(1) $[p].\cdot.f(p) \equiv: f(Vr).p. \vee .f(Fl). \sim (p) ::$ (**Th. 47**)

(2) $f(f(Fl)) \equiv: f(Vr).f(Fl). \vee .f(Fl). \sim (f(Fl)).\cdot.$ (1)

(3) $f(f(Fl)) \equiv .f(Vr).f(Fl) :$ (2)

Ts **(Th. 53**, Hp, 3)

Th. 66. $[f] .\!\cdot\!. \vartheta\rho\{f\} \supset: [\exists p].f(p). \equiv f(f(Vr))$ **(Th. 47, Th. 55)**

The proof is similar to that of the preceding theorem.

Th. 67. $[f] : \vartheta\rho\{f\} \supset .[p].f(f(Fl)) \supset f(p) \supset f(f(Vr))$

(Th. 65, Th. 66)*

Th. 68. $[f] : [p].f(p). \equiv f(f(Fl)) :\supset: [p,f] :\sim (p) \supset .f(p) \equiv f(Fl)$

Com.

$Hp :\supset:::$

(1) $[p,f] ::$

(a) $[\exists g] ::$

(α) $[r] .\!\cdot\!. g(r) \equiv :\sim (r). \supset .f(r) \equiv f(Fl) ::$[21]

(β) $g(Fl) \equiv :\sim (Fl) \supset .f(Fl) \equiv f(Fl) .\!\cdot\!.$ (α)

(γ) $\sim (Fl) \supset .f(Fl) \equiv f(Fl) :$

(δ) $g(Fl) .\!\cdot\!.$ (β, γ)

(ε) $g(g(Fl)) \equiv :\sim (g(Fl)) \supset .f(g(Fl)) \equiv f(Fl) .\!\cdot\!.$ (α)

(ζ) $\sim (g(Fl)) \supset .f(g(Fl)) \equiv f(Fl) :$ (δ)

(η) $g(g(Fl)) :$ (ε, ζ)

(ϑ) $[r].g(r) :$ (Hp, η)

(ι) $g(p) ::$ (ϑ)

(b) $\sim (p) \supset .f(p) \equiv f(Fl) :::$ (a-α, ι)

Ts (1-b)

Th. 69. $[f] : [\exists p].f(p). \equiv f(f(Vr)) :\supset: [f] : [p].f(p). \equiv f(\sim (f(Vr)))$

Com.

$Hp :\supset .\!\cdot\!.$

(1) $[f] : [\exists p]. \sim (f(p)). \equiv \sim (f(\sim (f(Vr)))) .\!\cdot\!.$

Ts (1)

* [Ed. Note: In the Polish original, the references to theorems 65 and 66 are absent.]

Th. 70. $[f] : [p].f(p). \equiv f(\sim (f(Vr))) :\supset: [p, f] : p \supset .f(p) \equiv f(Vr)$

The proof, analogous to that of **Th. 68**, does not present any difficulties.

Th. 71. $[f] : [\exists p].f(p). \equiv f(f(Vr)) :\supset: [p, f] : p \supset .f(p) \equiv f(Vr)$
(**Th. 69, Th. 70**)

Th. 72. $[f] : [p].f(p). \equiv f(f(Fl)) :\supset: [p, q, f] :\sim (p). \sim (q). \supset .$
$f(p) \equiv f(q)$

Com.

$Hp :\supset .\cdot.$

(1) $[p, q, f] :\sim (p). \sim (q). \supset .$

(a) $\sim (p). \sim (q).$

(b) $f(p) \equiv f(Fl).$ (**Th. 68**, Hp, a)

(c) $f(q) \equiv f(Fl).$ (**Th. 68**, Hp, a)

(d) $f(p) \equiv f(q).\cdot.$ (b, c)

Ts (1-d)

Th. 73. $[f] : [\exists p].f(p). \equiv f(f(Vr)) :\supset: [p, q, f] : p.q. \supset .f(p) \equiv f(q)$
(**Th. 71**)

The proof is quite similar to that of the preceding theorem. **Th. 72** demonstrates that the law of substitution for false sentences follows from sentence (K), while **Th. 73** shows that the law of substitution for true sentences follows from sentence (L).

Th. 74. $[f] : [p].f(p). \equiv f(f(Fl)) : [\exists p].f(p). \equiv f(f(Vr)) :\supset Sb$
(**Th. 72, Th. 73, Th. 15**)

Th. 75. $Sb \equiv: [f] : [p].f(p). \equiv f(f(Fl)) : [\exists p].f(p). \equiv f(f(Vr))$
(**Th. 65, Th. 66, Def. 5, Th. 74**)

Th. 76. $Sb \equiv .[p, f].f(f(Fl)) \supset f(p) \supset f(f(Vr))$
(**Th. 75, Th. 63, Th. 64**)

By putting together the main results obtained in Sections 3–6, we can conclude that the following sentences are equivalent:

(A) $[p, q, f] : p \equiv q.f(p). \supset f(q)$ (the law of substitution),

(B) $[f] : [p].f(p) \equiv vr(p). \vee .[p].f(p) \equiv as(p). \vee .[p].f(p) \equiv$
$\sim (p). \vee .[p].f(p) \equiv fl(p)$ (the law of the number of functions),

(C) $[p, f] .\cdot. f(p) \equiv: f(Vr).p. \vee .f(Fl). \sim (p)$ (the law of development),

(D) $[f] : [p].f(p). \equiv .f(Vr).f(Fl)$
(the first theorem on the lower bound of a function),

(E) $[f] : [\exists p].f(p). \equiv .f(Vr) \vee f(Fl)$
(the first theorem on the upper bound of a function),

(M) $[p, f].f(f(Fl)) \supset f(p) \supset f(f(Vr))$
(the second theorem on the bounds of a function).

ANNEX

To conclude the present work, I would like to briefly present the results, analogous to those presented in Sections 3–6, which were obtained when investigating functions of more than one argument or functions whose arguments are not sentences.

I. Functions of Several Arguments

One can state and prove a number of theorems, similar in their intuitive content to those which I presented in the preceding sections, for functions of more than one argument; furthermore, the proofs do not differ in any significant way.

For example, if we limit ourselves to functions of two arguments and if, by the analogy to **Def. 4**, we shall call such a function f a truth-function, provided that it satisfies the condition:

(a_1) $[p, q, r, s] : p \equiv q.r \equiv s.f(p, r). \supset f(q, s)$,

then we can easily prove that the following conditions are the necessary conditions for a given function f to be a truth-function:

(b_1) $[p, q].f(p, q). \vee :$
$[p, q] : f(p, q) \equiv .p \vee q : \vee : [p, q] : f(p, q) \equiv . \sim (p) \vee q : \vee : [p, q] :$

$f(p,q) \equiv .p\vee \sim (q) : \vee : [p,q] : f(p,q) \equiv . \sim (p)\vee \sim (q) : \vee .$
$[p,q].f(p,q) \equiv p. \vee .[p,q].f(p,q) \equiv q.\vee : [p,q] : f(p,q) \equiv .p \equiv q : \vee : [p,q] : f(p,q) \equiv .p \equiv\sim (q) : \vee : [p,q].f(p,q) \equiv\sim (q). \vee .$
$[p,q].f(p,q) \equiv\sim (p).\vee : [p,q] : f(p,q) \equiv .p.q : \vee : [p,q] : f(p,q) \equiv . \sim (p).q : \vee : [p,q] : f(p,q) \equiv .p. \sim (q) : \vee : [p,q] : f(p,q) \equiv . \sim (p). \sim (q) : \vee .$
$[p,q]. \sim (f(p,q)),$

(c_1) $[p,q].\cdot.f(p,q) \equiv: f(Vr,Vr).p.q. \vee .f(Vr,Fl).p. \sim (q). \vee . f(Fl,Vr). \sim (p).q. \vee .f(Fl,Fl). \sim (p). \sim (q),$

(d_1) $[p,q].f(p,q). \equiv .f(Vr,Vr).f(Vr,Fl).f(Fl,Vr).f(Fl,Fl),$

(e_1) $[\exists p,q].f(p,q). \equiv .f(Vr,Vr) \vee f(Vr,Fl) \vee f(Fl,Vr) \vee f(Fl,Fl).$

Conditions (a_1)–(e_1) are the counterparts of conditions (a)–(e) listed above. I am not stating the counterparts of conditions (k)–(m), since they would be quite complicated, and I do not find them interesting.

Conditions (d_1) and (e_1) are only the necessary conditions for a function f to be a truth-function; the remaining conditions are both necessary and sufficient. Sentences (A_1)–(E_1), stating that every function f satisfies conditions (a_1)–(e_1), are equivalent; we could name them in the way analogous to that of sentences (A)–(E), that is: the *law of substitution (for functions of two arguments)*, the *law of the number of functions of two arguments*,[22] etc.

It is worth noting that sentences (A_1)–(E_1) are equivalent not only among themselves but also equivalent to the above stated sentences (A)–(E). In particular, it is easy to prove the equivalence of sentences (D) and (D_1); **Th. 57** may serve as one of the lemmas in the proof of such a theorem.

In fact, without leaving the domain of functions of two arguments, it is not difficult to construct yet another series of sentences, equivalent to sentence (A) – the law of substitution – and, as far as their content is concerned, to some extent intermediate between the sentences of the preceding sections and those which have just been mentioned.

Here are some examples:

(A′) $[p,q,r,f] : p \equiv q.f(p,r). \supset f(q,r),$

(D′) $[p,f] : [q].f(p,q). \equiv .f(p,Vr).f(p,Fl).$

II. Functions, Whose Arguments Are Not Sentences

Many issues emerge, analogous to those resolved above, in the study of functions whose arguments are not only sentences but also logical functions.

Let us restrict ourselves to the consideration of only one particular case: to functions of one argument whose arguments are not sentences but those functions whose properties we studied in Sections 3–6.

By analogy with the terminology of Russell and Whitehead,[23] we shall call such a function ϕ *extensional* if it satisfies the condition:

$$(\mathrm{a}_2)\ [f, g] : = \{f, g\}.\phi\{f\}. \supset \phi\{g\}.$$

By **Th. 38**, $\vartheta\rho$ is an example of such a function. The sentence:

$$(\mathrm{A}_2)\ [f, g, \phi] : = \{f, g\}.\phi\{f\}. \supset \phi\{g\},$$

as it seems, cannot be proved on the ground of any of the known systems of logistic, even after the addition of the law of substitution.

It is very easy to give a series of conditions, either only necessary or necessary and sufficient, for a given function ϕ to be extensional – conditions, which correspond to conditions (b)–(f) from the previous sections; it is equally easy to formulate sentences, equivalent to (A_2), which correspond to sentences (B)–(F). The proofs of the corresponding theorems, however, require the acceptance of the sentence (A) – the law of substitution – as a hypothesis; they do not differ significantly from the proofs of Sections 3–6, but the role analogous to that of **Th. 25** is played now by the law of the number of functions:

$$[f]. = \{f, vr\} \vee = \{f, as\} \vee = \{f, \sim\} \vee = \{f, fl\},$$

which follows, as we know, from the accepted hypothesis.

I shall give here the main sentences equivalent to sentence (A_2):

$$(\mathrm{C}_2)\ [f, \phi].\because.\phi\{f\} \equiv: \phi\{vr\}. = \{f, vr\}. \vee .\phi\{as\}. \\ = \{f, as\}. \vee .\phi\{\sim\}. = \{f, \sim\}. \vee .\phi\{fl\}. = \{f, fl\},$$

$$(\mathrm{D}_2)\ [\phi] : [f].\phi\{f\}. \equiv .\phi\{vr\}.\phi\{as\}.\phi\{\sim\}.\phi\{fl\},$$

$$(\mathrm{E}_2)\ [\phi] : [\exists f].\phi\{f\}. \equiv .\phi\{vr\} \vee \phi\{as\} \vee \phi\{\sim\} \vee \phi\{fl\},$$

$$(\mathrm{N})\ [\phi].\because.[\exists g] : [f].\phi\{f\} \equiv g(f(Vr), f(Fl)).$$

Sentence (N) does not have an exact counterpart in the preceding series of sentences. From this sentence one can easily derive the sentence (B_2) – the law of the number of functions, which I am not stating here for technical reasons, as it must contain sixteen logical components, and as its formulation would be very laborious without the help of auxiliary definitions.

NOTES

[1] Volume IV of the journal *Fundamenta Mathematicae* (Warszawa, 1922) contains my paper 'Sur le terme primitif de la Logistique', the content of which covers, more or less, Sections 1–2 of the present paper; the subsequent sections of this paper (Sections 3–6) correspond roughly to the paper 'Sur les truth-functions au sens de MM. Russell et Whitehead', which will appear in volume V of the journal mentioned above.*

[2] The possibility of constructing different theories of logical types was already anticipated by the inventor of the first of them – Russell. Cf. A.N. Whitehead and B. Russell, *Principia Mathematica*, Cambridge 1910, Vol. I, p. vii.

[3] One way in which Leśniewski's theory of types affected the lay-out of this work is that for functions, whose arguments are not sentences, I am using distinct parentheses. Cf. **Def. 4** in Section 2 and **Def. 6** in Section 3.

[4] I am using the term 'quantifier' in the sense of Peirce ([cf.] 'On the Algebra of Logic', *American Journal of Mathematics* VII, 1885, p. 197), who denotes with this term the symbols 'Π' (universal quantifier) and 'Σ' (particular quantifier), representing abbreviations of the expressions: 'for every signification of terms ...' and 'for some signification of terms ...'.**

[5] See B. Russell, *The Principles of Mathematics*, Cambridge, 1903, pp. 16–18.

[6] See H.M. Sheffer, 'A set of five postulates for Boolean algebras with application to logical constants', *Transactions of the American Mathematical Society* 14, 1913. See also B. Russell, *Introduction to Mathematical Philosophy*, London–New York, 1920, p. 144 and following.

[7] In his work, written approximately at the same time as the work of Sheffer mentioned above but independently from it, Łukasiewicz shows that, in addition to the signs of implication and rejection, other logistic terms possess the required property: each of them can be adopted in the system of logistic as the sole primitive term. The sign of equivalence is not among them.

[8] Russell and Whitehead chose the opposite approach in the work cited above: by giving definitions the form '$A = B\ Df$', they use a symbol which occurs neither in axioms nor theorems of their system.

[9] In this work I adopt the symbolism of Russell and Whitehead with some minor modifications; in particular, instead of expressions of the form 'ϕx' I write: '$\phi(x)$'.

[10] I am using the symbols 'Vr' and 'Fl' instead of '1' and '0' – the terms occurring, for instance, in E. Schröder's *Vorlesungen über die Algebra der Logik*, I Band, Leipzig 1890, p. 188. See also L. Couturat, [... *L'algèbre de la logique*, Paris, 1905, p. 18.]

[11] In the proofs of theorems which have the form of a conditional sentence, the term '*Hp*' replaces the antecedent (hypothesis) while the term '*Ts*' the consequent (thesis) of a theorem.‡

[12] To pass from (4) to (5), I am using the following theorem communicated to me by Łukasiewicz:

$$[p, q, r] .\because . p \equiv .q \equiv r :\equiv: p \equiv q. \equiv r.$$

I will not give a proof, quite easy in fact, of this theorem which expresses an interesting property of equivalence (this theorem is analogous to the laws of associativity for logical addition and multiplication).

[13] A.N. Whitehead and B. Russell, op. cit., pp. 120–121. The Polish term *funkcja prawdziwościowa* was used for the first time by E. Żyliński in his lecture (*O przedstawialności funkcyj prawdziwościowych jednych przez drugie*) delivered during the first Polish Philosophical Congress in Lwów, May [... 1923].

[14] Cf. A.N. Whitehead and B. Russell, op. cit., pp. 120–121.

---

* [Ed. Note: Tarski's paper 'Sur le terme primitif de la Logistique' appeared in 1923.]

** [Ed. Note: Peirce gives the interpretation of the expressions '$\Pi_i x_i$' and '$\Sigma_i x_i$' on pp. 194–195 of his paper.]

‡ [Ed. Note: In addition, the term 'Com' is the abbreviation of the word 'Commentary'.]

15 These results of Leśniewski have not yet been published.
16 Cf. A.N. Whitehead and B. Russell, op. cit., pp. 98–101.
17 Cf. L. Couturat, op. cit., p. 31; E. Schröder, op. cit., p. 409, 44_+.
18 Cf. L. Couturat, op. cit., p. 38; E. Schröder, op. cit., p. 427, 48_+.
19 Łukasiewicz called attention to the fact that the attribution of properties (d) and (e) to every function, or analogous properties to functions of several arguments, and, therefore, the acceptance of sentences (D) and (E) as true, allows considerable simplifications during the construction of the system of logistic. Cf. J. Łukasiewicz, 'Logika dwuwartościowa', Księga pamiątkowa ku uczczeniu Kazimierza Twardowskiego, Lwów 1921, p. 199.*
20 Cf. L. Couturat, op. cit., p. 39 (Remarque).**
21 The auxiliary definition, which I am using in the present proof, may seem superfluous. I have adopted this approach, however, for a better clarity of the proof.
22 In his work, mentioned by me in Note 7, Łukasiewicz proves the laws of the number of functions – sentences (B) and (B_1) – basing his arguments on some assumptions of intuitive nature and concerning functions. Symbolic counterparts of these assumptions could be, as it seems, the sentences:

$$\text{(P)}\quad [p, f] : f(p) \equiv f(Vr) . \vee . f(p) \equiv f(Fl),$$

$$\text{(P}_1\text{)}\quad [p, q, f] : f(p,q) \equiv f(Vr, Vr) . \vee . f(p,q) \equiv f(Vr, Fl) . \vee . f(p,q) \equiv f(Fl, Vr) . \vee . f(p,q) \equiv f(Fl, Fl),$$

both of which are equivalent, as it can be easily verified, to the law of substitution.
23 Cf. A.N. Whitehead and B. Russell, op. cit., p. 22.

---

* [Ed. Note: Cf. J. Łukasiewicz, 'Two-Valued Logic', *Przegląd Filozoficzny* XXIII (1921), pp. 189–205.]
** [Ed. Note: Tarski refers here to Remarque 1°.]

BOLESŁAW SOBOCIŃSKI

AN INVESTIGATION OF PROTOTHETIC *

INTRODUCTION

This paper was already published in Polish, in 1939, as an offprint from the first volume of *Collectanea Logica*. I am publishing it for the second time, in English, for the following reasons.

In 1937, in Poland, at the suggestion of Jan Łukasiewicz, we founded *Collectanea Logica*, a periodical devoted to logic, its history, and its applications. It was to be issued as one large volume each year, and would be international in character, containing papers written in Polish, English, French, German, Italian, and Latin. J. Łukasiewicz was the editor of *Collectanea Logica*, and I was its managing editor. The organization of the periodical and the technical difficulties caused by the typesetting of complicated logical symbolism were the reasons why the printing of the first volume, begun in 1938, advanced rather slowly. Nevertheless, on the first of September 1939, the first part of the volume, which would have had 500 pages, was printed; the second part was already collected and in proof. Moreover, the first five papers from the first part were already published as offprints. At the siege of Warsaw, in September 1939, the printing-house of the periodical was completely burned down, together with all the prepared type-blocks and offprints. All the final proofs of the first volume, most of the prepared offprints, and the archives of the periodical escaped in my flat. All was destroyed in August 1944 during the Warsaw Uprising. In that way, the desire of Polish logicians to create an important international periodical devoted to logic and related sciences, comparable with *Fundamenta Mathematicae* (a well-known Polish mathematical publication), could not be realized. Even if the type-blocks for the first volume could be reproduced, it would be impossible to realize this plan at the present time, because of the political situation in Poland, and the dispersion of the periodical's organizers caused by the war. It will be best, perhaps, to give a short description of what would have been published there, as most

* Bibliographical Note. This paper appeared in print in *Cahiers de l'Institut d'Études Polonaises en Belgique* 5, 1–44. Bruxelles, 1949.

of these works and their results have not yet appeared elsewhere; moreover, even these papers which escaped as offprints are not generally available.

Volume I of *Collectanea Logica* contained the following papers:

I. Stanisław Leśniewski, Einleitende Bemerkungen zur Fortsetzung meiner Mitteilung u.d.T. "Grundzüge eines neuen Systems der Grundlagen der Mathematik" [, pp. 1–60].
II. Stanisław Leśniewski, Grundzüge eines neuen Systems der Grundlagen der Mathematik, §12 [, pp. 61–144].
III. Jan Łukasiewicz, Der Äquivalenzenkalkül [, pp. 145–169].
IV. Bolesław Sobociński, Z badań nad prototetyką (An Investigation of Protothetic) [, 171–177].
V. Bolesław Sobociński, Aksjomatyzacja konjunkcyjno-negacyjnej teorii dedukcji (Axiomatization of a Conjunctive–Negative Calculus of Propositions) [, 179–193].
VI. Inocenty M. Bocheński, La logique de Théophraste [, 195–304].
VII. Bolesław Sobociński, O aksjomatykach prototetyki (On the Axiom-Systems of Protothetic).
VIII. Stanisław Leśniewski, O pewnym jedynym aksjomacie prototetyki (On a Certain Single Axiom of Protothetic).
IX. Bolesław Sobociński, O różnych systemach prototetyki (On Different Systems of Protothetic).
X. Jerzy Słupecki, Przyczynek do prototetyki (A Contribution to Protothetic).
XI. Bolesław Sobociński, Uwagi w związku z pracą p. J. Słupeckiego: "Przyczynek do prototetyki" (Remarks Concerning the Paper of J. Słupecki: "A Contribution to Protothetic").*

Of these eleven papers, the first five were published as offprints,[1] and only a few of I and II (published in 1938) were sent to some people and institutions by St. Leśniewski shortly before his death. Offprints of the papers III, IV, and V were published just before the war, and could be distributed by their authors only to very few people. Most of the copies of I–V were destroyed. Luckily, one of the proofs of paper VI escaped the destruction, so that its author could reconstruct and publish it in 1947.[2] The manuscripts and proofs of VII–XI were completely destroyed.

Of the papers mentioned, only that of I. Bocheński was concerned with the history of Greek logic. I will not give any account of it here, since it has been

* [Ed. Note: English translations of papers I, III, and IV appeared in McCall (1967); II was reviewed by Quine in the *Journal of Symbolic Logic* 5 (1940), p. 84; V appeared in English in the *Journal of Computing Systems* 1 (1954), 229–242; VI appeared in *Collectanea Friburgensia*, Nouvelle Série Fasc. 32, Fribourg en Suisse, 1947. Most of the results of papers VII, IX, and XI are probably included in Sobociński (1954) and Sobociński (1960) in the Protothetic Bibliography.]

published in the meantime. Papers III and V were devoted to the calculus of propositions.

In his paper III,[3] J. Łukasiewicz provided a historical account of the axiomatization of the equivalential calculus of propositions,[4] and proved that each of the eleven-letter theses:

(1) $EEpqEErqEpr$[5]

(2) $EEpqEEprErq$

(3) $EEpqEErpEqr$

can be taken as the sole axiom of this theory. He also proved that no other eleven-letter thesis and no shorter one has the property in question*; in other words, each of the theses mentioned above can be considered as a shortest single axiom of the theory.[6]

In my paper V,[7] I showed that the complete system of the calculus of propositions can be based on the primitive terms 'conjunction' and 'negation', i.e. 'K' and 'N', and on two rules:

(a) the usual rule of substitution of the calculus of propositions,

and

(b) the rule of detachment adjusted to the above primitive terms:

> If we have in the system two theses of the forms '$NK\alpha N\beta$' and 'α', then we can add to the system the thesis of the form 'β'.[8]

Finally, we have four independent axioms of this theory, viz.:

(1) $NKNKNprKNKNqrKNKpqr$

(2) $NKNpKpq$

(3) $NKNpKqKrp$

(4) $NKNKpqNNKpNNq$.

---

* [Ed. Note: This remark by Sobociński is not exactly correct. While Łukasiewicz indeed proved that any single axiom for the equivalential calculus must consist of at least eleven symbols, he did not claim, however, that the three axioms discovered by him are the only eleven-letter single axioms for the calculus. In fact, there are other such axioms; for more discussion see Kalman, J.A., 'A Shortest Single Axiom for the Classical Equivalential Calculus', *Notre Dame Journal of Formal Logic* XIX (1978), pp. 141–144.]

This system is worth noting since all of its theses are valid in Heyting's intuitionistic logic and because of certain connections between this system and the Boolean algebra of solids.[9,10]

All the other papers were concerned with protothetic. Protothetic is a deductive theory constructed by the late Stanisław Leśniewski.[11] As we know, he based the whole system of contemporary mathematics on three deductive theories, protothetic, ontology, and mereology, which he conceived and constructed.[12] [. . .]*

I will neither describe the characteristic features of these theories nor comment on the theoretical basis of St. Leśniewski's system. I will only state that:

1. The precision and the conciseness of the formalization, symbolism, and the formulation of the rules of procedure for the above theories are unparalleled among the known deductive systems.
2. The principles on which these theories are based differ in many respects from those that were usually accepted before the discovery of the Russell Antinomy. They allow, nevertheless, to formulate and prove all the theorems of classical logic.
3. There are straightforward proofs that these theories and the entire system are consistent, and, consequently, that no known logical antinomy can be reconstructed in them.[13]

St. Leśniewski presents the main features of his system of the foundations of mathematics in his works: Leśniewski (1927, 1929) and (1930).[14] He gives there the theoretical foundations of his three deductive theories, the history of their origin, the description of their symbolism, first axiom-systems, and concise formulation of rules of procedure of protothetic and ontology.[15] These rules are expressed in the symbolism of a metasystem. In the treatise Leśniewski (1927), a large number of theorems are derived from the axioms of mereology. However, St. Leśniewski gives no systematic proofs of theorems valid in protothetic or ontology[16]; he presents neither the proof of completeness of protothetic nor the consistency proofs for any of these theories or his entire system.

St. Leśniewski's papers I and II, which were published as offprints from volume I of *Collectanea Logica*, are the continuations of Leśniewski (1929). The first of them contains supplements and explanations of subjects described in Leśniewski (1929), an account of the construction of the so-called computative system of protothetic, and the history of the successive simplifications of the axiom-system of protothetic based on 'equivalence' as the only primitive term. This historical account ends in the year 1926, when St. Leśniewski proved that the thesis:

---

* [Ed. Note: The following four paragraphs of the rough description of protothetic, ontology, and mereology are omitted here.]

(A) $[fpqrst]{::}p \equiv q. \equiv{::} [g]{.\cdot.}f(pf(p[u].u)). \equiv {.\cdot.}[u].f(qu). \equiv: g(r \equiv s. \equiv tq). \equiv .g(s \equiv t. \equiv rp)$[17]

can be taken as the sole axiom of this system.

In the second paper, St. Leśniewski showed how all the theses of the calculus of propositions can be deduced from the original axioms ⌊of protothetic⌋ with the help of the concisely formulated rules of procedure.[18]

These original axioms are:

I. $[pqr]{.\cdot.}p \equiv r. \equiv .q \equiv p :\equiv .r \equiv q$

II. $[pqr]{.\cdot.}p \equiv .q \equiv r :\equiv: p \equiv q. \equiv r$

III. $[gp]{::}[f] :: g(pp). \equiv {.\cdot.}[r] : f(rr). \equiv .g(pp) :\equiv: [r] : f(rr). \equiv . g(p \equiv .[q].qp) ::\equiv .[q].g(qp).$

My papers IV, VII, and IX presented the results of several years of research on protothetic. In the first one, Sobociński (1939a), I gave a number of interesting theses of protothetic which were unknown at that time (the full translation of the paper is given in this publication). The subject-matter of this paper is related to papers Tarski (1923a, 1923b, 1924), in which it is proved that conjunction can be defined in terms of equivalence. To understand the theorems given in my treatise, and appreciate their significance, it is not necessary to be acquainted with protothetic. It is enough to be familiar with elementary calculus of propositions, properties of the universal quantifiers, and variable functors of propositional arguments.

My two large papers (VII and IX) were concerned with subjects closely related to the inner structure of protothetic. I established a number of connections and dependencies between different theses in this hitherto little investigated theory.[19] In the first of them, I gave an account of my successive results concerning the simplification of the axiom of protothetic, based on the primitive term 'equivalence', results which I obtained in the years 1937–1938. My next important result concerning the structure of protothetic was closely related to the above treatises as well as to the following theorem established by St. Leśniewski:

an axiom-system of protothetic yields a complete system, when, using its rules of procedure, one can prove:

(1) all theses of the complete equivalential calculus of propositions,

(2) the following laws of the calculus of propositions (the so-called four laws of the logical product):

(a) $1.1 \equiv 1$[20]

(b) $0.0 \equiv 0$

(c) $0.1 \equiv 0$

(d) $1.0 \equiv 0$,

(3) that if $F(p)$ is a meaningful formula in this system and if the formulas $F(0)$ and $F(1)$ can be proved in the system, then one can add to the system, as its thesis, the formula $[p].F(p)$ (the so-called law of bivalence).[21]

In 1938, in connection with my inquires regarding the axiomatic basis of protothetic, I proved that the condition (2) of St. Leśniewski's theorem is superfluous because it results from the conditions (1) and (3). Moreover, in the condition (1), it is possible to replace the theory of equivalential calculus of propositions by some of its fragments.

These inquiries resulted in the proof that the thesis:

(B) $[fpq]\therefore\because f(pf(p[u].u)).\equiv{::}\,[r]::f(p\equiv q.\equiv qr).\equiv\therefore p\equiv q.\equiv:$
$r\equiv q.\equiv .p\equiv r,$[22]

much shorter than the above mentioned thesis (A), can be taken as the sole axiom of protothetic.

In the second paper (IX), I presented the solutions to a number of problems posed by St. Leśniewski or by myself, regarding the structure of different systems of protothetic. All these systems are equivalent to the one generally used*; they differ only as far as the primitive terms or the rules of procedure are concerned.[23]

In 1938, St. Leśniewski, starting from the results presented in my paper VII, obtained a new and shorter single axiom of protothetic, based on the term 'equivalence', showing that it can be the thesis:

(C) $[pq]::p\equiv q.\equiv{::}[f]\therefore f(qf(q[u].u)).\equiv{::}\,[r]::f(pr).\equiv\therefore r\equiv:$
$q\equiv .r\equiv p.$[24]

Mortal illness prevented him from writing a paper on this subject. This result, which I edited, was to be published as paper VIII.

The short notes (X and XI) by Dr. Jerzy Słupecki and myself would have contained a certain contribution to the investigations reported in paper VII, and a certain suggestion of mine based on it.

The study of protothetic advanced considerably after 1939. For instance, at the end of 1945, using rather complicated and difficult deductions, I proved that the following thesis:

(D) $[pq]::p\equiv q.\equiv\therefore[f]\therefore f(pf(p[u].u)).\equiv:[r]:f(qr).\equiv .q\equiv p$

can be taken as the sole axiom of protothetic, based on the primitive term 'equivalence'. It is the shortest known axiom of this theory and probably it

* [Ed. Note: Sobociński presumably refers to the system $\mathfrak{S}_5$.]

is the shortest of all such axioms.[25] This result is closely connected with my deductions which were to be published in papers VII and IX.

AN INVESTIGATION OF PROTOTHETIC*

While investigating various problems of protothetic, I observed that a number of theorems which, as far as I could ascertain, remained unknown at that time (December 1935), were theses of protothetic.[26] As this finding is closely associated with the theorem of Dr. Alfred Tarski concerning the definability, in protothetic, of conjunction in terms of equivalence,[27] I have decided to publish the theses discovered by myself.

Editorial and space-saving considerations compel me to use a slightly modified form of the symbolism of Peano and Russell instead of the original symbolism of protothetic. This notation should be easy to follow for anyone familiar with the usual symbolism of Peano and Russell and with the contributions Leśniewski (1929, 1930), Sobociński (1934), and Tarski (1923a, 1923b, 1924). However, to avoid any misunderstanding, all the more important theorems which occur in this contribution will be given in endnotes in the original symbolism of protothetic.[28] For the reasons already mentioned I have also decided to present the discovered theorems without proofs, for I assume that every reader familiar with elementary logic can easily prove them by the truth table method.[29]

As we know, Alfred Tarski proved the following two theses of protothetic:

$$A.\ [pq] :: p.q. \equiv .\therefore [f]. \therefore p \equiv : [r].p \equiv f(r). \equiv .[r].q \equiv f(r)$$

$$B.\ [pq]. \therefore p.q. \equiv : [f] : p \equiv .f(p) \equiv f(q),$$ [30]

each of which can be accepted in protothetic as the definition of the connective '.' by means of the connective '≡'. In their efforts to establish and simplify the axiom system of protothetic, Professor Stanisław Leśniewski and Dr. Mordchaj Wajsberg discovered numerous modifications of theses A and B; but these modifications, which, with one exception, remained unpublished, keep the characteristic features of Tarski's theorem.[31] On the other hand, although the theorems I have discovered are based on Tarski's idea that in protothetic conjunction can be defined by means of equivalence, the universal quantifier, and variable propositional functors, they differ significantly from theorems A and B as well as from all their known modifications. The theorems discovered by myself are as follows:

---

* Bibliographical Note. This part of the paper is reprinted from *Polish Logic 1920–1939*, McCall, S., editor. Clarendon Press, Oxford, 1967, by permission of Oxford University Press.

$C1.$ $[pq] : p.q. \equiv .[f].f(pq) \equiv f(q1)$[32]

$C2.$ $[pq] : p.q. \equiv .[f].f(pq) \equiv f(1p)$

$C3.$ $[pq] : p.q. \equiv .[f].f(qp) \equiv f(p1)$

$C4.$ $[pq] : p.q. \equiv .[f].f(qp) \equiv f(1q)$

$C5.$ $[pq] : p.q. \equiv .[f].f(pp) \equiv f(q1)$

$C6.$ $[pq] : p.q. \equiv .[f].f(pp) \equiv f(1q)$

$C7.$ $[pq] : p.q. \equiv .[f].f(qq) \equiv f(p1)$

$C8.$ $[pq] : p.q. \equiv .[f].f(qq) \equiv f(1p)$

$C9.$ $[pq] : p.q. \equiv .[f].f(pq) \equiv f(11)$[33]

$C10.$ $[pq] : p.q. \equiv .[f].f(qp) \equiv f(11)$

$C11.$ $[pq] : p.q. \equiv .[f].f(1p) \equiv f(q1)$[34]

$C12.$ $[pq] : p.q. \equiv .[f].f(1q) \equiv f(p1).$

It should be noted that Theorems $C1$–$C12$ can be written in a modified form by substituting everywhere for the sign '1' one of the expressions: '$p \equiv p$', '$q \equiv q$', '$p \equiv q$', or '$q \equiv p$'. It can be seen that the fundamental difference between Theorems A and B on the one hand and Theorems $C1$–$C12$ on the other is twofold. First, in Theorems $C1$–$C12$ the variable functor 'f' is a propositional functor of two arguments instead of one. Second, while in each of Theorems $C1$–$C12$ the expression in the scope of the inside quantifier is an equivalence, neither of whose arguments is an equivalence, in Tarski's theorems one of the arguments of the equivalence occurring after the inside quantifier is also an equivalence.

Having discovered Theorems $C1$–$C12$, I noted that in protothetic we are able to construct in a similar fashion theorems which within this system can be accepted as the definitions of three other propositional functors of two arguments, namely the functors defined by means of the expressions '$\sim (p \vee q)$', '$\sim (p \supset q)$', and '$\sim (p \vee \sim q)$'.

If the symbolic expression '$p \mathbin{\text{ó}} q$', equivalent to '$\sim (p \vee q)$', is introduced, so that

$$[pq] : p \mathbin{\text{ó}} q. \equiv . \sim (p \vee q)^{35}$$

becomes a valid theorem, the following theses of protothetic can be established:

$D1.\quad [pq] : p \mathbin{\acute{\circ}} q. \equiv .[f].f(pq) \equiv f(q0)$[36]

$D2.\quad [pq] : p \mathbin{\acute{\circ}} q. \equiv .[f].f(0q) \equiv f(p0)$

$D3.\quad [pq] : p \mathbin{\acute{\circ}} q. \equiv .[f].f(0p) \equiv f(q0)$

$D4.\quad [pq] : p \mathbin{\acute{\circ}} q. \equiv .[f].f(qp) \equiv f(00)$

$D5.\quad [pq] : p \mathbin{\acute{\circ}} q. \equiv .[f].f(pq) \equiv f(00)$

$D6.\quad [pq] : p \mathbin{\acute{\circ}} q. \equiv .[f].f(qq) \equiv f(0p)$

$D7.\quad [pq] : p \mathbin{\acute{\circ}} q. \equiv .[f].f(qq) \equiv f(p0)$

$D8.\quad [pq] : p \mathbin{\acute{\circ}} q. \equiv .[f].f(pp) \equiv f(0q)$

$D9.\quad [pq] : p \mathbin{\acute{\circ}} q. \equiv .[f].f(pp) \equiv f(q0)$

$D10.\ [pq] : p \mathbin{\acute{\circ}} q. \equiv .[f].f(qp) \equiv f(0q)$

$D11.\ [pq] : p \mathbin{\acute{\circ}} q. \equiv .[f].f(qp) \equiv f(p0)$

$D12.\ [pq] : p \mathbin{\acute{\circ}} q. \equiv .[f].f(pq) \equiv f(0p).$

If we regard the new symbolic expression '$p \multimap q$' as equivalent to '$\sim (p \supset q)$' and accept the proposition

$[pq] : p \multimap q. \equiv . \sim (p \supset q)$[37]

the following theorems become theses of protothetic:

$E1.\ [pq] : p \multimap q. \equiv .[f].f(pq) \equiv f(10)$[38]

$E2.\ [pq] : p \multimap q. \equiv .[f].f(qp) \equiv f(01)$

$E3.\ [pq] : p \multimap q. \equiv .[f].f(1q) \equiv f(p0)$

$E4.\ [pq] : p \multimap q. \equiv .[f].f(0p) \equiv f(q1).$

If the new symbolic expression '$p \multimapinv q$', equivalent to '$\sim (p \vee \sim q)$', and the proposition

$[pq] : p \multimapinv q. \equiv . \sim (p \vee \sim q)$[39]

are accepted, the following theses should be included in protothetic:

$F1.\ [pq] : p \multimap q. \equiv .[f].f(pq) \equiv f(01)$[40]

$F2.\ [pq] : p \multimap q. \equiv .[f].f(qp) \equiv f(10)$

$F3.\ [pq] : p \multimap q. \equiv .[f].f(1p) \equiv f(q0)$

$F4.\ [pq] : p \multimap q. \equiv .[f].f(0q) \equiv f(p1).$

It should be observed that a modified form of Theorems $E1$–$F4$ can be obtained by the substitution of either '$p \equiv p$' or '$q \equiv q$' for '1'.

NOTES

1 In the Bibliographical Abbreviations, these papers are referred to as Leśniewski (1938a, 1938b), Łukasiewicz (1939) and Sobociński (1939a, 1939b).

2 Cf. Bocheński (1947).

3 Cf. Łukasiewicz (1939).

4 It is a calculus of propositions based on the single primitive term 'equivalence'. The first axiom-system of this theory and the proof of its completeness was given by St. Leśniewski; cf. Leśniewski (1929, 1938a) and Łukasiewicz (1939).

5 In this paper, I write theses of the calculus of propositions in the known symbolism of J. Łukasiewicz. An account of this notation is given in the papers: Łukasiewicz (1939, 1948), Łukasiewicz-Tarski (1930). Cf. also Feys (1944), pages 193–196. I will only mention that the symbol 'E' corresponds to '$\equiv$', and, further, that the symbols 'K' and 'N' correspond to '.' (or '&') and '$\sim$', respectively.

6 Curiously, even the thesis:

$$EEpqEEqrErp$$

does not possess this property.

7 Cf. Sobociński (1939b).

8 It corresponds exactly to the usual rule of detachment formulated in terms of implication:

If expressions of the form '$C\alpha\beta$' and 'α' are accepted as theses, then one can add to the system the thesis of the form 'β'.

This correspondence is a simple consequence of the well-known equivalence of the expressions '$C\alpha\beta$' and '$NK\alpha N\beta$'.

9 The first fact was noted by J. Łukasiewicz, the second observation was made by Dr. Adolf Lindenbaum, a lecturer of mathematics at the University of Warsaw, who was killed by the Germans in Wilno, in 1941.

As far as I could ascertain, I was the first to investigate the conjunction–negation calculus of propositions. All expositions of the calculus of propositions which I am familiar with, contain at least the thesis '$NKpNp$' or '$NKNpp$' of this theory. (Cf. Whitehead–Russell, 1925.)

10 My papers Sobociński (1939a, 1939b) were reviewed by Heinrich Scholz (a professor at the University of Münster, Germany) in Zentralblatt für Mathematik in 1940 and 1941. [Cf. Band 23, pp. 97–98 and p. 289 – Ed.]

11 Stanisław Leśniewski, born March 18, 1886, died May 13, 1939; professor of the philosophy of mathematics at the University of Warsaw. [Concerning Leśniewski's date of birth, see

Luschei (1962), note 16 – Ed.]

[12] The names of these theories are of Greek origin. Protothetic is a theory on which other theories must be based; its name derives from the word '$\pi\rho\hat{\omega}\tau o\varsigma$'. Ontology contains only one primitive term, the copula 'is' (Latin 'est', Greek '$\dot{\epsilon}\sigma\tau\acute{\iota}$'); its name derives from the participle of the word '$\epsilon\hat{\iota}\nu\alpha\iota$', which is '$\breve{o}\nu$'. 'Mereology' derives from the word '$\mu\acute{\epsilon}\rho o\varsigma$' – a part. It is obvious that the name 'ontology', as used by St. Leśniewski, is only a homonym of 'ontology' which is a part of Metaphysics (cf. Leśniewski, 1927, Chapter XI).

In St. Leśniewski's system, protothetic is not based on any earlier theory. When constructing proofs in other deductive theories, we must use only theses of protothetic (for instance in ontology), only theses of protothetic and ontology (for instance in mereology or in arithmetic), only theses of all three theories (for instance in certain systems of geometry), and so on. In this way a certain hierarchy of deductive theories is created in which protothetic takes the first place, ontology the second, mereology, arithmetic, and theories of groups the third, some systems of geometry the fourth, and so on.

[13] A number of loose but important remarks explaining certain properties of St. Leśniewski's system [of the foundations of mathematics] can be found in the works Tarski (1933, 1936).

[14] In Chapters I–III of Leśniewski (1927), St. Leśniewski outlined his criticism of the earlier systems of the foundations of mathematics; in Chapters IV–X he presented his system of mereology; in Chapter XI he presented the theoretical foundations of ontology. In the paper Leśniewski (1929), one can find the description of the origin and the development of protothetic and concisely formulated rules of procedure of this theory. The paper Leśniewski (1930) presents the fundamentals of ontology, its axiomatization, and its rules of procedure. The question of rules of procedure of St. Leśniewski's system, difficult at first sight, was very clearly explained by him in his paper Leśniewski (1931).

[15] In St. Leśniewski's system all theories based on protothetic and ontology (viz. of the third and higher orders, cf. note 12) have their rules of procedure exactly analogous to the rules of ontology. For this reason mereology does not require a separate formulation of such rules.

[16] There is only one paper in which one can find some of the theorems of ontology deduced from its axiom, namely my treatise Sobociński (1934). Cf. also Sobociński (1949).

[17] Cf. Leśniewski (1929, 1938a). In the original symbolism of protothetic this thesis has the form:

$$\llcorner fpqrst \lrcorner^{\ulcorner} \phi \left(\phi(pq) \llcorner g \lrcorner^{\ulcorner} \phi \left(f\left(pf(p\llcorner u \lrcorner^{\ulcorner} u^{\urcorner})\right) \phi \left(\llcorner u \lrcorner^{\ulcorner} f(qu)^{\urcorner} \phi \left(g \left(\phi \left(\phi(rs)t \right) q \right) g \left(\phi \left(\phi(st)r \right) p \right) \right) \right) \right)^{\urcorner} \right)^{\urcorner}$$

and is composed of 82 signs. Thesis A, as well as further theses, are written using a slightly modified symbolism of Peano–Russell.

[18] Leśniewski deduced Łukasiewicz's well-known axioms of the calculus of propositions:

(1) $CCpqCCqrCpr$

(2) $CCNppp$

(3) $CpCNpq$

(cf. for instance Feys, 1944, p. 27). Then he offered a simple proof that every thesis which can be deduced from theses (1)–(3) with the help of the usual rules of procedure of the calculus of propositions, can be also inferred from them using the rules of procedure of protothetic.

[19] The present knowledge of the inner structure of protothetic is much less advanced than our knowledge of propositions, which we attained thanks to the work carried out by J. Łukasiewicz, by his school, and by other investigators.

[20] For the sake of clarity, I use the symbol '1' for '$[u].u \equiv u$', and the symbol '0' for '$[u].u$' (see, for instance, Leśniewski, 1929, pp. 12–13).

[21] In view of condition 3, if, for instance, we have already in the system the theses:
(a) $0 \supset 0$
(b) $1 \supset 1$,
then we can add to this system the thesis:
(c) $[p].p \supset p$.
This law has its origin in the bivalence of protothetic, which means that in protothetic, as well as in the ordinary calculus of propositions, each proposition is either true or false.

[22] In the original symbolism of protothetic this thesis has the form:

$$\llcorner fpq \lrcorner^{\ulcorner} \phi \Big(f \big(pf(p \llcorner u \lrcorner^{\ulcorner} u^{\urcorner}) \big) \llcorner r \lrcorner^{\ulcorner} \phi \Big(f \big(\phi \big(\phi(pq)q \big) r \big) \phi \big(\phi(pq)\phi \big(\phi(rq)\phi(pr) \big) \big) \Big)^{\urcorner} \Big)^{\urcorner}$$

and is composed of 66 signs. This thesis, in contrast with the above thesis A, is not organic.

For the meaning of the term 'organic', cf. Łukasiewicz–Tarski (1930), p. 8 (Satz 9), where the definition of the term 'single axiom', continuously used through this paper, is also given (cf. op. cit., p. 7, Satz 8). Cf. also Łukasiewicz (1939).

[23] In his papers Leśniewski (1929, 1938a), St. Leśniewski frequently writes about systems of protothetic that can be constructed using primitive terms other than 'equivalence', and, in this context, about an appropriate formulation of the rules of procedure.

[24] In the original symbolism of protothetic this organic thesis has the form:

$$\llcorner pq \lrcorner^{\ulcorner} \phi \Big(\phi(pq) \llcorner f \lrcorner^{\ulcorner} \phi \Big(f \big(qf(q \llcorner u \lrcorner^{\ulcorner} u^{\urcorner}) \big) \llcorner r \lrcorner^{\ulcorner} \phi \Big(f(pr) \phi \big(r\phi \big(q\phi(rp) \big) \big) \Big)^{\urcorner} \Big)^{\urcorner} \Big)^{\urcorner}$$

and is composed of 62 signs.

[25] In the original symbolism of protothetic this organic thesis has the form:

$$\llcorner pq \lrcorner^{\ulcorner} \phi \Big(\phi(pq) \llcorner f \lrcorner^{\ulcorner} \phi \Big(f \big(pf(p \llcorner u \lrcorner^{\ulcorner} u^{\urcorner}) \llcorner r \lrcorner^{\ulcorner} \phi \big(f(qr)\phi(qp) \big)^{\urcorner} \big) \Big)^{\urcorner} \Big)^{\urcorner}$$

and is composed of 54 signs.

[26] For information about protothetic see Leśniewski (1927, 1929, 1930). The expression 'thesis' as used in this paper means 'theorem valid in protothetic'.

[27] See Tarski (1923a, 1923b, 1924), and also Leśniewski (1929), pp. 11 ff.

[28] Cf. Sobociński (1934), p. 155, footnote 2.

[29] See Tarski (1923b), pp. 75 ff., and Tarski (1924).

[30] See Tarski (1923b), pp. 73 and 75, and Tarski (1923a), pp. 199–200.

[31] The thesis which was published is to be found in Leśniewski (1929), p. 32.

[32] It should be observed that if in the system of protothetic which Leśniewski called $\mathfrak{S}_5$ one of the theses given here were to be accepted as a definition, then, in accordance with the directive accepted in system $\mathfrak{S}_5$, the *definiens* should be written on the left-hand side and the *definiendum* on the right-hand side of the equivalence. See Leśniewski (1929), pp. 54 and 76, point 1 of the directive, and p. 70, terminological explanation T.E. XLIV.

In the original symbolism of protothetic the thesis corresponding to $C1$ is written as follows:

$$\llcorner pq \lrcorner^{\ulcorner} \text{⏀} \left(\text{⫯}(pq) \llcorner f \lrcorner^{\ulcorner} \text{⏀} \left(f(pq) f \left(q \llcorner u \lrcorner^{\ulcorner} \text{⏀}(uu)^{\urcorner} \right) \right)^{\urcorner} \right)^{\urcorner}.$$

[33] In the original notation of protothetic the thesis corresponding to $C9$ is written as follows:

$$\llcorner pq \lrcorner^{\ulcorner} \text{⏀} \left(\text{⫯}(pq) \llcorner f \lrcorner^{\ulcorner} \text{⏀} \left(f(pq) f \left(\llcorner u \lrcorner^{\ulcorner} \text{⏀}(uu)^{\urcorner} \llcorner u \lrcorner^{\ulcorner} \text{⏀}(uu)^{\urcorner} \right) \right)^{\urcorner} \right)^{\urcorner}.$$

[34] In the original notation of protothetic the thesis corresponding to $C11$ is written as follows:

$$\llcorner pq \lrcorner^{\ulcorner} \text{⏀} \left(\text{⫯}(pq) \llcorner f \lrcorner^{\ulcorner} \text{⏀} \left(f(\llcorner u \lrcorner^{\ulcorner} \text{⏀}(uu)^{\urcorner} p) f \left(q \llcorner u \lrcorner^{\ulcorner} \text{⏀}(uu)^{\urcorner} \right) \right)^{\urcorner} \right)^{\urcorner}.$$

[35] The sign '⫰' is taken from the original notation of protothetic. The functor '⫰' is one of the two propositional functors of two arguments, each of which is sufficient to define in the theory of deduction all the remaining functors. This proposition has been proved by Henry M. Sheffer; see Sheffer (1913). Moreover, Professor Eustachy Żyliński has proved that out of sixteen possible binary connectives only these two propositional functors of two arguments have the characteristic in question. See Żyliński (1925, 1927).

The truth-functional behaviour of this functor can be described by means of the matrix method in the following way:

⫰	0	1
0	1	0
1	0	0

For the matrix method see Łukasiewicz–Tarski (1930), §1, Definition 3, p. 4.

[36] In the original symbolism of protothetic the thesis corresponding to $D1$ is written as follows:

$$\llcorner pq \lrcorner^{\ulcorner} \text{⏀} \left(\text{⫰}(pq) \llcorner f \lrcorner^{\ulcorner} \text{⏀} \left(f(pq) f \left(q \llcorner u \lrcorner^{\ulcorner} u^{\urcorner} \right) \right)^{\urcorner} \right)^{\urcorner}.$$

[37] The sign '⊸' is taken from the original notation of protothetic. With the aid of the matrix method the truth table of '⊸' can be presented as follows:

⊸	0	1
0	0	0
1	1	0

[38] The thesis corresponding to $E1$ is written in the original symbolism of protothetic as follows:

$$\llcorner pq \lrcorner \ulcorner \phi \left(-\!\circ(pq) \llcorner f \lrcorner \ulcorner \phi \left(f(pq) f \left(\llcorner u \lrcorner \ulcorner \phi(uu) \urcorner \llcorner u \lrcorner \ulcorner u \urcorner \right) \right) \urcorner \right) \urcorner .$$

[39] The sign '$\circ\!-$' is taken from the original notation of protothetic. The truth table of '$\circ\!-$', presented be means of the matrix method, is as follows:

$\circ\!-$	0	1
0	0	1
1	0	0

[40] In the original symbolism of protothetic the thesis corresponding to $F1$ is written as follows:

$$\llcorner pq \lrcorner \ulcorner \phi \left(\circ\!- (pq) \llcorner f \lrcorner \ulcorner \phi \left(f(pq) f \left(\llcorner u \lrcorner \ulcorner u \urcorner \llcorner u \lrcorner \ulcorner \phi(uu) \urcorner \right) \right) \urcorner \right) \urcorner .$$

REFERENCES

Bocheński, Inocenty M. (1947). 'La logique de Théophraste', *Collectanea Friburgensia*. Nouvelle Série Fasc. 32, Fribourg en Suisse, 195–304.

Feys, Robert (1944). *Logistiek. Geformaliseerde Logica*, Antwerp (Dutch).

Leśniewski, Stanisław (1927). 'O podstawach matematyki (On the Foundations of Mathematics)', *Przegląd Filozoficzny* XXX—XXXIV [: XXX (1927), 164–206; XXXI (1928), 261–291; XXXII (1929), 60–101; XXXIII (1930), 77–105; XXXIV (1931), 142–170].

Leśniewski, Stanisław (1929). 'Grundzüge eines neuen Systems der Grundlagen der Mathematik', *Fundamenta Mathematicae* XIV, 1–81.

Leśniewski, Stanisław (1930). Über die Grundlagen der Ontologie, *Comptes rendus des Séances de la Société des Sciences et des Lettres de Varsovie, Classe III* 23, 111–132.

Leśniewski, Stanisław (1931). 'Über Definitionen in der sogenannten Theorie der Deduktion', *Comptes rendus des Séances de la Société des Sciences et des Lettres de Varsovie, Classe III* 24, 289–309.

Leśniewski, Stanisław (1938a). 'Einleitende Bemerkungen zur Fortsetzung meiner Mitteilung u.d.T. "Grundzüge eines neuen Systems der Grundlagen der Mathematik"', *Collectanea Logica* I (offprint), 1–60.

Leśniewski, Stanisław (1938b). 'Grundzüge eines neuen Systems der Grundlagen der Mathematik, §12', *Collectanea Logica* I (offprint), 61–144.

Łukasiewicz, Jan (1939). 'Der Äquivalenzenkalkül', *Collectanea Logica* I (offprint), 145–169.

Łukasiewicz, Jan (1948). 'The Shortest Axiom of the Implicational Calculus of Propositions', *Proceedings of the Royal Irish Academy* 52, Section A, no. 3, 25–33.

Łukasiewicz, Jan and Tarski, Alfred (1930). 'Untersuchungen über den Aussagenkalkül', *Comptes rendus des Séances de la Société des Sciences et des Lettres de Varsovie, Classe III* 23, 30–50.

Sheffer, Henry Maurice (1913). 'A Set of Five Independent Postulates for Boolean Algebras, with Application to Logical Constants', *Transactions of the American Mathematical Society* 14, 481–488.

Sobociński, Bolesław (1934). 'O kolejnych uproszczeniach aksjomatyki "Ontologji" prof. St. Leśniewskiego (On the Successive Simplifications of the Axiom-System of Leśniewski's Ontology)', *Fragmenty Filozoficzne* 1, Volume in commemoration of fifteen years' teaching at the University of Warsaw by Professor Tadeusz Kotarbiński, 143–160.

Sobociński, Bolesław (1939a). 'Z badań nad prototetyką (An Investigation of Protothetic)', *Collectanea Logica* I (offprint), 171–177.

Sobociński, Bolesław (1939b). 'Aksjomatyzacja konjunkcyjno-negacyjnej teorii dedukcji (Axiomatization of the Conjunctive-Negative Theory of Deduction)', *Collectanea Logica* I (offprint).

Sobociński, Bolesław (1949). 'L'analyse de l'antinomie Russellienne par Leśniewski', *Methods* 1, 94–107, 220–228, 308–316, and 2 (1950), 237–257.

Tajtelbaum, Alfred (1923a). 'Sur le terme primitif de la Logistique', *Fundamenta Mathematicae* IV, 196–200.

Tajtelbaum-Tarski, Alfred (1923b). 'O wyrazie pierwotnym logistyki (On the Primitive Term of Logistic)', *Przegląd Filozoficzny* XXVI, 68–89. [Cf. paper III in this volume – Ed.]

Tajtelbaum-Tarski, Alfred (1924). 'Sur les truth-functions au sens de MM. Russell et Whitehead', *Fundamenta Mathematicae* V, 59–74.

Tarski, Alfred (1933). *Pojęcie Prawdy w Językach Nauk Dedukcyjnych* (The Concept of Truth in the Languages of the Deductive Sciences)', *Prace Towarzystwa Naukowego Warszawskiego, Wydział III Nauk Matematyczno-Fizycznych* 34, Warszawa, vii + 116 pp.

Tarski, Alfred (1936). 'Der Wahrheitsbegriff in den formalisierten Sprachen', *Studia Philosophica* 1, 261–405.

Whitehead, Alfred North and Russell, Bertrand (1925). *Principia Mathematica*, ed. 2, Vol. I, Cambridge.

Żyliński, Eustachy (1925). 'Some Remarks Concerning the Theory of Deduction', *Fundamenta Mathematicae* VII, 203–209.

Żyliński, Eustachy (1927). 'O przedstawialności funkcji prawdziwościowych jednych przez drugie (On the Representation of Truth-Functions by Other Truth-Functions)', *Przegląd Filozoficzny* XXX.

JERZY SŁUPECKI

ST. LEŚNIEWSKI'S PROTOTHETICS *

FOREWORD**

In this paper I am discussing one of the three logical systems formulated by the late Professor Stanisław Leśniewski, Ph.D., namely, that which he at first named *logistics* and later *protothetics*[1]; on the two other systems, i.e. *ontology* and *mereology* I am working at present. As Professor Leśniewski's manuscripts were destroyed during the Warsaw Uprising in 1944, the main source for this paper constitute the notes taken by students on his lectures, in the first place by the late Mr. Jerzy Billich of the Warsaw University.[‡] The initiative of collecting those notes which escaped destruction and of arranging systematically the result of Professor Leśniewski's research work, was taken by Professor Tadeusz Kotarbiński, who asked me as one of Professor Leśniewski's students to work on these notes. Professor Leśniewski left a great part of the results of his research work unpublished. Those which he did publish are often fragmentary and thus do not represent a complete picture of the systems he created. Besides, some of his papers are styled in an extremely difficult manner, while some others are at present almost unavailable. These are the reasons that Professor Leśniewski's works are not much known, even in this country, and therefore subject to falling into oblivion, notwithstanding

* Bibliographical Note. This paper originally appeared in *Studia Logica* I (1953), 44–112, by permission of the Institute of Philosophy and Sociology of the Polish Academy of Sciences.

** In the period between the two world wars, the work of Stanisław Leśniewski (1886–1939) had considerable influence on the development of logic in Poland. Many logicians accepted Leśniewski's conceptions enthusiastically whereas mathematicians adopted mostly a critical attitude. Leśniewski himself published but little, and it is therefore difficult to form an opinion of his logical systems on the basis of his few printed works.

With a view to enabling students of logic to become acquainted with, and to evaluate the works of Leśniewski, the editors of *Studia Logica* have decided to publish an extensive report on his lifework written by Professor Słupecki and compiled from notes taken of Leśniewski's lectures. This report has been divided into parts of which the first is given below and the remaining parts will appear in subsequent volumes of *Studia Logica – The Editors*.

‡ [Ed. Note: These lecture notes were published in 1988 by Kluwer Academic Publishers (see Leśniewski (1988) in the Protothetic Bibliography).]

the fact that according to unanimous opinion of those who know them, they are of lasting value.

This paper does not bear the character of a historical study on Professor Leśniewski's system as the notes on which it is based were not written by himself. Moreover, the exposition of protothetics given here deviates in many points from that contained in those notes. This seemed to be indicated, in the first place, for didactic reasons, my aim being to make this paper comprehensible also to non-specialists,[2] although it has, of course, been taken for granted that the reader is familiar with the fundamental branches of mathematical logic.

I have taken account of all the results contained in the notes with the exception of a few which I consider either unessential or indeed obsolete. In order to make the exposition of protothetics both concise and complete I was bound to amplify considerably the material found in the notes. However, all the results given here and not contained in the notes were undoubtedly known to Professor Leśniewski. The paper includes no results of my own but a considerable part of the proofs has been entirely worked out by myself. I mention this so that I alone should be held responsible for any inaccuracies and errors which may have occurred.

This paper is not written with that exactitude which Professor Leśniewski always observed in his lectures and publications. This reservation, however, refers only to the meta-logical considerations whereas in the proofs of theorems of the system I have strictly followed the proofs found in the notes.

The lectures "On Certain Problems of Protothetics", the notes of which form the basis of this paper, were delivered by Professor Leśniewski at the Warsaw University during the academic year 1932–1933. The subject-matter of this paper is discussed in a more detailed way in Section 2.

1. SEMANTIC CATEGORIES OF EXPRESSIONS OF PROTOTHETICS

I shall outline here the theory of the division of expressions appearing in Leśniewski's logical systems. The classes of that division are called semantic categories. The notes[3] contain a remark stating that this division is analogous to the division of expressions of everyday language into parts of sentences. It is also connected by certain analogies with the theory of types. Just as the latter, the division of expressions into semantic categories prevents logical antinomies. From a formal point of view, the theory of the division of logical expressions into semantic categories is simpler than the ramified theory of types, but on the other hand it is richer than the so-called simple theory of types. The essential difference between these two theories consists in the fact that the theory of types systematizes such objects as individuals, classes and relations, whereas the theory of semantic categories systematizes expressions of logic, i.e., elements of language.[4]

A precise discussion of Leśniewski's theory requires a fixed symbolism and regular reference to such symbolism.[5] Here, however, we shall describe the theory in general and that description will be independent of the symbolism used in various systems of logic. This will somewhat impair the precision of our remarks, but on the other hand will make them more intuitive. In this way, too, we shall show a way of applying Leśniewski's theory to non-logical languages.

Among logical expressions we distinguish two basic semantic categories: the category of *propositions* and the category of *names*. The former includes not only propositions in a narrower sense, i.e., expressions in which there appear no free variables, but also all the expressions containing free variables which change into propositions if proper constant terms are substituted for these variables. Thus in particular, the variables of the propositional calculus belong to the category of propositions. Analogous remarks refer to the expressions belonging to the semantic category of names.

All logical expressions which do not belong to one of these basic categories and which are neither quantifiers nor parentheses, are given the common name of *functors*. In Leśniewski's theory, no semantic category is assigned to quantifiers and to parentheses.[6] The constants of propositional calculus (the symbols of implication, conjunction, negation, etc.) as well as the symbols a, i, e, o of Aristotle's syllogistics can be quoted as examples of functors. Every functor together with its arguments forms an expression which belongs either to the category of propositions or to the category of names. In the former case it is called *proposition-forming*, in the latter, *name-forming*.*

The semantic category of a functor, variable or constant, depends on whether it is a proposition-forming or a name-forming functor, on the number of its arguments, and on the semantic categories of these arguments. Thus, for example, two functors of two arguments belong to the same semantic category if and only if both of them are either proposition-forming or name-forming; if their first arguments belong to the same semantic category; and if their second arguments belong to the same semantic category.

Further remarks will refer to the semantic categories of the expressions of protothetics. These categories include the category of propositions and the categories of those proposition-forming functors of which all the arguments belong to the semantic category of the expressions of protothetics. Thus, the categories of the functors of the propositional calculus (regardless of the number of their arguments), and the categories of those functors the arguments of which are functors of the propositional calculus or propositions, etc., belong to the semantic categories of the expressions of protothetics.

The notion of the semantic categories of the expressions of protothetics enables us to characterize the theorems of that system.

* [Ed. Note: For a clarification of this part of the exposition of Leśniewski's theory of semantic categories, see Lejewski (1956) in the Protothetic Bibliography.]

The theorems of protothetics may contain: constant or variable expressions belonging to any semantic category of the expressions of protothetics, quantifiers binding variables belonging to those categories, and parentheses. No other expressions appear in the theorems of protothetics.

We must also introduce the notion of the order of a functor belonging to a semantic category of the expressions of protothetics.[7]

Every functor whose all arguments, regardless of their number, belong to the category of propositions, is a functor of the first order.

Every functor of which one argument at least is a functor of the $n-1$-st order and none of the arguments is a functor of an order higher than $n-1$ is a functor of the n-th order (where $n > 1$).

2. PROBLEMS DISCUSSED IN THIS PAPER

This paper deals with three systems of protothetics. The primitive term of the first two is implication, of the third – equivalence. The systems having implication as their primitive term will be called S and S_1, and the system having equivalence as its primitive term will be called S_2. The merit of the third system consists in the fact that definitions, which in Leśniewski's systems are always theorems, can in this system be formulated by means of the primitive term, in a way which is generally (e.g., in mathematics) used to formulate those definitions which are theorems. For this reason the research work conducted by Leśniewski and his students in the last years before World War II was concentrated mostly on this system. However, the systems having implication as their primitive term are, from a formal point of view, simpler than S_2, and that is why they have been worked out in more detail than the latter. Consequently, we devote to them no less attention than to S_2.

It is obvious that in S and S_1 we are able to define equivalence.

Thus the expression defining that term can in these systems be used as a scheme of definitions. Since this expression, however, has a rather complicated structure, we shall note definitions in the form of two implications: simple and converse. This method of noting definitions will enable us to base the systems S and S_1 on a simple and easily comprehensible axiom.

In the notes we find a remark stating that protothetics is a generalization of the propositional calculus. This generalization can be achieved gradually. The simplest generalization consists in adding to the expressions of the propositional calculus quantifiers binding propositional variables. A system of the propositional calculus enriched by quantifiers and having implication as its primitive term is discussed in Section 5.

A further generalization of the propositional calculus is the system which Leśniewski called a slightly enlarged theory of deduction and which in this paper is referred to as elementary protothetics. In the theorems of that system appear constant and variable functors of the first semantic order only, and also quantifiers binding propositional variables and functorial variables.[8]

In Sections 6 and 8 the characteristic theorems of elementary protothetics are discussed. Since variables representing functors do appear in these theorems, we can express by means of them those properties of the theorems of the propositional calculus which are usually spoken of in the meta-system. The only axiom of that system of elementary protothetics which has implication as its primitive term is given in Section 9.

In order to pass from elementary protothetics to systems the theorems of which contain expressions belonging to all those semantic categories which we can define starting from the category of propositions, we must add new rules of inference to those used in elementary protothetics. This role can be played by both the rule of verification and the rule of extensionality. These rules are discussed in Sections 10 and 12. By choosing one or the other we obtain two different but equivalent systems of protothetics, both of them having implication as their primitive term. The proof of the equivalence of these two systems is given in Section 12. Since one and the same theorem is the axiom of both these systems, to prove their equivalence it suffices to show that the rules of inference of one of these systems may be derived from the rules of inference and the theorems of the other system. We are proving also (see Section 13) that given any proposition noted in terms of protothetics either that proposition or its negation is a theorem of both these systems. Hence we are proving that both these systems are *complete*.

The last section deals with that system of protothetics which uses equivalence as its primitive term. We are showing here: (1) that conjunction and implication may be defined in terms of equivalence, (2) that the proposition which is the only axiom of S and of S_1 is a theorem of S_2, and (3) that the rules of inference of S and of S_1 may be derived from the rules of inference and the theorems of S_2.

The system of axioms of S_2, as given in this paper, consists of four propositions. Leśniewski and his students were building theorems equivalent to this system of axioms. The problem of finding such theorems, though important for formal reasons, is of but little interest from the intuitive point of view. That is why we are not considering it in this paper.

3. THE SYMBOLISM USED BY LEŚNIEWSKI

Leśniewski in his works used either a symbolism of his own[9] or a symbolism modelled on that of Peano and Russell.[10] In Leśniewski's original symbolism, as in the symbolism without parentheses of Łukasiewicz, functors always precede their arguments. However, parentheses form a part of Leśniewski's symbolism and are playing there an essential role. Namely, in conformity with the rules of that symbolism – the arguments of two functors are put in parentheses of the same form if and only if these functors belong to the same semantic category or to the same order, provided that all the arguments of both these functors belong to the same semantic category. Thus arguments of the functors of implication, conjunction and negation are put in parentheses of the same form. Expressions governed by a quantifier are in Leśniewski's symbolism put in parentheses of a special form.

For technical reasons, however, in some of his works Leśniewski did not use his original symbolism but one modelled on that of Peano and Russell replacing, among other things, parentheses by sets of dots. Yet the arguments of functors of semantic orders higher than the first were always put by him in parentheses having that form which is characteristic of the semantic category of these functors. Further, in his symbolism the functors of an order not lower than the second always precede their arguments.

We shall use here that symbolism using, however, parentheses in the place of sets of dots so that in the symbolism used in this paper the dot will appear only as the sign of conjunction. An expression governed by a quantifier will be put in brackets.

(In Leśniewski's systems, theorems do not contain free variables. However, in the case of variables bound by quantifiers standing at the beginning of a theorem the same rules of substitution are valid as in the case of free variables. For this reason the variables bound by quantifiers standing at the beginning of a theorem will be called free, whereas variables governed by inner quantifiers will be given the name of bound variables.

In Leśniewski's symoblism, variables have no special forms of their own. A symbol appearing in a theorem is a variable if it is equiform to any symbol contained in the brackets of a quantifier which is a part of the theorem in question.)

4. RULES OF INFERENCE IN PROTOTHETICS

The respective rules of inference of the three systems of protothetics discussed in this paper are not identical. They have, however, many common features. Moreover, the formulation of certain rules in Leśniewski's systems differs obviously from the formulation of analogous rules in logical systems of other authors.

The fact that definitions are theorems of a given system is characteristic of all the systems of Leśniewski. As mentioned above, the definitions occurring in the system of protothetics are written either in the form of an equivalence or in the form of two implications. If at a certain stage of building a system of protothetics we have all theorems of the propositional calculus at our disposal, then it makes no difference which of these two methods of writing definitions we use, since two implications which jointly define a certain term and the equivalence which could be used to define this term are inferentially equivalent. The structure of definitions which can be added to the systems of protothetics is restricted by certain rules. According to one of them, for example, all symbols and only those which occur in the definiendum as free variables appear in the definiens too as free variables. These rules will not be discussed here in detail since they do not differ essentially from the rules to which the notation of definitions is subjected in all systems of logic. It should be added that in Leśniewski's systems definitions make it possible to introduce into a system expressions belonging to new semantic categories, not represented in that system before. This is the only method of introducing into a system expressions belonging to new semantic categories, not represented in the axioms.

Since in Leśniewski's systems definitions are theorems, their role is not limited to the shortening of expressions, but they are indispensable for proving numerous theorems. Thus in his systems definitions do not satisfy the conditions of *non-creativity*; they satisfy, however, the condition of *translatability* as proved by Lemma 10 in Section 13.

In Leśniewski's systems, the rule of substitution does not permit to replace complex expressions by other expressions, but only to replace free variables, representing functors of any semantic category, by constants or variables of the same semantic category. In spite of this limitation, the rule of substitution used in his systems, taken together with the rule of definition, makes it possible to obtain in those systems all the theorems which result from an accepted theorem on the strength of functional substitution.[11] This will be illustrated by the following example:

From the axiom of complete induction:

(a) $[f]\{f(0) \cdot [x]\{f(x) \supset f(\operatorname{seq} x)\} \supset [y]\{f(y)\}\}$

we obtain by applying functional substitution the proposition

(b) $(0+1 > 0) \cdot [x]\{(x+1 > x) \supset (\operatorname{seq} x + 1 > \operatorname{seq} x)\} \supset [y]\{y+1 > y\}.$

We now introduce into the system of arithmetic the following definition:

(c) $[x]\{f^*(x) \equiv (x+1 > x)\}.$

By substituting in (a) the functor f^*, defined above, for the variable f, we obtain

(d) $f^*(0) \cdot [x]\{f^*(x) \supset f^*(\operatorname{seq} x)\} \supset [y]\{f^*(y)\}.$

By substituting in (c) 0 for the variable x and applying the following theorem of the propositional calculus:

$[p, q, r, s]\{(p \equiv q) \cdot (p \cdot r \supset s) \supset (q \cdot r \supset s)\},$

we obtain from (d) the proposition

(e) $(0+1>0)\cdot[x]\{f^*(x)\supset f^*(\mathrm{seq}\,x)\}\supset[y]\{f^*(y)\}$.

By eliminating in an analogous way the expressions $f^*(x)$, $f^*(\mathrm{seq}\,x)$ and $f^*(y)$ from (e) we obtain (b).

By substituting in any theorem certain expressions for its free variables we can increase or reduce the number of non-equiform free variables of that theorem. In the former case we write the added variables within the brackets of the general quantifier standing at the beginning of the theorem; in the latter case we omit within the brackets of that quantifier those variables which after the substitution cease to appear in the theorem as free variables. We proceed in the same way when by adding a quantifier to the antecedent or to the consequent of a theorem we have reduced the number of its non-equiform free variables.

The common trait of the rules of detachment, used in those systems of protothetics which have implication as their primitive term and in that system which has equivalence as its primitive term, consists in the fact that these rules can be applied only to those implications or equivalences which are not preceded by quantifiers. This again will be illustrated by an example.

Let the proposition $[p]\{\alpha\}$ be a theorem of S, S_1 and S_2; the proposition $[p]\{\alpha\supset\beta\}$ a theorem of S and S_1[12]; and the proposition $[p]\{\alpha\equiv\beta\}$ a theorem of S_2.[13]

A secondary rule of S and S_1 permits to distribute a general quantifier preceding an implication; a primary rule of S_2 permits to distribute a general quantifier preceding an equivalence. Hence $[p]\{\alpha\}\supset[p]\{\beta\}$ is a theorem of S and of S_1, and $[p]\{\alpha\}\equiv[p]\{\beta\}$ is a theorem of S_2. The rule of detachment may be applied to both these propositions.

An exact statement of the rules of inference for protothetics is given in Leśniewski (1929).

5. PROPOSITIONAL CALCULUS WITH QUANTIFIERS

Implication is the sole primitive term of the system with which we shall deal here.[14] Negation will be defined by means of that term and the general quantifier. Since equivalence is not at our disposal, we write this definition in the form of two implications:

D1. (α) $[p]\{\sim(p)\supset(p\supset[p]\{p\})\}$,

(β) $[p]\{(p\supset[p]\{p\})\supset\sim(p)\}$.

The consequent $[p]\{p\}$ of the implication $p\supset[p]\{p\}$, which is the definiens of D1 is, of course, a false proposition. Consequently, this implication is true if its antecedent is false, and false if its antecedent is true. Thus the expression which defines negation satisfies the conditions characteristic of that function. Having defined negation we can with its help, and with the help of implication, define all constant terms of the propositional calculus.

The following propositions:

A1. $[p, q, r]\{(p \supset q) \supset ((q \supset r) \supset (p \supset r))\}$;

A2. $[p, q]\{q \supset (p \supset q)\}$;

A3. $[p, q]\{((p \supset q) \supset p) \supset p\}$

are the axioms of the system.

These are the axioms (in the literature of the subject they are quoted as the axioms of Tarski–Bernays) of the propositional calculus which has the sign of implication as its only constant term.[15]

In proving the theorems of the system under discussion the following rules of inference may be used: (1) the rule of substitution, with restrictions analogous to those adhered to in the functional calculus; (2) the rule of detachment; (3) two rules for using the general quantifier: the first allows to add the general quantifier to the antecedent of an implication,[16] the second to add it to the consequent of an implication, provided that the antecedent does not contain a free variable having the same form as the variable bound by this quantifier. The last two rules are analogous to the rules for using the quantifier in the functional calculus. In building the system we are applying also the rule permitting to add definitions.

THEOREM. *All theorems of the propositional calculus which has implication and negation as its primitive terms result from the axioms* A1–3.

Proof. To prove this theorem it suffices to show that the propositions:

A1′. $[p]\{(\sim(p) \supset p) \supset p\}$;

A2′. $[p, q]\{p \supset (\sim(p) \supset q)\}$

result from these axioms, since these propositions, together with A1, form the system of axioms in Łukasiewicz's propositional calculus.

Since the following four propositions are theorems of the propositional calculus which has implication as its primitive term, they result from A1–3:

T1. $[p, q, r]\{((q \supset p) \supset p) \supset ((q \supset r) \supset ((r \supset p) \supset p))\}$;

T2. $[p]\{p \supset p\}$;

T3. $[p,q,r]\{(r \supset q) \supset ((p \supset r) \supset (p \supset q))\}$;

T4. $[p,q,r]\{(p \supset (q \supset r)) \supset (q \supset (p \supset r))\}$.

Further theorems, which along with implication include negation, require proofs. These proofs will not be given here in a complete form. In simpler cases we limit ourselves to noting – on the right of a theorem which is being introduced into the system – the numbers of those theorems from which it results. In more complicated cases we add brief explanations.

T5. $[p]\{(\sim(p) \supset p) \supset p\}$. (T1; A3; D1, β)

To obtain T5 we substitute in T1 for the variables q and r the expressions $p \supset [p]\{p\}$ and $\sim(p)$, respectively, and in A3 we substitute $[p]\{p\}$ for q. Next, we apply twice the rule of detachment. The theorem thus obtained is equiform to A1′.

T6. $[q]\{[p]\{p\} \supset q\}$. (T2)

We have applied here the rule of adding the quantifier to the antecedent and the rule of substitution.

T7. $[p,q]\{(p \supset [p]\{p\}) \supset (p \supset q)\}$. (T3; T6)

T8. $[p,q]\{\sim(p) \supset (p \supset q)\}$. (A1; D1, α; T7)

T9. $[p,q]\{p \supset (\sim(p) \supset q)\}$. (T4; T8)

The theorem thus obtained is equiform to A2′.

We have thus proved that the axioms of the propositional calculus which has implication and negation as its primitive terms result from A1–3.

6. CHARACTERISTIC THEOREMS OF ELEMENTARY PROTOTHETICS

We shall now discuss the intuitive meaning of several characteristic theorems of elementary protothetics.[17] To formulate them we shall need the terms defined in the following definitions having the forms of equivalences:

D1. $[p]\{as\,(p) \equiv p\}$;

D2. $[p]\{vr\,(p) \equiv (p \equiv p)\}$;

D3. $[p]\{fl\,(p) \equiv (p \equiv \sim(p))\}$;

D4. $1 \equiv [p]\{p \equiv p\}$;

D5. $0 \equiv [p]\{p\}$.

D1–3 define functors of the propositional calculus. The other two definitions do not belong to that calculus and define constants whose semantic category is that of propositions. The symbols: *as*, *vr*, *fl* are abbreviations of the terms: *assertio*, *verum*, and *falsum*.[18]

The following theorem is called *the law of extensionality*:

(a) $[f, p, q]\{(p \equiv q) \supset (f(p) \equiv f(q))\}$.

In conformity with this theorem we conclude that two propositional functions are equivalent if their arguments are equivalent. Propositional functions which satisfy (a) are called *truth-functions*. Thus the law of extensionality states that all functions of one argument of elementary protothetics are truth-functions.

(b) $[f]\{[p]\{f(p) \equiv as\,(p)\} \vee [p]\{f(p) \equiv vr\,(p)\} \vee [p]\{f(p) \equiv fl\,(p)\} \vee$
$[p]\{f(p) \equiv \sim(p)\}\}$.

This law is called *the law of the quantity of functions*; it states that any propositional function of one argument is equivalent to one of the four functions: $as\,(p)$, $vr\,(p)$, $fl\,(p)$ or $\sim(p)$, so that these functions exhaust all the alternatives of the propositional functions of one argument.

(c) $[f, p]\{f(p) \equiv (f(1) \cdot p \vee f(0) \cdot \sim(p))\}$.

This is *the law of development*; it states that any propositional function $f(p)$ after the substitution for p of any true proposition is equivalent to the expression $f(1)$, and after the substitution of any false proposition, to the expression $f(0)$.

(d) $[f, p]\{f(1) \cdot f(0) \supset f(p)\}$.

This is *the law of verification*; it states that any function of one argument of elementary protothetics which is satisfied by the expressions 1 and 0 is satisfied by any proposition.

(e) $[f]\{[p]\{f(p)\} \equiv f(1) \cdot f(0)\}$.

This theorem will be called *the law on the limit of a function*.[19] It can be used for defining the general quantifier by means of conjunction and of the terms 1 and 0.

(f) $[f, q]\{[p]\{f(p)\} \equiv f(q) \cdot f(\sim(q))\}$.

This theorem is called *generalized law on the limit of a function*. It states that a propositional function of one propositional argument is satisfied by any proposition if and only if it is satisfied by any two contradictory propositions.

Theorems (a)–(f) describe the properties of functions of one argument of elementary protothetics. They can be generalized so as to include functions of more arguments. The theorems which are generalizations of the propositions (a)–(f) will be discussed in Section 8.

7. PROOF OF THE EQUIVALENCE OF PROPOSITIONS (a)–(f) FROM SECTION 6

Two propositions α and β are said to be *equivalent* in a certain system, if the proposition $\alpha \equiv \beta$ can be obtained from theorems of that system by means of its rules of inference strengthened as follows. The rules of substitution, of adding definitions to the system and of the use of quantifiers are extended so as to cover expressions of semantic categories which are not represented in the axioms of the system.

We shall show that the propositions (a)–(f) from Section 6 are equivalent[20] in the system discussed in Section 5. When proving equivalences built of any two propositions from among (a)–(f), Section 6, we may refer to any theorems of that system and use its rules of inference strengthened as specified above. Consequently, in the proofs given here we do not refer to the specific axioms of any system of elementary protothetics; it is also for this reason that the axiom of that system of elementary protothetics which has implication as its primitive term will be given only in Section 9.

First of all, we shall prove those theorems which are easily obtainable from D1–5, Section 6.

T1. $[p, q]\{vr(p) \equiv vr(q)\}$. (D2, Section 6)

T2. $[p, q]\{fl(p) \equiv fl(q)\}$. (D3, Section 6)

T3. $[p, q]\{(p \equiv q) \supset (as(p) \equiv as(q))\}$. (D1, Section 6)

The proofs of T1–3 are based, respectively, on the following theorems of the propositional calculus:

$$[p, q, r, s]\{(r \equiv (p \equiv p)) \cdot (s \equiv (q \equiv q)) \supset (r \equiv s)\};$$

$$[p, q, r, s]\{(r \equiv (p \equiv \sim(p))) \cdot (s \equiv (q \equiv \sim(q))) \supset (r \equiv s)\};$$

$$[p, q, r, s]\{(r \equiv p) \cdot (s \equiv q) \supset ((p \equiv q) \supset (r \equiv s))\}.$$

T4. $[p]\{vr\,(p)\}$. (D2, Section 6)

T5. $[p]\{\sim(fl\,(p))\}$. (D3, Section 6)

T6. 1. (D4, Section 6)

T7. $\sim(0)$. (D5, Section 6)

In the proof of T7 we are applying the proposition $\sim([p]\{p\})$ which is a theorem of the propositional calculus with quantifiers.

T8. $\sim(1 \equiv 0)$. (T6; T7)

T9. $[p]\{(p \equiv 1) \vee (p \equiv 0)\}$. (T6; T7)

Subsequent theorems will be needed to show the equivalence of the law of extensionality (a, Section 6) and the law of the quantity of functions (b, Section 6). Their proofs are rather elaborate. If a theorem is an implication, its proof will be noted in *suppositional* form.

In a suppositional proof the antecedent of the theorem to be proved is written first; if this antecedent is a conjunction, we write all its factors in a column. These factors are called *assumptions of the proof.* Other expressions are added to the proof if they may be obtained by applying the rule of detachment from expressions appearing earlier in the proof, from theorems and from definitions of the system.

The proof is completed when an expression equiform to the consequent of the theorem to be proved has been added. All expressions of the proof are written in a column and numbered on the left, with the exception of the last expression of the proof, which is given no number. At the right of every expression of the proof which is not an assumption, we give in parentheses the numbers of the expressions of the proof or of the theorems and of definitions of the system from which this expression results. We do not, however, give references to those theorems of the propositional calculus with quantifiers which are indispensable for the proof. We also do not specify the substitutions performed in the course of the proof; in more complicated cases we do so in a commentary added to the proof. The dots which follow the expressions of proofs are symbols of conjunction. The last assumption is followed by the symbol of implication.

T10. $[f, q, r]\{[p]\{f(p) \equiv vr\,(p)\} \supset (f(q) \equiv f(r))\}$.

Proof. (1) $[p]\{f(p) \equiv vr(p)\} \supset$

(2) $f(q) \equiv vr(q)$. (1)

(3) $f(r) \equiv vr(r)$. (1)

$f(q) \equiv f(r)$ (T1; 2; 3)

We shall explain in more detail how the expression (2) is obtained from the expression (1). The following proposition is a result of substitution applied to T2, Section 5:

$$[p]\{(f(p) \equiv vr(p)) \supset (f(p) \equiv vr(p))\}.$$

By applying first the rule of adding the quantifier to the antecedent of an implication, and subsequently the rule of substitution, we obtain:

(a) $[q]\{[p]\{f(p) \equiv vr(p)\} \supset (f(q) \equiv vr(q))\}$.

By detaching (1) from the expression obtained from (a) by cancelling the quantifier standing at its beginning we obtain (2). In further proofs we shall often obtain expressions in an analogous way but without giving detailed explanations. We must draw attention to the fact that (2) was not obtained from (1) by substitution, although the variable q appears in (2) in those places where the variable p appeared in (1).[21] In similar cases we shall say that a variable has been *replaced* by another variable or expression.

On the example of the proof of T10 we shall explain how a suppositional proof can be transformed into a complete one. For this purpose let us note that we can obtain the proposition

(b) $[r]\{[p]\{f(p) \equiv vr(p)\} \supset (f(r) \equiv vr(r))\}$

in a way similar to that in which we obtained the proposition (a).

Let us further note that the proposition

(c) $[f, q, r]\{(f(q) \equiv vr(q)) \cdot (f(r) \equiv vr(r)) \supset (f(q) \equiv f(r))\}$

is obtained by applying the rule of detachment to T1 and to a proposition which is a result of substitution applied to the following theorem of the propositional calculus:

$$[p, q, r, s]\{(p \equiv q) \supset ((r \equiv p) \cdot (s \equiv q) \supset (r \equiv s))\}.$$

By substituting in the following theorem of the propositional calculus:

$$[p, q, r, s]\{(p \supset q) \supset ((p \supset r) \supset ((q \cdot r \supset s) \supset (p \supset s)))\}$$

the expressions (1), (2), (3) and the last expression of the proof of T10 for the variables p, q, r and s, respectively, and by detaching successively the propositions (a), (b) and (c), we obtain T10.[22]

T11. $[f, q, r]\{[p] f(p) \equiv fl(p)\} \supset (f(q) \equiv f(r))\}$.

Proof. (1) $[p]\{f(p) \equiv fl(p)\} \supset$

(2) $f(q) \equiv fl(q)$. (1)

(3) $f(r) \equiv fl(r)$. (1)

$f(q) \equiv f(r)$ {T2; 2; 3}

T12. $[f, q, r]\{[p]\{f(p) \equiv as(p)\} \cdot (q \equiv r) \supset (f(q) \equiv f(r))\}$.

Proof. (1) $[p]\{f(p) \equiv as(p)\}$.

(2) $q \equiv r \supset$

(3) $f(q) \equiv as(q)$. (1)

(4) $f(r) \equiv as(r)$. (1)

(5) $as(q) \equiv as(r)$. (T3; 2)

$f(q) \equiv f(r)$ (3; 4; 5)

T13. $[f, q, r]\{[p]\{f(p) \equiv \sim(p)\} \cdot (q \equiv r) \supset (f(q) \equiv f(r))\}$.

Proof. (1) $[p]\{f(p) \equiv \sim(p)\}$.

(2) $q \equiv r \supset$

(3) $f(q) \equiv \sim(q)$ (1)

(4) $f(r) \equiv \sim(r)$. (1)

(5) $\sim(q) \equiv \sim(r)$. (2)

$f(q) \equiv f(r)$ (3; 4; 5)

T14. $[f, r, s, t]\{[p, q]\{(p \equiv q) \supset (f(p) \equiv f(q))\} \cdot f(r) \cdot \sim(r) \cdot f(s) \cdot s \supset f(t)\}$.

Proof. (1) $[p, q]\{(p \equiv q) \supset (f(p) \equiv f(q))\}$.

(2) $f(r)$.

(3) $\sim(r)$.

(4) $f(s)$.

(5) $s \supset$

(6) $s \equiv \sim(r)$. (3; 5)

(7) $f(\sim(r))$. (1; 6; 4)

(8) $(r \equiv t) \supset f(t)$. (1; 2)

(9) $(\sim(r) \equiv t) \supset f(t)$. (1; 7)

$f(t)$ (8; 9)

In the last line of the proof we refer to the following theorem of the propositional calculus:

$$[p, r, t]\{((r \equiv t) \supset p) \cdot ((\sim(r) \equiv t) \supset p) \supset p\}.$$

In conformity with T14, any truth-function which is satisfied by a certain false proposition and by a certain true proposition is satisfied by any proposition.

T15. $[f, r, s, t]\{[p, q]\{(p \equiv q) \supset (f(p) \equiv f(q))\} \cdot \sim(f(r)) \cdot r \cdot \sim(f(s)) \cdot \sim(s) \supset \sim(f(t))\}$.

Proof. (1) $[p, q]\{(p \equiv q) \supset (f(p) \equiv f(q))\}$.

(2) $\sim(f(r))$.

(3) r.

(4) $\sim f(s))$.

(5) $\sim s \supset$

(6) $r \equiv \sim(s)$. (3; 5)

(7) $\sim(f(\sim(s)))$. (1; 6; 2)

(8) $(s \equiv t) \supset \sim(f(t))$. (1; 4)

(9) $(\sim(s) \equiv t) \supset \sim(f(t))$. (1; 7)

$\sim(f(t))$ (8; 9)

In conformity with T15, any truth-function which is not satisfied by a certain true proposition nor by a certain false proposition, is not satisfied by any proposition.

T16. $[f, s, t]\{[p, q]\{(p \equiv q) \supset (f(p) \equiv f(q))\} \cdot f(s) \cdot \sim(s) \cdot \sim(t) \supset f(t)\}$.

Proof. (1) $[p, q]\{(p \equiv q) \supset (f(p) \equiv f(q))\}$.

(2) $f(s)$.

(3) $\sim(s)$.

(4) $\sim(t) \supset$

(5) $s \equiv t$. (3; 4)

$f(t)$ (1; 5; 2)

T17. $[f, s, t]\{[p, q]\{(p \equiv q) \supset (f(p) \equiv f(q))\} \supset \sim(f(s) \cdot \sim(s) \cdot \sim(t) \cdot \sim(f(t)))\}$. (T16)

In proving T17 we base ourselves on the following theorem of the propositional calculus:

$$[p, q, r, s]\{(p \cdot q \cdot r \supset s) \equiv \sim(p \cdot q \cdot r \cdot \sim(s))\}.$$

T18. $[f, s, t]\{[p, q]\{(p \equiv q) \supset (f(p) \equiv f(q))\} \cdot \sim(f(s)) \cdot s \cdot t \supset \sim(f(t))\}$.

Proof. (1) $[p, q]\{(p \equiv q) \supset (f(p) \equiv f(q))\}$.

(2) $\sim(f(s))$.

(3) s.

(4) $t \supset$

(5) $s \equiv t.$ (3; 4)

$\sim(f(s))$ (1; 5; 2)

T19. $[f,s,t]\{[p,q]\{(p \equiv q) \supset (f(p) \equiv f(q))\} \supset \sim(\sim(f(s)) \cdot s \cdot t \cdot f(t))\}.$ (T18)

According to T17 and T19, no truth-function assumes different logical values when applied to two propositions having the same logical value.

T20. $[f,r,s,t,u]\{[p,q]\{(p \equiv q) \supset (f(p) \equiv f(q))\} \cdot \sim(f(s) \equiv as\,(s)) \cdot \sim(f(r) \equiv vr\,(r)) \cdot \sim(f(t) \equiv \sim(t)) \supset (f(u) \equiv fl\,(u))\}.$

Proof. (1) $[p,q]\{(p \equiv q) \supset (f(p) \equiv f(q))\}.$

(2) $\sim(f(s) \equiv as\,(s)).$

(3) $\sim(f(r) \equiv vr\,(r)).$

(4) $\sim(f(t) \equiv \sim(t)) \supset$

(5) $\sim(f(s) \equiv s).$ (2; D1, Section 6)

(6) $f(s) \cdot \sim(s) \vee \sim(f(s)) \cdot s.$ (5)

(7) $f(t) \cdot t \vee \sim(f(t)) \cdot \sim(t)$ (4)

(8) $f(s) \cdot \sim(s) \cdot f(t) \cdot t \vee f(s) \cdot \sim(s) \cdot \sim(f(t)) \cdot \sim(t) \vee \sim(f(s)) \cdot s \cdot f(t) \cdot t \vee \sim(f(s)) \cdot s \cdot \sim(f(t)) \cdot \sim(t).$ (6; 7)

(9) $\sim(f(r)).$ (3; T4)

(10) $\sim(f(s) \cdot \sim(s) \cdot \sim(f(t)) \cdot \sim(t)).$ (T17; 1)

(11) $\sim(\sim(f(s)) \cdot s \cdot f(t) \cdot t).$ (T19; 1)

(12) $f(s) \cdot \sim(s) \cdot f(t) \cdot t \supset f(r).$ (T14; 1)

(13) $\sim(f(s) \cdot \sim(s) \cdot f(t) \cdot t).$ (12; 9)

(14) $\sim(f(s)) \cdot s \cdot \sim(f(t)) \cdot \sim(t).$ (8; 13; 10; 11)

$$(15)\ \sim(f(s)) \cdot s \cdot \sim(f(t)) \cdot \sim(t) \supset \sim(f(u)). \qquad \text{(T15; 1)}$$

$$(16)\ \sim(f(u)). \qquad \text{(15; 14)}$$

$$f(u) \equiv fl\,(u) \qquad \text{(16; T5)}$$

In obtaining the expressions (6), (7) and (8) we based ourselves respectively on the following theorems of the propositional calculus:

$$[p, q]\{\sim(p \equiv q) \supset (p \cdot \sim(q) \vee \sim(p) \cdot q)\};$$

$$[p, q]\{\sim(p \equiv \sim(q)) \supset (p \cdot q \vee \sim(p) \cdot \sim(q))\};$$

$$[p, q, r, s]\{(p \cdot \sim(q) \vee \sim(p) \cdot q) \cdot (r \cdot s \vee \sim(r) \cdot \sim(s)) \supset$$
$$(p \cdot \sim(q) \cdot r \cdot s \vee p \cdot \sim(q) \cdot \sim(r) \cdot \sim(s) \vee$$
$$\sim(p) \cdot q \cdot r \cdot s \vee \sim(p) \cdot q \cdot \sim(r) \cdot \sim(s))\}.$$

T21. $[f, r, s, t, u]\{[p, q]\{(p \equiv q) \supset (f(p) \equiv f(q))\} \supset ((f(s) \equiv as\,(s)) \vee$
$(f(r) \equiv vr\,(r)) \vee (f(t) \equiv \sim(t)) \vee (f(u) \equiv fl\,(u)))\}.$ (T20)

T22. $[f]\{[p, q]\{(p \equiv q) \supset (f(p) \equiv f(q))\} \supset ([p]\{f(p) \equiv as\,(p)\} \vee$
$[p]\{f(p) \equiv vr\,(p)\} \vee [p]\{f(p) \equiv fl\,(p)\} \vee [p]\{f(p) \equiv \sim(p)\})\}.$ (T21)

Here we have applied four times the following theorem of elementary protothetics:

$$[f, g]\{[p, q]\{f(p) \vee g(q)\} \supset ([p]\{f(p)\} \vee [q]\{g(q)\})\}.$$

This theorem, analogous to a theorem of the functional calculus, is obtained from theorems of the propositional calculus by applying the rules for using the general quantifier.

T23. $[f]\{([p]\{f(p) \equiv as\,(p)\} \vee [p]\{f(p) \equiv vr\,(p)\} \vee [p]\{f(p) \equiv fl\,(p)\} \vee$
$[p]\{f(p) \equiv \sim(p)\}) \supset [p, q]\{(p \equiv q) \supset (f(p) \equiv f(q))\}\}.$
(T12; T10; T11; T13)

T24. $[f, p, q]\{(p \equiv q) \supset (f(p) \equiv f(q))\} \equiv [f]\{[p]\{f(p) \equiv as\,(p)\} \vee$
$[p]\{f(p) \equiv vr\,(p)\} \vee [p]\{f(p) \equiv fl\,(p)\} \vee [p]\{f(p) \equiv \sim(p)\}\}.$
(T22; T23)

We have thus proved the equivalence of the law of extensionality and the law of the quantity of functions. We shall prove in turn the equivalence of the law of extensionality and the law of development (c, Section 6).

T25. $[f, r]\{[p, q]\{(p \equiv q) \supset (f(p) \equiv f(q))\} \cdot f(1) \cdot r \supset f(r)\}$.

Proof. (1) $[p, q]\{(p \equiv q) \supset (f(p) \equiv f(q))\}$.

(2) $f(1)$.

(3) r.

(4) $1 \equiv r$. (T6; 3)

$f(r)$ (1; 4; 2)

T26. $[f, r]\{[p, q]\{(p \equiv q) \supset (f(p) \equiv f(q))\} \cdot f(0) \cdot \sim(r) \supset f(r)\}$.

Proof. (1) $[p, q]\{(p \equiv q) \supset (f(p) \equiv f(q))\}$.

(2) $f(0)$.

(3) $\sim(r) \supset$

(4) $0 \equiv r$. (T7; 3)

$f(r)$ (1; 4; 2)

According to T25, any truth-function which is satisfied by 1 is satisfied by any true proposition; according to T26, any truth-function which is satisfied by 0 is satisfied by any false proposition.

T27. $[f, r]\{[p, q]\{(p \equiv q) \supset (f(p) \equiv f(q))\} \cdot (f(1) \cdot r \vee$
$f(0) \cdot \sim(r)) \supset f(r)\}$. (T25; T26)

T28. $[f, r]\{[p, q]\{(p \equiv q) \supset (f(p) \equiv f(q))\} \cdot f(r) \supset (f(1) \cdot r \vee f(0) \cdot \sim(r))\}$.

Proof. (1) $[p, q]\{(p \equiv q) \supset (f(p) \equiv f(q))\}$.

(2) $f(r) \supset$

(3) $r \supset f(1)$. (1; T6; 2)

(4) $\sim(r) \supset f(0)$. (1; T7; 2)

$$f(1) \cdot r \vee f(0) \cdot \sim(r) \qquad (3; 4)$$

In the last line of the proof we have applied the following theorem of the propositional calculus:

$$[p, q, r]\{(r \supset p) \cdot (\sim(r) \supset q) \supset (p \cdot r \vee q \cdot \sim(r))\}.$$

T29. $[f, r]\{[p, q]\{(p \equiv q) \supset (f(p) \equiv f(q))\} \supset (f(r) \equiv (f(1) \cdot r \vee f(0) \cdot \sim(r)))\}$. (T27; T28)

T30. $[f, p, q]\{[r]\{f(r) \equiv (f(1) \cdot r \vee f(0) \cdot \sim(r))\} \cdot (p \equiv q) \supset (f(p) \equiv f(q))\}$.

Proof. (1) $[r]\{f(r) \equiv (f(1) \cdot r \vee f(0) \cdot \sim(r))\}$.

(2) $p \equiv q \supset$

(3) $f(p) \equiv (f(1) \cdot p \vee f(0) \cdot \sim(p))$. (1)

(4) $f(q) \equiv (f(1) \cdot q \vee f(0) \cdot \sim(q))$. (1)

(5) $(f(1) \cdot p \vee f(0) \cdot \sim(p)) \equiv (f(1) \cdot q \vee f(0) \cdot \sim(q))$. (2)

$f(p) \equiv f(q)$ (3; 4; 5)

T31. $[f, p, q]\{(p \equiv q) \supset (f(p) \equiv f(q))\} \equiv [f, p]\{f(p) \equiv (f(1) \cdot p \vee f(0) \cdot \sim(p))\}$. (T29; T30)

We have thus shown the equivalence of the law of extensionality and the law of development. Subsequent theorems will be necessary to prove the equivalence of the law of extensionality and the law of verification (d, Section 6).

T32. $[f, r]\{[p, q]\{(p \equiv q) \supset (f(p) \equiv f(q))\} \cdot f(1) \cdot f(0) \supset f(r)\}$.

Proof. (1) $[p, q]\{(p \equiv q) \supset (f(p) \equiv f(q))\}$.

(2) $f(1)$.

(3) $f(0) \supset$

(4) $(r \equiv 1) \supset f(r)$. (1; 2)

(5) $(r \equiv 0) \supset f(r)$. (1; 3)

$f(r)$ (T9; 4; 5)

In the last line of the proof we have applied the following theorem of the propositional calculus:

$$[p, q]\{(p \vee q) \cdot (p \supset r) \cdot (q \supset r) \supset r\}.$$

We introduce an auxiliary definition necessary for the proof of further theorems:

D1. $[p, q]\{\Phi\langle q\rangle(p) \equiv [f]\{(p \equiv q) \supset (f(p) \equiv f(q))\}\}$.

Definitions of this type are characteristic for protothetics. The functor $\Phi\langle q\rangle$, defined in D1, is a proposition-forming functor of one propositional argument p. This functor depends on the propositional parameter q, which is placed in parentheses differing in shape from all parentheses used so far. Propositional parameters of functors of a propositional argument will always be placed in parentheses of that shape.

T33. $[p, q]\{\Phi\langle p\rangle(q) \equiv \Phi\langle q\rangle(p)\}$. (D1)

To obtain this theorem we have substituted in D1 the variables q and p for the variables p and q, respectively, thus changing not only the argument of the functor defined in D1 but its parameter as well.

T34. $\Phi\langle 1\rangle(1)$. (D1)

T35. $\Phi\langle 0\rangle(0)$. (D1)

T36. $\Phi\langle 1\rangle(0)$. (D1; T8)

T37. $\Phi\langle 0\rangle(1)$. (D1; T8)

T38. $[p]\{\Phi\langle p\rangle(p)\}$. (D1)

T39. $[p]\{\Phi\langle p\rangle(\sim(p))\}$. (D1)

T40. $[f, q, r]\{[f, p]\{f(1) \cdot f(0) \supset f(p)\} \cdot (q \equiv r) \supset (f(q) \equiv f(r))\}$.

Proof. (1) $[f, p]\{f(1) \cdot f(0) \supset f(p)\}$.

(2) $q \equiv r \supset$

(3) $\Phi\langle 1\rangle(r)$. (1; T34; T36)

(4) $\Phi\langle r\rangle(1)$. (T33; 3)

(5) $\Phi\langle 0\rangle(r)$. (1; T37; T35)

(6) $\Phi\langle r\rangle(0)$. (T33; 5)

(7) $\Phi\langle r\rangle(q)$. (1; 4; 6)

$f(q) \equiv f(r)$ (D1; 7; 2)

The expression (3) has been obtained from (1) by replacing in the latter the variable f by $\Phi\langle 1\rangle$ and by performing twice the operation of detachment. The expressions (5) and (7) have been obtained in an analogous way.

T41. $[f,p,q]\{(p \equiv q) \supset (f(p) \equiv f(q))\} \equiv [f,p]\{(f(1) \cdot f(0) \supset f(p)\}$.
(T32; T40)

T41 states the equivalence of the laws of extensionality and verification.

T42. $[f]\{[p]\{f(p)\} \supset f(1) \cdot f(0)\}$.

Proof. (1) $[p]\{f(p)\} \supset$

(2) $f(1)$. (1)

(3) $f(0)$. (1)

$f(1) \cdot f(0)$ (2; 3)

T43. $[f,p,q]\{(p \equiv q) \supset (f(p) \equiv f(q))\} \equiv [f]\{[p]\{f(p)\} \equiv f(1) \cdot f(0)\}$.
(T41; T42)

In the proof we have applied the following theorem of the propositional calculus:

$$[p,q,r]\{(p \equiv (q \supset r)) \cdot (r \supset q) \supset (p \equiv (q \equiv r))\}.$$

T43 states the equivalence of the law of extensionality and the theorem on the limit of a function. Our aim now is to show the equivalence of the law of extensionality and the generalized theorem on the limit of a function (f, Section 6).

T44. $[f,r,s]\{[f,p,q]\{f(q)\cdot f(\sim(q))\supset f(p)\}\cdot(r\equiv s)\supset(f(r)\equiv f(s))\}$.

Proof. (1) $[f,p,q]\{f(q)\cdot f(\sim(q))\supset f(p)\}$.

(2) $r\equiv s\supset$

(3) $\Phi\langle s\rangle(r)$. (1; T38; T39)

$f(r)\equiv f(s)$ (D1; 3; 2)

T45. $[f,r,s]\{[p,q]\{(p\equiv q)\supset(f(p)\equiv f(q))\}\cdot f(r)\cdot f(\sim(r))\supset f(s)\}$.

Proof. (1) $[p,q]\{(p\equiv q)\supset(f(r)\equiv f(q))\}$.

(2) $f(r)$.

(3) $f(\sim(r))\supset$

(4) $(r\equiv s)\supset f(s)$. (1; 2)

(5) $(\sim(r)\equiv s)\supset f(s)$. (1; 3)

$f(s)$ (4; 5)

In the proof we have applied the theorem of propositional calculus quoted in the remark explaining the last line of the proof of T14.

T46. $[f,p,q]\{(p\equiv q)\supset(f(p)\equiv f(q))\}\equiv$
$[f,p,q]\{f(q)\cdot f(\sim(q))\supset f(p)\}$. (T44; T45)

T47. $[f,q]\{[p]\{f(p)\}\supset f(q)\cdot f(\sim(q))\}$.

Proof. (1) $[p]\{f(p)\}\supset$

(2) $f(q)$. (1)

(3) $f(\sim(q))$. (1)

$f(q)\cdot f(\sim(q))$ (2; 3)

T48. $[f,p,q]\{(p\equiv q)\supset(f(p)\equiv f(q))\}\equiv[f,q]\{[p]\{f(p)\}\equiv$
$f(q)\cdot f(\sim(q))\}$. (T46; T47)

In the proof we have applied the theorem of the propositional calculus quoted in the remark explaining the proof of T43.

T48 states the equivalence of the law of extensionality and the generalized theorem on the limit of a function. We have thus proved that in the system discussed in Section 5 the law of extensionality is equivalent to any of the propositions (b)–(f), Section 6, and so we have proved that in that system all the propositions (a)–(f), Section 6, are equivalent.

8. PROPERTIES OF PROPOSITIONAL FUNCTIONS OF MORE THAN ONE ARGUMENT

All the functions which appear in theorems discussed in Sections 6 and 7 are functions of one propositional argument. Analogous theorems are valid for functions of more than one argument. We shall now give theorems whose intuitive meaning is quite close to the intuitive meaning of the propositions (a)–(f), Section 6, and in which functions of two arguments appear.[23]

The law of extensionality for functions of two arguments:

$$(\mathrm{a}')\ [f,p,q,r,s]\{(p \equiv r)\cdot(q \equiv s) \supset (f(p,q) \equiv f(r,s))\}.$$

Let

$$f_1, f_2, \ldots, f_{16}$$

be a sequence of all different functors of two arguments belonging to the propositional calculus. The scheme of definition for the expressions of that sequence has the form:

$$[p,q]\{f_i(p,q) \equiv (f_j(p)\cdot q \vee f_k(p)\cdot \sim(q))\}$$
$$i = 1,\ldots,16; \quad j,k = 1,2,3,4,$$

where f_1, f_2, f_3, f_4 are *as*, *vr*, *fl*, and $\sim$, respectively.

The law of the quantity of functions of two arguments has the form:

$$(\mathrm{b}')\ [f]\{[p,q]\{f(p,q) \equiv f_1(p,q)\} \vee [p,q]\{f(p,q) \equiv f_2(p,q)\} \vee \ldots \vee [p,q]\{f(p,q) \equiv f_{16}(p,q)\}\}.$$

The law of development of functions of two arguments:

$$(\mathrm{c}')\ [f,p,q]\{f(p,q) \equiv (f(1,1)\cdot p\cdot q \vee f(1,0)\cdot p\cdot \sim(q) \vee f(0,1)\cdot \sim(p)\cdot q \vee f(0,0)\cdot \sim(p)\cdot \sim(q))\}.$$

The law of verification of functions of two arguments:

(d′) $[f,p,q]\{f(1,1)\cdot f(1,0)\cdot f(0,1)\cdot f(0,0)\supset f(p,q)\}$.

The theorem on the limit of functions of two arguments:

(e′) $[f]\{[p,q]\{f(p,q)\}\equiv f(1,1)\cdot f(1,0)\cdot f(0,1)\cdot f(0,0)\}$.

The generalized theorem on the limit of functions of two arguments:

(f′) $[f,r,s]\{[p,q]\{f(p,q)\}\equiv f(r,s)\cdot f(r,\sim(s))\cdot f(\sim(r),s)\cdot f(\sim(r),\sim(s))\}$.

Like the propositions (a)–(f), Section 6, the propositions (a′)–(f′) are equivalent in the system discussed in Section 5, and the proofs of those equivalences are analogous to the proofs given in Section 7; moreover, each proposition from among (a′)–(f′) is equivalent in that system to its corresponding proposition from among (a)–(f), Section 6.

We shall now, by way of example, give the proofs of the equivalence of the propositions (a) and (a′), and (d) and (d′).* We shall need for that purpose the following auxiliary definitions:

D1. $[f,p,q]\{\chi\langle f,p\rangle(q)\equiv f(p,q)\}$.

D2. $[f,p,q]\{\psi\langle f,p\rangle(q)\equiv f(q,p)\}$.[24]

T1. $[f,r,s,t,u]\{[f,p,q]\{(p\equiv q)\supset(f(p)\equiv(f(q))\}\cdot(r\equiv t)\cdot(s\equiv u)\supset(f(r,s)\equiv f(t,u))\}$.

Proof. (1) $[f,p,q]\{(p\equiv q)\supset(f(p)\equiv f(q))\}$.

(2) $r\equiv t$.

(3) $s\equiv u\supset$

(4) $\psi\langle f,s\rangle(r)\equiv\psi\langle f,s\rangle(t)$. (1; 2)

(5) $f(r,s)\equiv f(t,s)$. (4; D2)

(6) $\chi\langle f,t\rangle(s)\equiv\chi\langle f,t\rangle(u)$. (1; 3)

(7) $f(t,s)\equiv f(t,u)$. (6; D1)

---

* [Ed. Note: The equivalence of (c) and (c′), (e) and (e′), and of (f) and (f′) is demonstrated in Chikawa (1967); cf. the Protothetic Bibliography.]

$f(r,s) \equiv f(t,u)$ (5; 7)

D3. $[f,p,q]\{\omega\!\Leftarrow\! f\!\Rightarrow\!(p,q) \equiv (f(p) \equiv vr(q))\}$.

D3 includes the expression $vr(q)$ in order that the same free variables might appear both in the definiendum and in the definiens.

T2. $[f,p,q]\{\omega\!\Leftarrow\! f\!\Rightarrow\!(p,q) \equiv f(p)\}$. (D3; T4, Section 7)

T3. $[f,p,q]\{[f,r,s,t,u]\{(r \equiv t)\cdot(s \equiv u) \supset (f(r,s) \equiv f(t,u))\}\cdot$
$(p \equiv q) \supset (f(p) \equiv f(q))\}$.

Proof. (1) $[f,r,s,t,u]\{(r \equiv t)\cdot(s \equiv u) \supset (f(r,s) \equiv f(t,u))\}$.

(2) $p \equiv q \supset$

(3) $\omega\!\Leftarrow\! f\!\Rightarrow\!(p,p) \equiv \omega\!\Leftarrow\! f\!\Rightarrow\!(q,q)$. (1; 2)

$f(p) \equiv f(q)$ (3; T2)

To obtain (3) we have replaced in (1)[25] the variables r and s by the variable p, the variables t and u by the variable q, and the variable f by the expression $\omega\!\Leftarrow\! f\!\Rightarrow$.

T4. $[f,p,q]\{(p \equiv q) \supset (f(p) \equiv f(q))\} \equiv [f,p,q,r,s]\{(p \equiv r)\cdot(q \equiv s) \supset$
$(f(p,q) \equiv f(r,s))\}$. (T1; T3)

We have thus proved the equivalence of the propositions (a) and (a′); we shall in turn prove the equivalence of (d) and (d′).

T5. $[f,q,r]\{[f,p]\{f(1)\cdot f(0) \supset f(p)\}\cdot f(q,1)\cdot f(q,0) \supset \psi\!\leftarrow\! f,r\!\rightarrow\!(q)\}$.

Proof. (1) $[f,p]\{f(1)\cdot f(0) \supset f(p)\}$.

(2) $f(q,1)$.

(3) $f(q,0) \supset$

(4) $\chi\!\leftarrow\! f,q\!\rightarrow\!(1)$. (D1; 2)

(5) $\chi\!\leftarrow\! f,q\!\rightarrow\!(0)$. (D1; 3)

(6) $\chi\langle\!\!-f, q-\!\!\rangle(r)$. (1; 4; 5)

(7) $f(q, r)$. (D1; 6)

$\psi\langle\!\!-f, r-\!\!\rangle(q)$ (D2; 7)

T6. $[f, q, r]\{[f, p]\{f(1) \cdot f(0) \supset f(p)\} \cdot f(1, 1) \cdot f(1, 0) \cdot f(0, 1) \cdot f(0, 0) \supset f(q, r)\}$.

Proof. (1) $[f, p]\{f(1) \cdot f(0) \supset f(p)\}$.

(2) $f(1, 1)$.

(3) $f(1, 0)$.

(4) $f(0, 1)$.

(5) $f(0, 0) \supset$

(6) $\psi\langle\!\!-f, r-\!\!\rangle(1)$. (T5; 1; 2; 3)

(7) $\psi\langle\!\!-f, r-\!\!\rangle(0)$. (T5; 1; 4; 5)

(8) $\psi\langle\!\!-f, r-\!\!\rangle(q)$. (1; 6; 7)

$f(q, r)$ (D2; 8)

T7. $[f, r]\{[f, p, q]\{f(1, 1) \cdot f(1, 0) \cdot f(0, 1) \cdot f(0, 0) \supset f(p, q)\} \cdot f(1) \cdot f(0) \supset f(r)\}$.

Proof. (1) $[f, p, q]\{f(1, 1) \cdot f(1, 0) \cdot f(0, 1) \cdot f(0, 0) \supset f(p, q)\}$.

(2) $f(1)$.

(3) $f(0) \supset$

(4) $\omega\langle\!\!=f=\!\!\rangle(1, 1)$. (T2; 2)

(5) $\omega\langle\!\!=f=\!\!\rangle(1, 0)$. (T2; 2)

(6) $\omega\langle\!\!=f=\!\!\rangle(0, 1)$. (T2; 3)

(7) $\omega\!\!\leftleftarrows\! f \!\rightrightarrows\!(0,0)$. (T2; 3)

(8) $\omega\!\!\leftleftarrows\! f \!\rightrightarrows\!(r,r)$. (1; 4; 5; 6; 7)

$f(r)$ (T2; 8)

T8. $[f,p]\{f(1)\cdot f(0) \supset f(p)\} \equiv [f,p,q]\{f(1,1)\cdot f(1,0)\cdot f(0,1)\cdot f(0,0) \supset f(p,q)\}$. (T6, T7)

T8 asserts the equivalence of the propositions (d) and (d′).

The following two propositions are the respective laws of extensionality for functions of n and of $n+1$ arguments:

(g) $[p_1,\ldots,p_n,q_1,\ldots,q_n]\{(p_1 \equiv q_1)\cdot\ldots\cdot(p_n \equiv q_n) \supset (f(p_1,\ldots,p_n) \equiv f(q_1,\ldots q_n))\}$;

(h) $[p_1,\ldots,p_{n+1},q_1,\ldots,q_{n+1}]\{(p_1 \equiv q_1)\cdot\ldots\cdot(p_{n+1} \equiv q_{n+1}) \supset (f(p_1,\ldots,p_{n+1}) \equiv f(q_1,\ldots q_{n+1}))\}$.

The propositions (g) and (h) are equivalent. When proving this we shall use the following definitions analogous to D1–3:

D4. $[f,p_1,\ldots,p_{n+1}]\{\chi\!\leftarrow\! f,p_1\!\rightarrow\!(p_2,\ldots,p_{n+1}) \equiv f(p_1,\ldots,p_{n+1})\}$;

D5. $[f,p_1,\ldots,p_{n+1}]\{\psi\!\leftarrow\! f,p_{n+1}\!\rightarrow\!(p_1,\ldots,p_n) \equiv f(p_1,\ldots,p_{n+1})\}$;

D6. $[f,p_1,\ldots,p_{n+1}]\{\omega\!\!\leftleftarrows\! f \!\rightrightarrows\!(p_1,\ldots,p_{n+1}) \equiv (f(p_1,\ldots,p_n) \equiv vr(p_{n+1}))\}$.

T9. $[f,r_1,\ldots,r_{n+1},s_1,\ldots,s_{n+1}]\{[f,p_1,\ldots,p_n,q_1,\ldots,q_n]\{(p_1 \equiv q_1)\cdot\ldots\cdot(p_n \equiv q_n) \supset (f(p_1,\ldots,p_n) \equiv f(q_1,\ldots,q_n))\}\cdot(r_1 \equiv s_1)\cdot\ldots\cdot(r_{n+1} \equiv s_{n+1}) \supset (f(r_1,\ldots,r_{n+1}) \equiv f(s_1,\ldots,s_{n+1}))\}$.

Proof. (1) $[f,p_1,\ldots,p_n,q_1,\ldots,q_n]\{(p_1 \equiv q_1)\cdot\ldots\cdot(p_n \equiv q_n) \supset (f(p_1,\ldots,p_n) \equiv f(q_1,\ldots,q_n))\}$.

(2) $(r_1 \equiv s_1)\cdot\ldots\cdot(r_{n+1} \equiv s_{n+1}) \supset$

(3) $\psi\!\leftarrow\! f,r_{n+1}\!\rightarrow\!(r_1,\ldots,r_n) \equiv \psi\!\leftarrow\! f,r_{n+1}\!\rightarrow\!(s_1,\ldots,s_n)$. (1; 2)

(4) $f(r_1, \ldots, r_{n+1}) \equiv f(s_1, \ldots, s_n, r_{n+1}).$ (3; D5)

(5) $\chi\dashleftarrow f, s_1 \dashrightarrow(s_2, \ldots, s_n, r_{n+1}) \equiv \chi\dashleftarrow f, s_1 \dashrightarrow(s_2, \ldots, s_{n+1}).$ (1; 2)

(6) $f(s_1, \ldots, s_n, r_{n+1}) \equiv f(s_1, \ldots, s_{n+1}).$ (5; D4)

$f(r_1, \ldots, r_{n+1}) \equiv f(s_1, \ldots, s_{n+1})$ (4; 6)

T10. $[f, p_1, \ldots, p_{n+1}]\{\omega\Leftarrow f \Rightarrow(p_1, \ldots, p_{n+1}) \equiv f(p_1, \ldots, p_n)\}.$ (D6; T4, Section 7)

This theorem is analogous to T2.

T11. $[f, p_1, \ldots, p_n, q_1, \ldots, q_n]\{[f, r_1, \ldots, r_{n+1}, s_1, \ldots, s_{n+1}]$
$\{(r_1 \equiv s_1)\cdot\ldots\cdot(r_{n+1} \equiv s_{n+1}) \supset (f(r_1, \ldots, r_{n+1}) \equiv f(s_1, \ldots, s_{n+1}))\}\cdot$
$(p_1 \equiv q_1) \cdot \ldots \cdot (p_n \equiv q_n) \supset (f(p_1, \ldots, p_n) \equiv f(q_1, \ldots, q_n))\}.$

Proof. (1) $[f, r_1, \ldots, r_{n+1}, s_1, \ldots, s_{n+1}]\{(r_1 \equiv s_1) \cdot \ldots \cdot$
$(r_{n+1} \equiv s_{n+1}) \supset (f(r_1, \ldots, r_{n+1}) \equiv f(s_1, \ldots, s_{n+1}))\}.$

(2) $(p_1 \equiv q_1) \cdot \ldots \cdot (p_n \equiv q_n) \supset$

(3) $\omega\Leftarrow f \Rightarrow(p_1, \ldots, p_n, p_n) \equiv \omega\Leftarrow f \Rightarrow(q_1, \ldots, q_n, q_n)$ (1; 2)

$f(p_1, \ldots, p_n) \equiv f(q_1, \ldots, q_n)$ (3; T10)

T12. $[f, p_1, \ldots, p_n, q_1, \ldots, q_n]\{(p_1 \equiv q_1) \cdot \ldots \cdot (p_n \equiv q_n) \supset$
$(f(p_1, \ldots, p_n) \equiv f(q_1, \ldots, q_n))\} \equiv$
$[f, p_1, \ldots, p_{n+1}, q_1, \ldots, q_{n+1}]\{(p_1 \equiv q_1) \cdot \ldots \cdot$
$(p_{n+1} \equiv q_{n+1}) \supset (f(p_1, \ldots, p_{n+1}) \equiv f(q_1, \ldots, q_{n+1}))\}.$ (T9; T11)

We have thus proved the equivalence of propositions (g) and (h), and the proof was obviously analogous to the proof of the equivalence of (a) and (a′).

We now formulate the theorem which we have proved only for certain cases; the proofs for the remaining cases, however, are analogous:

THEOREM. *The laws of extensionality, quantity of functions, development and verification as well as both theorems on the limit of a function, formulated for functors of any number of propositional arguments, are equivalent in the system discussed in Section 5.*

9. THE AXIOM OF THE SYSTEM OF ELEMENTARY PROTOTHETICS WHICH HAS IMPLICATION AS ITS PRIMITIVE TERM

The notes contain the following remark: a set of axioms of any system of elementary protothetics should satisfy two conditions:

1. all theorems of the propositional calculus should result from these axioms (since elementary protothetics is a generalization of the propositional calculus with quantifiers);
2. the following proposition:

A1. $[f, q]\{f([p]\{p \supset p\}) \supset (f([p]\{p\}) \supset f(q))\}$

should result from these axioms.

In the systems of elementary protothetics which have implication as their primitive term the valid rules of inference are those of the system discussed in Section 5, strengthened as specified at the beginning of Section 7.

THEOREM 1. *All theorems of the propositional calculus may be deduced from* A1 *by applying the rules of inference used in the systems of elementary protothetics which have implication as their primitive term.*[26]

Proof. From the theorem formulated in Section 5 it follows that in order to prove Theorem 1 it suffices to show that A1–3, Section 5, may be deduced from A1.

The definitions needed when proving this will be noted in the form of two implications.

D1. (α) $[p]\{as(p) \supset p\}$;

(β) $[p]\{p \supset as(p)\}$.

T1. $[p]\{[p]\{p\} \supset as(p)\}$. (D1, β)

We have applied here the rule of adding the general quantifier to the antecedent of an implication.

When noting further theorems we shall use the following abbreviations:

instead of $[p]\{p \supset p\}$	we shall write 1,
instead of $[p]\{p\}$	we shall write 0.

The abbreviation of T1 has the form:

T1′. $[p]\{0 \supset as\,(p)\}$.

In certain cases, however, it will be convenient to give the full form of a theorem along with its abbreviation.

Using the above abbreviations we shall give A1 the following form:

A1′. $[f, q]\{f(1) \supset (f(0) \supset f(q))\}$.[27]

T2. $0 \supset as\,(1)$. (T1)

T3. $1 \supset as\,(1)$. (D1, β)

D2. (α) $[q]\{\alpha(q) \supset (q \supset as\,(q))\}$;
(β) $[q]\{(q \supset as\,(1)) \supset \alpha(q)\}$.

T4. $\alpha(1)$. (D2, β; T3)

T5. $\alpha(0)$. (D2, β; T2)

T6. $[q]\{\alpha(q)\}$. (A1′; T4; T5)

T7. $[q]\{q \supset as\,(1)\}$. (D2, α; T6)

T8. $as\,(1)$. (T7)

T9. 1. (D1, α; T8)

When written without abbreviations, this theorem has the form:

T9′. $[p]\{p \supset p\}$. (T9)

T10. $[p]\{[p]\{p\} \supset p\}$. (T9′)

T10′. $[p]\{0 \supset p\}$. (T10)

T11. $0 \supset (1 \supset 0)$. (T10′)

T12. $[p]\{[p]\{p \supset p\} \supset (p \supset p)\}$. (T9′)

T12′. $[p]\{1 \supset (p \supset p)\}$. (T12)

T13. $1 \supset (1 \supset 1)$. (T12′)

D3. (α) $[q]\{\beta(q) \supset (q \supset (1 \supset q))\}$;
(β) $[q]\{(q \supset (1 \supset q)) \supset \beta(q)\}$.

T14. $\beta(1)$. (D3, β; T13)

T15. $\beta(0)$. (D3, β; T11)

T16. $[q]\{\beta(q)\}$. (A1′; T14; T15)

T17. $[q]\{q \supset (1 \supset q)\}$. (D3, α; T16)

T18. $[p]\{1 \supset (0 \supset p)\}$. (T17; T10′)

T19. $1 \supset (0 \supset 1)$. (T18)

T20. $0 \supset (0 \supset 0)$. (T10′)

D4. (α) $[q]\{\gamma(q) \supset (q \supset (0 \supset q))\}$;
(β) $[q]\{(q \supset (0 \supset q)) \supset \gamma(q)\}$.

T21. $\gamma(1)$. (D4, β; T19)

T22. $\gamma(0)$. (D4, β; T20)

T23. $[q]\{\gamma(q)\}$. (A1′; T21; T22)

T24. $[q]\{q \supset (0 \supset q)\}$. (D4, α; T23)

D5. (α) $[p, q]\{\delta\langle q\rangle(p) \supset (q \supset (p \supset q))\}$;
(β) $[p, q]\{(q \supset (p \supset q)) \supset \delta\langle q\rangle(p)\}$.

T25. $[q]\{\delta\langle q\rangle(1)\}$. (D5, β; T17)

T26. $[q]\{\delta\langle q\rangle(0)\}$. (D5, β; T24)

T27. $[p, q]\{\delta\langle q\rangle(p)\}$. (A1′; T25; T26)

T28. $[p, q]\{q \supset (p \supset q)\}$. (D5, α; T27)

The theorem obtained is equiform to A2, Section 5. Our aim is to obtain a theorem which would be equiform to A1, Section 5.

T29. $[p,q]\{q \supset (p \supset p)\}$. (T28; T9′)

T29′. $[q]\{q \supset 1\}$. (T29)

T30. $[p,q,r]\{r \supset (q \supset (p \supset p))\}$. (T28; T29)

T31. $[p,q,r]\{r \supset (q \supset (p \supset q))\}$. (T28)

T32. $(0 \supset 0) \supset ((0 \supset 0) \supset (0 \supset 0))$. (T28)

T33. $(1 \supset 0) \supset ((0 \supset 0) \supset (1 \supset 0))$. (T28)

T34. $(0 \supset 1) \supset ((1 \supset 0) \supset (0 \supset 0))$. (T30)

T35. $(0 \supset 0) \supset ((0 \supset 1) \supset (0 \supset 1))$. (T29)

T36. $(1 \supset 1) \supset ((1 \supset 0) \supset (1 \supset 0))$. (T29)

T37. $(1 \supset 0) \supset ((0 \supset 1) \supset (1 \supset 1))$. (T30)

T38. $(0 \supset 1) \supset ((1 \supset 1) \supset (0 \supset 1))$. (T28)

T39. $(1 \supset 1) \supset ((1 \supset 1) \supset (1 \supset 1))$. (T28)

T33–38 are all the possible expressions which may be obtained from A1, Section 5, by substituting for the variables the expressions 1 and 0. Thus they are the *zero-one tests* of that axiom. By using several times A1′ and the definitions analogous to D1–4, we easily obtain:

T40. $[p,q,r]\{(p \supset q) \supset ((q \supset r) \supset (p \supset r))\}$.

This theorem is equiform to A1, Section 5.

All the zero-one tests of the following proposition:

T41. $[p,q,r]\{(p \supset (q \supset r)) \supset (q \supset (p \supset r))\}$

are results of substitution applied to T9′, T30 and T31. Hence T41 may be deduced from A1.

T42. $[p,q]\{p \supset ((p \supset q) \supset q)\}$. (T41; T9′)

T43. $[p]\{((0 \supset p) \supset q) \supset q\}$. (T42; T10′)

All the zero-one tests of A3, Section 5, are results of substitution applied to T43 and T29, so that this axiom results from A1. We have proved that all the propositions A1–3, Section 5, result from A1, and hence we conclude that Theorem 1 is true.

DEFINITION. The system of elementary protothetics which has implication as its sole primitive term, which has the proposition A1 as its sole axiom, and in which the rules of inference of the system discussed in Section 5, strengthened as specified at the beginning of Section 7, are valid, shall be called *system E*.

It follows easily from Theorem 1 and the theorem formulated in Section 8 that all the theorems discussed in Sections 6 and 8 are theorems of E.[28] The following theorem, the proof of which we omit, is also true:

THEOREM 2. *The system E is a complete system.*

10. THE SYSTEM OF PROTOTHETICS IN WHICH THE RULE OF VERIFICATION IS VALID

There exist propositions which contain only expressions belonging to the semantic categories characteristic of protothetics but which do not belong to the system E[29] and which, as it seems, do not result from its theorems in spite of the fact that from the intuitive point of view we would undoubtedly include those propositions in the class of true propositions.

The following propositions can be given as examples:

(1) $[f, g, \varphi]\{[p, q]\{f(p, q) \equiv g(p, q)\} \supset (\varphi\{f\} \equiv \varphi\{g\})\}$;

(2) $[f, \varphi]\{\varphi\{as\} \cdot \varphi\{vr\} \cdot \varphi\{fl\} \cdot \varphi\{\sim\} \supset \varphi\{f\}\}$.[30]

In order to enlarge the system E so that it should include as its theorems the propositions (1) and (2), as well as all those propositions of protothetics which are undoubtedly true, it is necessary either to add new axioms to the axiom of the system E or to adopt new rules of inference. The first method, as it seems, would not suffice to solve the problem in question since there are always propositions which contain expressions of semantic orders higher than the semantic order of any expression contained in the axioms; such propositions, consequently, do not result from those axioms. On the contrary, the other method makes it possible to solve our problem.

The rule which was at first adopted by Leśniewski was the *rule of verification*, which is a generalization of a rule formulated in Łukasiewicz (1921). Łukasiewicz's rule allows to add to the system of the propositional calculus any proposition which fulfills the following condition: both expressions resulting from that proposition if we substitute in it for one of its propositional variables the respective values 1 and 0, are theorems of that system. The generalized rule refers not only to propositional variables, but to variables of all those semantic categories which are proper to protothetics.

The system which has implication as its sole primitive term, of which the system E is a part, and in which all the rules of inference of E, strengthened as specified at the beginning of Section 7, as well as the rule of verification are valid, is historically the first system of protothetics. It will be called *system* S. The proof of its completeness will be given in Section 12.

In order to be able to formulate precisely the rule of verification we shall define the notion of a *verifier* of a variable of a given semantic category.[31] This definition will have an inductive form, the induction referring to the semantic order of the variable in question.

The verifiers of a propositional variable are the expressions 1 and 0.[32]

By way of example, we shall in a more detailed way discuss the verifiers of a variable f, which represents functors of one propositional argument. There are four such verifiers: f_1, f_2, f_3, f_4. We characterize them assuming that the following propositions are true:

(3) $f_1(1), f_1(0)$;

(4) $f_2(1), \sim(f_2(0))$;

(5) $\sim(f_3(1)), f_3(0)$;

(6) $\sim(f_4(1)), \sim(f_4(0))$.

Each pair of propositions (3)–(6) shall be called the *characteristic* of that verifier which appears in it. The definitions of those verifiers will be given the following form:

D1. $[p]\{f_1(p) \equiv ((p \equiv 1) \vee (p \equiv 0))\}$;

D2. $[p]\{f_2(p) \equiv (p \equiv 1)\}$;

D3. $[p]\{f_3(p) \equiv (p \equiv 0)\}$;

D4. $[p]\{f_4(p) \equiv \sim((p \equiv 1) \vee (p \equiv 0))\}$.

When comparing the definitions of the verifiers f_1, f_2 and f_3 with their characteristics we note that the expression $p \equiv 1$ is the definiens or, in case of this definiens being an alternative, a part of the definiens of any of those definitions, if and only if the proposition having 1 as its argument is not negated in the characteristic of the verifier which is being defined. An analogous remark may be made with respect to the expression $p \equiv 0$. As regards the verifier f_4 let us note that its characteristic contains only negated propositions, and its definition has to the right of the symbol of equivalence the negation of the alternative of the expressions $p \equiv 1$ and $p \equiv 0$.

When giving a general definition of a verifier we must speak of expressions of any semantic category. Consequently, we are not allowed to use parentheses which determine the semantic category of functors. We shall use, however, ordinary parentheses, but we must bear in mind that the parenthesized expressions must not necessarily belong to the semantic category of propositions.

Before discussing the verifiers of a variable belonging to any semantic category, we shall give a scheme for definitions of the *equivalence of two functors* belonging to an arbitrarily chosen, but one and the same for both functors, semantic category:

D5. $[\varphi, \psi]\{(\varphi \equiv \psi) \equiv [\xi_1, \ldots, \xi_n]\{\varphi(\xi_1, \ldots, \xi_n) \equiv \psi(\xi_1, \ldots, \xi_n)\}\}$.

No assumptions are made here concerning the semantic category of the variables $\xi_1, \ldots, \xi_n$. The definition of the equivalence of two functors belonging to a given semantic category can be written in the correct notation. For example, the definition of the equivalence of two functors of one propositional argument has the following form:

$$[f, g]\{\equiv \langle f, g\rangle \equiv [p]\{f(p) \equiv g(p)\}\}.$$

The functor defined by D5 will be called the *sign of interfunctorial equivalence*.

It can easily be noticed that the functors $f_1, \ldots, f_4$, defined by D1–4, are equivalent to the functors *vr*, *as*, $\sim$ and *fl*, respectively.

Further remarks in this section will be limited to functors of two arguments. The generalization of those remarks, so that they might be applied to functors of any number of arguments, does not involve any essential difficulties.

Let φ be a variable of any semantic order representing functors of two arguments. Let f and g be the arguments of the functor φ. Let the respective verifiers of the variables φ, f and g form the sequences:

(7) $\varphi_1, \ldots, \varphi_l$;

(8) $f_1, \ldots, f_m$;

(9) $g_1, \ldots, g_n$.

The characteristic of an arbitrarily chosen verifier φ_i belonging to the sequence (7) is a system of $m.n$ propositions, anyone of which has the form

(10) $\varphi_i(f_j, g_k)$ or $\sim(\varphi_i(f_j, g_k))$,

where f_j and g_k are any verifiers of the variables f and g, and for each pair of verifiers, the first of which belongs to (8) and the second to (9), exactly one of the propositions (10) belongs to the characteristic of the verifier φ_i. We are assuming that the propositions belonging to this characteristic are true.

The characteristic of any functor φ_i of two arguments f and g can be noted in the form of a table analogous to tables characterizing the properties of functors of the propositional calculus:

φ_i	g_1	g_2	$\dots$	g_n
f_1	.	0	$\dots$	.
f_2	.	.		
.	.			
.	.			
.	.			
f_m	1			

The first column of the table contains all the verifiers of the variable f, and the uppermost line, all the verifiers of the variable g. The symbols 1 and 0, appearing in the table by way of example, show that the propositions $\varphi_i(f_m, g_1)$ and $\sim(\varphi_i(f_1, g_2))$ are true.

We shall now discuss the form of the definition of the verifier φ_i. If its characteristic contains more than one non-negated proposition, then the definiens is an alternative of propositions of type:

(11) $(f \equiv f_j) \cdot (g \equiv g_k)$.

This alternative has as many components as there are non-negated propositions in the characteristic of φ_i, and to each of those propositions there corresponds a proposition of type (11) which contains the same verifiers belonging to the sequences (8) and (9) as does the given non-negated proposition of the characteristic.

If the characteristic of φ_i contains only one non-negated proposition, then the definiens is a proposition of type (11) and contains the same verifiers belonging to (8) and (9) as does the non-negated proposition of the characteristic.

If the characteristic of φ_i includes negated propositions only, then the definiens is the negation of the alternative of all propositions of type (11).

The verifier can be defined in the following abbreviated form:

The expressions 1 *and* 0 *are the verifiers of a propositional variable, and functors defined by definitions having the structure specified above are the verifiers of the functorial variable* φ. *Those definitions contain verifiers of* *arguments of the functor φ. Those arguments are of lower semantic orders than the variable φ.*

We shall now state two theorems on verifiers.

THEOREM 1. *The equivalence of two equiform verifiers and the negation of the equivalence of two non-equiform verifiers of one and the same variable result from the theorems of elementary protothetics on the strength of the rules of inference of that system, strengthened as specified at the beginning of Section 7.*

Proof. In the case of equiform verifiers, the definiens of D5 is a result of substitution applied to the proposition $[p]\{p \equiv p\}$, which is a theorem of elementary protothetics. Hence in that case Theorem 1 is true.

If the non-equiform verifiers are 1 and 0, Theorem 1 results from the fact that the expression $\sim(1 \equiv 0)$ is a theorem of elementary protothetics (T8, Section 7). Let us assume now that our theorem is true for the non-equiform verifiers of the variables f and g which are arguments of the variable functor φ. Let the terms of the sequence (7)–(9) be all the verifiers of the variables φ, f and g, respectively. Let φ_r and φ_s be any non-identical terms of (7). For the time being we assume that the characteristics of both these verifiers contain non-negated propositions.

From the structure of the definitions of verifiers we conclude that there exists an expression of type (11) which appears in the definiens of one of these verifiers and does not appear in the definiens of the other verifier.

We may assume that this expression appears in the definiens of the verifier φ_r, and that it contains the expressions f_j and g_k of the sequences (8) and (9). Hence, from the inductive hypothesis and from the fact that the theorem is true for equiform functors, it follows that the expressions $\varphi_r(f_j, g_k)$ and $\sim(\varphi_s(f_j, g_k))$, are theorems of the system S. Therefore, the expression

$$\sim(\varphi_r(f_j, g_k) \equiv \varphi_s(f_j, g_k))$$

is a theorem of that system, and consequently also the expression

$$\sim([f, g]\{\varphi_r(f, g) \equiv \varphi_s(f, g)\}).$$

From the above and from D5 we conclude that the expression $\sim(\varphi_r \equiv \varphi_s)$ is a theorem of the system S. In case of the characteristic of one of the verifiers φ_r and φ_s containing solely negated propositions, the above reasoning is not subject to any essential change.

THEOREM 2. *Every proposition occurring in the characteristic of a verifier of a functorial variable of any semantic category results from theorems of elementary protothetics on the strength of the rules of inference valid in that system, strengthened as specified at the beginning of Section 7.*

Proof. Taking into account the structure of the definitions of verifiers (of which D1–4 are examples) and Theorem 1 we obtain easily Theorem 2. For the proof the following propositions of the propositional calculus are required:

$$[p_1, \ldots, p_k]\{p_i \supset (p_1 \vee \ldots \vee p_k)\} \qquad i = 1, \ldots, k;$$

$$[p_1, \ldots, p_k]\{\sim(p_1) \cdot \ldots \cdot \sim(p_k) \supset \sim(p_1 \vee \ldots \vee p_k)\}.$$

We now formulate the *rule of verification*:

Let α be any proposition of protothetics. If every proposition obtainable from α by substituting in α for a certain of its free variables representing functors the verifiers of this variable is a theorem of the system S, then α is also a theorem of S.

The following is a generalization of the rule of verification.

THEOREM 3. *If every proposition obtainable from any proposition α belonging to protothetics by substituting in α for all or some of its free variables (representing either functors or propositions) their verifiers is a theorem of the system S, then the proposition α is also a theorem of that system.*

Proof. From the theorem given in Section 8 it follows that the rule of verification may be generalized so as to cover propositional variables. When continuing the proof we apply induction on the number of those free variables for which their verifiers have been substituted in α. For the proof certain auxiliary definitions are required.

11. THEOREMS WHICH ARE GENERALIZATIONS OF THE LAWS OF ELEMENTARY PROTOTHETICS

Each of the theorems of elementary protothetics discussed in Sections 6 and 8, with the exception of the law of development and the generalized theorem on the limit of a function, can be generalized so that in the generalized theorems functors of higher orders appear in place of functors of the first semantic order. The terms applied when referring to theorems of elementary protothetics will also be used for the corresponding generalized theorems.

As in the preceding section, the remarks made in this section will refer to functors of two arguments only. Let φ be a variable functor of two arguments, belonging to any semantic category. Let f and g be the arguments of φ. Let further the sequences (7)–(9), Section 10, be the sequences of all the verifiers of the variables φ, f and g, and let f', g' be variables of the same semantic category as f and g.

The law of extensionality:

(a) $[f, f', g, g', \varphi]\{(f \equiv f') \cdot (g \equiv g') \supset (\varphi(f, g) \equiv \varphi(f', g'))\}.$

The law of the quantity of functions:

(b) $[\varphi]\{[f,g]\{\varphi(f,g) \equiv \varphi_1(f,g)\} \vee \ldots \vee [f,g]\{\varphi(f,g) \equiv \varphi_l(f,g)\}\}$.

The law of verification:

(c) $[f,g,\varphi]\{\varphi(f_1,g_1) \cdot \varphi(f_1,g_2) \cdot \ldots \cdot \varphi(f_m,g_n) \supset \varphi(f,g)\}$.[33]

The theorem on the limit of a function:

(d) $[\varphi]\{[f,g]\{\varphi(f,g)\} \equiv \varphi(f_1,g_1) \cdot \varphi(f_1,g_2) \cdot \ldots \cdot (f_m,g_n)\}$.

Proposition (2), Section 10, is the correct notation of the law of verification formulated for those functors whose sole argument is a functor of one propositional argument. We also give the correct notation of the law of extensionality for these functors:

$$[f,f',\varphi]\{\equiv \langle f,f' \rangle \supset (\varphi(f') \equiv \varphi(f))\}.$$

The proofs of the theorems given below will be limited to the case of functors of two arguments. A generalization of those proofs does not involve any essential difficulties.

THEOREM 1. *The law of extensionality formulated for functors of any semantic category is a theorem of S.*

Proof. Let us substitute in the proposition (a) for the variables f, f', g and g' any of their verifiers. If the verifiers substituted for f and f' or the verifiers substituted for g and g' are not equiform, then on the strength of Theorem 1, Section 10, we conclude easily that the proposition obtained as a result of this substitution is a theorem of S. In the opposite case, the consequent of the proposition obtained as a result of this substitution is an equivalence of two equiform expressions and hence a theorem of S. From this and from Theorem 3, Section 10, we conclude that Theorem 1 is true.

THEOREM 2. *The law of verification, formulated for functors of any semantic category, is a theorem of S.*

Proof. By substituting in the proposition (c) for the variables f and g any of their verifiers we obtain a proposition which is a result of substitution applied to a theorem of the propositional calculus. From this and from Theorem 3, Section 10, we obtain Theorem 2.

THEOREM 3. *The theorem on the limit of a function, formulated for functors of any semantic category, is a theorem of S.*

Proof. It can easily be seen that the propositions (c) and (d) are equivalent in S. From this and from Theorem 2, we obtain Theorem 3.

In the next section we shall prove that also the law of the quantity of functions formulated for functors of any semantic category is a theorem of S.

12. THE SYSTEM OF PROTOTHETICS IN WHICH THE RULE OF EXTENSIONALITY IS VALID

We shall consider now a system of protothetics, equivalent to the system S, in which, however, instead of the rule of verification another rule, namely the so called rule of extensionality, is being used:[34]

Rule of extensionality: *Given any functor of one argument and of at least the second semantic order, the law of extensionality formulated for that functor is a theorem of that system.*

The system having implication as its only primitive term, including as its part the system E, and in which all the rules of inference of E, strengthened as specified at the beginning of Section 7, as well as the rule of extensionality, are valid will be called system S_1.

The following theorem is a generalization of the rule of extensionality.

THEOREM 1. *The law of extensionality formulated for functors of any semantic category is a theorem of* S_1.

Proof. From the theorem given in Section 8 we conclude that the laws of extensionality formulated for functors of the first semantic order are theorems of S_1. By reasoning in a way similar to that followed in the proofs of T1–4, Section 8, and T9–12, Section 8, we arrive at the conclusion that the laws of extensionality formulated for functors belonging to the same semantic order but having different numbers of arguments are equivalent in the system S_1. From this Theorem 1 results.

Our object now being the proof of the equivalence of the systems S and S_1, we shall need the following:

THEOREM 2. *The laws of verification* (c, Section 11) *and of the quantity of functions* (b, Section 11), *formulated for functors of any semantic category, are theorems of the system S.*

Proof. The proof will be carried out by induction on the semantic order of those functors for which these laws are formulated.

For functors of the first semantic order Theorem 2 results from the theorem given in Section 8.

Let n be any natural number ≥ 1, and let φ be a functor of the $n+1$-st semantic order. We assume that the theorem holds for functors belonging to orders not higher than n. We may also assume, without affecting thus our proof in an essential way, that φ is a functor of two arguments. Let f and g be the arguments of φ. The semantic order of both these variables is not higher than n. Let, further, the sequences (7), (8), (9), Section 10, be the respective sequences of verifiers of the variables φ, f, g. According to our inductive assumption the laws of the quantity of functions formulated as follows:

T1. $[f]\{(f \equiv f_1) \vee \ldots \vee (f \equiv f_m)\}$;

T2. $[g]\{(g \equiv g_1) \vee \ldots \vee (g \equiv g_n)\}$

are theorems of the system S.

From T1, T2 and the following theorem of the propositional calculus:

$$[p_1, \ldots, p_m, q_1, \ldots, q_n]\{(p_1 \vee \ldots \vee p_m) \cdot (q_1 \vee \ldots \vee q_n) \supset (p_1 \cdot q_1 \vee p_1 \cdot q_2 \vee \ldots \vee p_m \cdot q_n)\}$$ [35]

we may deduce:

T3. $[f, g]\{(f \equiv f_1) \cdot (g \equiv g_1) \vee (f \equiv f_1) \cdot (g \equiv g_2) \vee \ldots \vee (f \equiv f_m) \cdot (g \equiv g_n)\}$.

T4. $[f, g, \varphi]\{[f, f', g, g', \varphi]\{(f \equiv f') \cdot (g \equiv g') \supset (\varphi(f, g) \equiv \varphi(f', g'))\} \cdot \varphi(f_1, g_1) \cdot \varphi(f_1, g_2) \cdot \ldots \cdot \varphi(f_m, g_n) \supset \varphi(f, g)\}$.

Proof. (1) $[f, f', g, g', \varphi]\{(f \equiv f') \cdot (g \equiv g') \supset (\varphi(f, g) \equiv \varphi(f', g'))\}$.

(2) $\varphi(f_1, g_1) \cdot \varphi(f_1, g_2) \cdot \ldots \cdot \varphi(f_m, g_n) \supset$

(3) $((f \equiv f_1) \cdot (g \equiv g_1) \supset \varphi(f, g)) \cdot ((f \equiv f_1) \cdot (g \equiv g_2) \supset \varphi(f, g)) \cdot \ldots \cdot ((f \equiv f_m) \cdot (g \equiv g_n) \supset \varphi(f, g))$. (1; 2)

$\varphi(f, g)$ (T3; 3)

In the last line of the proof we have applied the following theorem of the propositional calculus:

$$[p_1, \ldots, p_{m \cdot n}, q]\{(p_1 \vee \ldots \vee p_{m \cdot n}) \cdot (p_1 \supset q) \cdot \ldots \cdot (p_{m \cdot n} \supset q) \supset q\}.$$

It follows from Theorem 1 that the following proposition is a theorem of S_1:

T5. $[f, f', g, g', \varphi]\{(f \equiv f') \cdot (g \equiv g') \supset (\varphi(f, g) \equiv \varphi(f', g'))\}$.

T6. $[f, g, \varphi]\{\varphi(f_1, g_1) \cdot \varphi(f_1, g_2) \cdot \ldots \cdot \varphi(f_m, g_n) \supset (f, g)\}$.[36] (T4; T5)

We have thus shown that from the assumption that Theorem 1 is true for functors of the n-th semantic order, it results that the law of verification formulated for a functor of the $n + 1$-st semantic order is a theorem of S_1.

We shall now prove that from the said assumption it follows that the law of the quantity of functions, formulated for a functor of the $n+1$-st semantic order, is also a theorem of S_1.

Let φ_k be any verifier of the variable φ, and let

$$(1)\ \alpha_1^k, \ldots, \alpha_{m \cdot n}^k$$

be the sequence of all propositions belonging to its characteristic. Let further

$$(2)\ \beta_1^k, \ldots, \beta_{m \cdot n}^k$$

be the sequence of expressions obtainable from the propositions (1) by substituting in them for the symbol φ_k the symbol φ.

It follows from Theorem 2, Section 10, that all the propositions

$$\text{T7. } \alpha_i^k, \qquad i = 1, \ldots, m \cdot n,$$

are theorems of the system S_1.

$$\text{T8. } [\varphi]\{\beta_1^k \cdot \ldots \cdot \beta_{m \cdot n}^k \supset (\beta_1^k \equiv \alpha_1^k) \cdot \ldots \cdot (\beta_{m \cdot n}^k \equiv \alpha_{m \cdot n}^k)\}.$$

Proof. (1) $\beta_1^k \cdot \ldots \cdot \beta_{m \cdot n}^k \supset$

$$(\beta_1^k \equiv \alpha_1^k) \cdot \ldots \cdot (\beta_{m \cdot n}^k \equiv \alpha_{m \cdot n}^k) \qquad \text{(T7; 1)}$$

We introduce the following auxiliary definition:

$$\text{D1. } [f, g, \varphi]\{\Phi\langle\varphi\rangle(f, g) \equiv (\varphi(f, g) \equiv \varphi_k(f, g))\}.$$

$$\text{T9. } [f, g, \varphi]\{(\beta_1^k \equiv \alpha_1^k) \cdot \ldots \cdot (\beta_{m \cdot n}^k \equiv \alpha_{m \cdot n}^k) \supset (\varphi(f, g) \equiv \varphi_k(f, g))\}.$$

(T6; D1)

We have substituted in T6 the functor $\Phi\langle\varphi\rangle$ for the variable φ. Let us also note that the factors of the conjunction which is the antecedent of that theorem are obtainable from D1 by substituting in D1 for the variables f and g their verifiers.

$$\text{T10. } [\varphi]\{\beta_1^k \cdot \ldots \cdot \beta_{m \cdot n}^k \supset (\varphi \equiv \varphi_k)\}. \qquad \text{(T8; T9; D5, Section 10)}$$

Analogous theorems can be proved for every verifier of the variable φ. Consequently, the following propositions

T11. $[\varphi]\{\beta_1^i \cdot \ldots \cdot \beta_{m \cdot n}^i \supset (\varphi \equiv \varphi_i)\}, \qquad i = 1, \ldots, l,$

are theorems of the system S_1, the propositions $\beta_1^i, \ldots, \beta_{m \cdot n}^i$ having been obtained from propositions belonging to the characteristic of the verifier φ_i in a way analogous to that in which the propositions (2) were obtained from the propositions (1). Further, let us note that it follows from the definition of the characteristics of the verifiers that an alternative in which the antecedents of all the propositions T11 appear as components, is a result of substitution applied to a theorem of the propositional calculus which is a generalization of the following theorem:

$$[p, q, r]\{p \cdot q \cdot r \vee \sim(p) \cdot q \cdot r \vee p \cdot \sim(q) \cdot r \vee p \cdot q \cdot \sim(r) \vee \sim(p) \cdot \sim(q) \cdot r \vee \sim(p) \cdot q \cdot \sim(r) \vee p \cdot \sim(q) \cdot \sim(r) \vee \sim(p) \cdot \sim(q) \cdot \sim(r)\}.$$

Hence the alternative of those antecedents is a theorem of S_1. From this, from T11 and from the following theorem of the propositional calculus:

$$[p_1, \ldots, p_l, q_1, \ldots, q_l]\{(p_1 \vee \ldots \vee p_l) \cdot (p_1 \supset q_1) \cdot \ldots \cdot (p_l \supset q_l) \supset (q_1 \vee \ldots \vee q_l)\}$$

it follows that the proposition:

$$[\varphi]\{(\varphi \equiv \varphi_1) \vee \ldots \vee (\varphi \equiv \varphi_l)\}$$

is a theorem of S_1. This theorem is the law of the quantity of functions, formulated for the functor φ.

Thus the inductive proof of Theorem 2 is completed.

THEOREM 3. *The systems S and S_1 are equivalent.*

Proof. A1, Section 9, is the common axiom of S and S_1. Consequently, to prove the theorem it suffices to show that the rules of inference of either of the two systems S and S_1 may be derived from that axiom and from the rules of inference of the other system. The rules of inference of S and S_1 are identical, except for the rules of verification and extensionality. From Theorem 1, Section 11 it follows that the rule of extensionality of S_1 may be derived from the axiom and from the rules of S. From Theorem 2 it follows that the rule of verification of S may be derived from the axiom and from the rules of S_1. Hence Theorem 3 is true.

THEOREM 4. *The law of the quantity of functions* (b, Section 11), *formulated for functors of any semantic category, is a theorem of the system S.*

Proof. This theorem is a direct consequence of Theorems 3 and 2.

13. THE PROOF OF COMPLETENESS OF THE SYSTEMS S AND $S1$

We will show now that given any proposition of protothetics α, α has the following property (hereafter called *property* W): either α or the negation of α is a theorem of the system S. Systems all of whose propositions have an analogous property will be called complete systems.

By proving the completeness of the system S we shall also prove the completeness of the system S_1, which is equivalent to the former.

LEMMA 1. *If the propositions α and α' are equivalent in S, and if α has the property W, then α' also has the property W.*

Proof. This lemma results from the fact that the propositions

$$[p, q]\{(p \equiv q) \supset (p \supset q)\}$$

and

$$[p, q]\{(p \equiv q) \supset (\sim(p) \supset \sim(q))\}$$

are theorems of S.

We say that the propositional expressions α and α' are *equivalent* in a certain system, if the expression, built of the symbol of equivalence and of those expressions, and preceded by general quantifiers binding all free variables of the expressions α and α' is a theorem of that system obtainable by applying its rules strengthened as specified at the beginning of Section 7.[37]

LEMMA 2. *Let β and β' be any two propositional expressions which are equivalent in S and are satisfying the following condition: every free variable appearing in one of these expressions is equiform to a certain free variable of the other expression.*

If $\alpha\beta$ is any expression of which a part is equiform to β and $\alpha\beta'$ – an expression which may be obtained from $\alpha\beta$ by substituting in it for any of its parts equiform to β the expression β', then the expressions $\alpha\beta$ and $\alpha\beta'$ are equivalent in S.

Proof. This lemma will be proved by induction on the number (n) of non-equiform variables of the expression β bound by those quantifiers of the expression $\alpha\beta$ which are not quantifiers of the expression β. These variables shall be called externally bound.

If $n = 0$, which means that the expression β contains no externally bound variables, the lemma results from the rule of adding definitions to the system and from the fact that the law of extensionality (a, Section 6) is a theorem of S.

Let now n be a natural number greater than 0. We assume that the lemma is true if the expression β contains less than n externally bound variables. We shall discuss two cases: (a) $\alpha\beta$ begins with a quantifier binding a variable of the expression β; and (b) $\alpha\beta$ does not begin with such a quantifier.

(a) Let $\alpha^*\beta$ be an expression obtained from $\alpha\beta$ by cancelling the variable which appears within the brackets of the quantifier standing at the beginning of that expression and is equiform to a certain free variable of the expression β; let $\alpha^*\beta'$ be an expression obtained from $\alpha\beta'$ in the same way as $\alpha^*\beta$ was obtained from $\alpha\beta$. According to the inductive assumption, the expressions $\alpha^*\beta$ and $\alpha^*\beta'$ are equivalent in S. In view of the rule allowing to distribute a quantifier preceding an equivalence[38] we conclude that also the expressions $\alpha\beta$ and $\alpha\beta'$ are equivalent in that system.

(b) Let us denote by $\gamma\beta$ any propositional expression which is a part of $\alpha\beta$ and which begins with a quantifier that binds the variable of the expression β and is not a part of that expression. We also assume that β is a part of $\gamma\beta$. It can easily be seen that in the case under discussion there exists an expression which possesses these properties. Let us denote by $\gamma\beta'$ an expression obtained from $\gamma\beta$ in the same way as $\alpha\beta'$ was obtained from $\alpha\beta$. On the basis of the inductive assumption or of part (a) of the proof we conclude that the expressions $\gamma\beta$ and $\gamma\beta'$ are equivalent in S. From this and from the inductive assumption it follows that the expressions $\alpha\beta$ and $\alpha\beta'$ are equivalent in that system.

Thus the inductive proof of the lemma has been completed.

LEMMA 3. *Every proposition α which contains the free or bound variable φ is equivalent in S to a proposition which contains neither the variable φ nor any variable or constant that is not a verifier of the variable φ or a functor of propositional calculus, and that differs in form from any variable or constant of the expression α.*[39]

Proof. This lemma may be easily obtained from Theorem 3, Section 2, from Lemma 11, and from the rule of adding definitions to the system.

LEMMA 4. *If the propositions α and β have the property W, then the negation of that proposition also has that property.*

Proof. If the negation of α is a theorem of S, Lemma 4 is obviously true. If α is a theorem of S, Lemma 4 is proved by the fact that the proposition

$$[p]\{p \supset \sim(\sim(p))\}$$

is a theorem of that system.

LEMMA 5. *If the propositions α and β have the property W, then the conditional sentence whose antecedent is equiform to α and whose consequent is equiform to β, also has the property W.*

Proof. According to the assumption of our lemma one of the four following cases holds:

(a) α and β are theorems of the system S;
(b) $\sim(\alpha)$ and β are theorems of the system S;
(c) α and $\sim(\beta)$ are theorems of the system S;

(d) $\sim(\alpha)$ and $\sim(\beta)$ are theorems of the system S.

In the cases (a) and (b) the lemma results from the fact that A2, Section 5, is a theorem of S. In the case (c) the lemma results from the fact that the expression

$$[p, q]\{p \supset (\sim (q) \supset \sim(p \supset q))\}$$

is a theorem of S. Finally, in the case (d) the lemma results from the fact that T8, Section 5, is a theorem of S.

Let φ be a functorial variable of any semantic category, and let $f, \ldots, u$ stand for the variable arguments of φ. Let further φ', f', …, u' stand for any verifiers of the variables φ, $f, \ldots, u$. The proposition $\varphi'(f', \ldots, u')$ shall be called an *elementary proposition*.

LEMMA 6. *Every elementary proposition has the property W.*

Proof. This lemma is a direct consequence of Theorem 2, Section 10.

LEMMA 7. *Let α be any elementary proposition containing either the symbol 1 or the symbol 0.*

Every proposition which may be obtained from α by substituting for all the symbols 1 and 0 in α any propositions which have the property W, also has the property W.

Proof. Let β be any proposition having the property W. Since the propositions 1 and $\sim(0)$ are theorems of the system S (cf. T6 and T7, Section 7), it can easily be concluded that either the proposition $1 \equiv \beta$ or the proposition $0 \equiv \beta$ is a theorem of that system.

From this, from the rule of adding definitions to the system and from Lemmata 1 and 2 it follows that Lemma 7 is true.

LEMMA 8. *Any proposition α of which every constant functor is either the functor of implication or the functor of negation or else a verifier of a certain variable and in which no free or bound variables appear has the property W.*

Proof. By induction on the number of symbols appearing in α:

If α contains only one symbol, then it is equiform either to the proposition 1 or to the proposition 0, both of which have the property W.

Let us now assume that α contains k symbols (where $k > 1$), and that the lemma holds for those propositions which contain less than k symbols.

We shall discuss four cases:

(a) the proposition α is a negation;
(b) the proposition α is an implication;
(c) the proposition α is an elementary proposition;
(d) the proposition α is a proposition which may be obtained from a certain elementary proposition by substituting in it the symbols 1 and 0 for any propositions.

From the inductive assumption and from Lemmata 4–7 we conclude that in each of the above cases the proposition α has the property W.

Thus the inductive proof of Lemma 8 has been completed.

LEMMA 9. *Every proposition α, which apart from functors of the propositional calculus contains only those constant functors which are verifiers of certain variables, has the property W.*

Proof. In view of Lemmata 3 and 1 we may assume that the proposition α does not include any variables; moreover, since every functor of the propositional calculus can be defined by means of implication and negation, and in view of Lemmata 2 and 1, we may assume that the sole functors of the propositional calculus which may appear in α are the functors of implication and negation.

From this and from Lemma 8 it follows that Lemma 9 is true.

LEMMA 10. *Let α stand for any propositional expression containing the free variables $f', \ldots, f^{(m)}$ of any semantic categories and not containing any other free variables. In a special case the expression α may not contain any free variables.*

We assume that the definition

$$\text{D1. } [f', \ldots, f^{(m)}]\{\Phi^*(f', \ldots, f^{(m)}) \equiv \alpha\}$$

is adopted in the system S.

Let β_{Φ^*} be any proposition containing the symbol Φ^*. β_{Φ^*} is *equivalent in S to a proposition which does not contain a functor equiform to Φ^* and whose every constant functor which is not a functor of the propositional calculus is equiform to a certain functor appearing in α or in β_{Φ^*}.*[40]

Proof. Let β_Φ be an expression from which β_{Φ^*} may be obtained by substituting in it the functor Φ^* for the variable Φ, and which does not contain any functor equiform to the functor Φ^*.

We add to S the following auxiliary definition:

$$\text{D2. } [\Phi]\{\Psi^*(\Phi) \equiv \beta_\Phi\}.$$

Since D5, Section 10, is a theorem of S, the expression

$$\text{T1. } [\Phi]\{\Phi \equiv \Phi\}$$

is also a theorem of that system.

From Theorem 1, Section 11, we conclude that also the following expression is a theorem of S:

T2. $[\Phi, \Phi', \Psi]\{(\Phi \equiv \Phi') \supset (\Psi(\Phi) \equiv \Psi(\Phi'))\}$.

T3. $[\Phi, \Phi', \Psi]\{\Psi(\Phi) \supset ((\Phi \equiv \Phi') \supset \Psi(\Phi'))\}$. (T2)

T4. $[\Phi, \Psi]\{[\Phi']\{(\Phi \equiv \Phi') \supset \Psi(\Phi')\} \supset \Psi(\Phi)\}$.

Proof. (1) $[\Phi']\{(\Phi \equiv \Phi') \supset \Psi(\Phi')\} \supset$

(2) $(\Phi \equiv \Phi) \supset \Psi(\Phi)$. (1)

$\Psi(\Phi)$ (2; T1)

T5. $[\Phi, \Psi]\{\Psi(\Phi) \equiv [\Phi']\{(\Phi \equiv \Phi') \supset \Psi(\Phi')\}\}$. (T3; T4)

T6. $[\Phi, \Psi]\{\Psi(\Phi) \equiv [\Phi']\{[f', \ldots, f^{(m)}]\{\Phi(f', \ldots, f^{(m)}) \equiv$
$\Phi'(f', \ldots, f^{(m)})\} \supset \Psi(\Phi')\}\}$. (T5; D5, Section 10)

T7. $\Psi^*(\Phi^*) \equiv [\Phi]\{[f', \ldots, f^{(m)}]\{\Phi^*(f', \ldots, f^{(m)}) \equiv$
$\Phi(f', \ldots, f^{(m)})\} \supset \Psi^*(\Phi)\}$. (T6)

T8. $\beta_{\Phi^*} \equiv [\Phi]\{[f', \ldots, f^{(m)}]\{\alpha \equiv \Phi(f', \ldots, f^{(m)})\} \supset \beta_\Phi\}$. (T7; D1; D2)

According to our assumptions, no functors equiform to Φ^* do appear in α or in β_Φ. Consequently, no functors having that property appear in the expression to the right of the symbol of equivalence in T8. Moreover, every constant functor appearing in the latter expression is either a functor of the propositional calculus or is equiform to one of the functors appearing in α or in β_{Φ^*}. From this and from the fact that T8 is a theorem of S it follows that Lemma 10 is true.

LEMMA 11. *Let α stand for any propositional expression which contains the free variables $f', \ldots, f^{(m)}$, $g', \ldots, g^{(n)}$ of any semantic categories and no other free variables.*
We assume that the definition

D3. $[f', \ldots, f^{(m)}, g', \ldots, g^{(n)}]\{\Phi^*\langle g', \ldots, g^{(n)}\rangle(f', \ldots, f^{(m)}) \equiv \alpha\}$

is adopted in S.

Let $g'^*, \ldots, g^{(n)*}$ be any expressions of the same semantic categories as the corresponding variables $g', \ldots, g^{(n)}$. Let β_{Φ^*} be any proposition containing the symbol $\Phi^*\langle g'^*, \ldots, g^{(n)*}\rangle$. *$\beta_{\Phi^*}$ is equivalent in S to a proposition of which every functor dependent on parameters is equiform to a certain functor* *of the proposition β_{Φ^*}, and which does not contain any functor equiform to the functor $\Phi^*\langle g'^*, \ldots, g^{(n)*}\rangle$.*

Proof. In view of Lemmata 3, 2 and 1 we may assume that $\Phi^*\langle g'^*, \ldots, g^{(n)*}\rangle$ does not contain free variables.

Let β_Φ stand for an expression from which β_{Φ^*} may be obtained by substituting in it the functor $\Phi^*\langle g'^*, \ldots, g^{(n)*}\rangle$ for the variable Φ, and which does not contain a functor equiform to the functor $\Phi^*\langle g'^*, \ldots, g^{(n)*}\rangle$.

We now introduce the following two auxiliary definitions:

D4. $[\Phi]\{\Psi^*(\Phi) \equiv \beta_\Phi\}$;

D5. $[f', \ldots, f^{(m)}, g', \ldots, g^{(n)}]\{\Phi^{**}(f', \ldots, f^{(m)}, g', \ldots, g^{(n)}) \equiv \alpha\}$.

T9. $\Psi^*(\Phi^*\langle g'^*, \ldots, g^{(n)*}\rangle) \equiv [\Phi]\{[f', \ldots, f^{(m)}]\{\Phi^*\langle g'^*, \ldots, g^{(n)*}\rangle (f', \ldots, f^{(m)}) \equiv \Phi(f', \ldots, f^{(m)})\} \supset \Psi^*(\Phi)\}$. (T6)

T10. $\beta_{\Phi^*} \equiv [\Phi]\{[f', \ldots, f^{(m)}]\{\Phi^{**}(f', \ldots, f^{(m)}, g'^*, \ldots, g^{(n)*}) \equiv \Phi(f', \ldots, f^{(m)})\} \supset \beta_\Phi\}$. (T9; D4; D3; D5)

Since the proposition to the right of the symbol of equivalence in T10 does not contain a functor equiform to the functor $\Phi^*\langle g'^*, \ldots, g^{(n)*}\rangle$, and since in that proposition every functor dependent on parameters is equiform to one of the functors appearing in β_{Φ^*}, Lemma 11 is true.

THEOREM 1. *The system S is a complete system.*

Proof. It follows from Lemma 11 that every proposition which contains functors dependent on parameters is equivalent in S to a proposition which does not contain such functors. From this and from Lemma 10 we conclude that in the system S every proposition is equivalent to a proposition every constant functor of which is either a functor of the propositional calculus or a verifier of a certain variable. From this and from Lemmata 9 and 1 it follows that every proposition of the system S has the property W. Thus Theorem 1 is proved.

THEOREM 2. *The system S_1 is a complete system.*

Proof. This theorem is a consequence of Theorem 1 and of Theorem 3, Section 11.

Let us also note that the completeness of the system S_1 can be proved exactly in the same way as the completeness of the system S.

14. THE SYSTEM OF PROTOTHETICS WHICH HAS EQUIVALENCE AS ITS PRIMITIVE TERM

We shall begin this section by proving that implication may be defined by using only the symbol of equivalence, the general quantifier and variables representing functors of one propositional argument.

First of all, let us note that implication can be defined as follows in terms of equivalence and conjunction:

(1) $[p,q]\{(p \supset q) \equiv (p \cdot q \equiv p)\}$.

Conjunction may be defined in two ways:

D1. $[p,q]\{p \cdot q \equiv [f]\{p \equiv (f(p) \equiv f(q))\}\}$.

(2) $[p,q]\{p \cdot q \equiv [f]\{p \equiv ([r]\{p \equiv f(r)\} \equiv [r]\{q \equiv f(r)\})\}\}$.[41]

The following proposition

D2. $[p,q]\{(p \supset q) \equiv ([f]\{p \equiv (f(p) \equiv f(q))\} \equiv p)\}$

is a synthesis of the propositions (1) and D1.

Thus equivalence can act as the sole primitive term of protothetics.

The antecedent of the proposition

(3) $[f,p,q]\{(p \equiv q) \supset (f(p) \equiv f(q))\} \supset [p,q]\{p \cdot q \equiv [f]\{p \equiv (f(p) \equiv f(q))\}\}$

is equiform to the law of extensionality (a, Section 6), and the consequent to D1.

The following theorem implies that both definitions of conjunction given above are correct. One of its consequences is also the theorem (discussed in Section 8) stating that D1 is a theorem of the system E.

THEOREM 1. *The propositions (2) and (3) may be derived from theorems of the system discussed in Section 5 by applying the rules of inference of that system strengthened as specified at the beginning of Section 7.*

Proof. We shall prove a number of theorems by applying the procedure described in the above theorem.[42]

T1. $[f,p,q]\{p \cdot q \supset (p \equiv ([r]\{p \equiv f(r)\} \equiv [r]\{q \equiv f(r)\}))\}$.

Proof. (1) p.

(2) $q \supset$

(3) $[r]\{p \equiv f(r)\} \equiv [r]\{q \equiv f(r)\}$. (1; 2)

$p \equiv ([r]\{p \equiv f(r)\} \equiv [r]\{q \equiv f(r)\})$ (1; 3)

T2. $[p]\{\sim(p \equiv as(\sim(p)))\}$. (D1, Section 6)

T3. $[p]\{\sim([r]\{p \equiv as(r)\})\}$. (T2)

T4. $[p, q]\{[r]\{p \equiv as(r)\} \equiv [r]\{q \equiv as(r)\}\}$. (T3)

T5. $[p, q]\{[f]\{p \equiv ([r]\{p \equiv f(r)\} \equiv [r]\{q \equiv f(r)\})\} \supset p\}$.

Proof. (1) $[f]\{p \equiv ([r]\{p \equiv f(r)\} \equiv [r]\{q \equiv f(r)\})\} \supset$

(2) $p \equiv ([r]\{p \equiv as(r)\} \equiv [r]\{q \equiv as(r)\})$. (1)

p (2; T4)

T6. $[p, q]\{(q \equiv vr(p)) \equiv q\}$. (T4; Section 7)

T7. $[q]\{[p]\{q \equiv vr(p)\} \equiv q\}$. (T6)

T8. $[p, q]\{[f]\{p \equiv ([r]\{p \equiv f(r)\} \equiv [r]\{q \equiv f(r)\})\} \supset q\}$.

Proof. (1) $[f]\{p \equiv ([r]\{p \equiv f(r)\} \equiv [r]\{q \equiv f(r)\})\} \supset$

(2) $p \equiv ([r]\{p \equiv vr(r)\} \equiv [r]\{q \equiv vr(r)\})$. (1)

(3) $p \equiv (p \equiv q)$. (2; T7)

q (3)

T9. $[p, q]\{p \cdot q \equiv [f]\{p \equiv ([r]\{p \equiv f(r)\} \equiv [r]\{q \equiv f(r)\})\}\}$.
(T1; T5; T8)

T10. $[p, q]\{[f]\{p \equiv (f(p) \equiv f(q))\} \supset q\}$.

Proof. (1) $[f]\{p \equiv (f(p) \equiv f(q))\} \supset$

(2) $p \equiv (as(p) \equiv as(q))$. (1)

(3) $p \equiv (p \equiv q)$. (2; D1, Section 6)

q (3)

T11. $[p, q]\{[f]\{p \equiv (f(p) \equiv f(q))\} \supset p\}$.

Proof. (1) $[f]\{p \equiv (f(p) \equiv f(q))\} \supset$

(2) $p \equiv (vr(p) \equiv vr(q))$. (1)

p (2; T1, Section 7)

T12. $[f, p, q]\{((p \equiv q) \supset (f(p) \equiv f(q))) \cdot p \cdot q \supset (p \equiv (f(p) \equiv f(q)))\}$.

Proof. (1) $(p \equiv q) \supset (f(p) \equiv f(q))$.

(2) $p\cdot$

(3) $q \supset$

(4) $p \equiv q\cdot$. (2; 3)

(5) $f(p) \equiv f(q)\cdot$. (1; 4)

$p \equiv (f(p) \equiv f(q))$ (2; 5)

T13. $[f, p, q]\{(p \equiv q) \supset (f(p) \equiv f(q))\} \supset [p, q]\{p \cdot q \equiv [f]\{p \equiv (f(p) \equiv f(q))\}\}$. (T10; T11; T12)

T9 and T13 are equiform to the propositions (2) and (3), respectively. Thus Theorem 1 has been proved.

The propositions:

A1. $[p, q, r]\{(p \equiv q) \equiv ((r \equiv q) \equiv (p \equiv r))\}$;

A2. $[p, q]\{(p \equiv q) \equiv [f]\{f(p) \equiv f(q)\}\}$;

A3. $[p, q]\{(p \equiv q) \equiv [f]\{(f(p) \equiv f(q)) \equiv (p \equiv q)\}\}$;

A4. $[f]\{f([p]\{p\}) \equiv (f([p]\{p\} \equiv [p]\{p\}) \equiv [q]\{f([p]\{p\}) \equiv f(q)\})\}$

have the following properties:

As shown by Łukasiewicz, all theorems of the propositional calculus which are expressed in terms of variables and the symbol of equivalence result from A1[43];

A2 corresponds to the law of extensionality (a, Section 6);

The structure of A3 is similar to that of A2, and the intuitive meaning of both axioms is similar;

A4 corresponds to A1, Section 9, the latter being the only axiom of the systems S and S_1.

DEFINITION. The system which has equivalence as its only primitive term, the propositions A1–4 as its axioms, and in which the rules of inference specified below are valid, shall be called the system S_2:

(a) the *rule of substitution*,

(b) the *rule of detachment*: if an equivalence of two propositions is a theorem of S_2, and the first of these propositions is also a theorem of that system, then the second proposition is a theorem of S_2;

(c) the *rule for the distribution of a general quantifier preceding an equivalence*: if α is a propositional expression consisting of two propositional expressions connected by the symbol of equivalence and if the proposition which may be obtained from α by adding at its beginning the general quantifier binding all the free variables of α is a theorem of S_2, then the equivalence of two propositions which may be obtained from the expressions the equivalence of which constitutes α by adding at their beginning general quantifiers binding all the free variables occurring in these expressions, is also a theorem of the system S_2.[44]

Let φ be a variable functor of one argument and of at least the second semantic order. The following proposition:

$$[f, g]\{(f \equiv g) \equiv [\varphi]\{\varphi(f) \equiv \varphi(g)\}\}$$

shall be called the *equivalential law of extensionality*. The symbol $\equiv$ occurring in the first parentheses of this law is the symbol of interfunctorial equivalence defined by D5, Section 10. Let us note that the functor of equivalence is the sole constant functor appearing in the definiens of D5, Section 10;

(d) the *rule of extensionality*: any equivalential law of extensionality is a theorem of S_2;

(e) the *rule of definition*: any correctly built definition is a theorem of S_2. The definitions of S_2 are, of course, equivalences.

We shall now prove that every theorem of S_1 is a theorem of S_2. To do so we must show that A1, Section 9, is a theorem of S_2, and that the rules of inference, valid in S_1, may be derived from the rules of inference and the axioms of S_2.

LEMMA 1. *The following propositions, which have the functor of equivalence as their sole constant functor, are theorems of S_2:*

T14. $[p]\{p \equiv p\}$;

T15. $[p,q]\{(p \equiv q) \equiv (q \equiv p)\}$;

T16. $[p,q]\{p \equiv (p \equiv (q \equiv q))\}$;

T17. $[p,q]\{p \equiv (q \equiv (p \equiv p))\}$;

T18. $[p,q]\{p \equiv (q \equiv (q \equiv p))\}$;

T19. $[p,q,r]\{(p \equiv q) \equiv ((q \equiv r) \equiv (p \equiv r))\}$;

T20. $[p,q,r]\{(p \equiv q) \equiv ((p \equiv r) \equiv (q \equiv r))\}$;

T21. $[p,q,r]\{(p \equiv q) \equiv (r \equiv ((p \equiv r) \equiv q))\}$;

T22. $[p,q,r,s]\{(p \equiv (q \equiv r)) \equiv ((s \equiv q) \equiv ((s \equiv r) \equiv p))\}$.

Proof. These theorems result from A1 since all theorems of the propositional calculus noted by means of variables and symbols of equivalence result, as mentioned above, from that axiom.[45]

We add to S_2 the following definitions: D1; D2; D1, Section 6; D5, Section 6; and

D3. $1 \equiv ([p]\{p\} \equiv [p]\{p\})$.

The following propositions:

(4) $1 \cdot 1 \equiv 1$;

(5) $0 \cdot 0 \equiv 0$;

(6) $1 \cdot 0 \equiv 0$;

(7) $0 \cdot 1 \equiv 0$

may serve to characterize conjunction. The set of these propositions corresponds to the zero-one matrix of that function. We shall now prove

LEMMA 2. *The propositions* (4)–(7) *are theorems of S_2.*

Proof. We shall prove further theorems of S_2.

T23. 1. (T15; D3; T14)

T24. $[p]\{1 \equiv (p \equiv p)\}$. (T16; T23)

T25. $[p]\{p \equiv (1 \equiv p)\}$. (T17; T23)

T26. $0 \equiv (0 \equiv 1)$. (T18; T23)

T27. $[p, q]\{[f]\{p \equiv (f(p) \equiv f(q))\} \equiv p \cdot q\}$. (T15; D1)

T28. $1 \cdot 1$. (T27; T24)

T29. $1 \cdot 1 \equiv 1$. (T17; T28; T23)

T30. $0 \equiv [f]\{0 \equiv (f(0) \equiv f(0))\}$.[46] (T16)

T31. $0 \cdot 0 \equiv 0$. (A1; D1; T30)

T32. $0 \equiv [f]\{f(1) \equiv f(0)\}$. (T19; T25; A2)

T33. $[f]\{f(1) \equiv f(0)\} \equiv [f]\{1 \equiv (f(1) \equiv f(0))\}$. (T25)

T34. $0 \equiv [f]\{1 \equiv (f(1) \equiv f(0))\}$. (T19; T32; T33)

T35. $1 \cdot 0 \equiv 0$. (A1; D1; T34)

T36. $0 \equiv [f]\{(f(0) \equiv f(1)) \equiv (0 \equiv 1)\}$. (T19; T26; A3)

T37. $[f]\{(f(0) \equiv f(1)) \equiv (0 \equiv 1)\} \equiv$
$[f]\{0 \equiv (f(0) \equiv f(1))\}$. (A1; T26)

T38. $0 \cdot 1 \equiv [f]\{(f(0) \equiv f(1)) \equiv (0 \equiv 1)\}$. (A1; D1; T37)

T39. $0 \cdot 1 \equiv 0$. (A1; T38; T36)

T29, T31, T35 and T39 are equiform to the propositions (4)–(7), respectively. Thus Lemma 2 has been proved.

LEMMA 3. *The proposition*

(8) $[f]\{f(0) \equiv (f(1) \equiv [q]\{f(0) \equiv f(q)\})\}$,

which corresponds to the law of verification (d, Section 6), *is a theorem of* S_2.

Proof. We add to the system two auxiliary definitions:

D4. $[r]\{\mu(r) \equiv [f]\{f(r) \equiv (f([p]\{p\} \equiv [p]\{p\}) \equiv [q]\{f(r) \equiv f(q)\})\}\}$;

D5. $[r]\{\nu(r) \equiv [f]\{f(0) \equiv (f(r) \equiv [q]\{f(0) \equiv f(q)\})\}\}$.

The following propositions are theorems of S_2.

T40. $[r]\{[f]\{f(r) \equiv (f([p]\{p\} \equiv [p]\{p\}) \equiv [q]\{f(r) \equiv f(q)\})\} \equiv \mu(r)\}$. (T15; D4)

T41. $\mu([p]\{p\})$. (T40; A4)

T42. $[p]\{p\} \equiv 0$. (T15; D5, Section 6)

T43. $\mu(0)$. (A2; T42; T41)

T44. $[f]\{f(0) \equiv (f([p]\{p\} \equiv [p]\{p\}) \equiv [q]\{f(0) \equiv f(q)\})\}$. (D4; T43)

T45. $[r]\{[f]\{f(0) \equiv (f(r) \equiv [q]\{f(0) \equiv f(q)\})\} \equiv \nu(r)\}$. (T15; D5)

T46. $\nu([p]\{p\} \equiv [p]\{p\})$. (T45; T44)

T47. $([p]\{p\} \equiv [p]\{p\}) \equiv 1$. (T15; D3)

T48. $\nu(1)$. (A2; T47; T46)

T49. $[f]\{f(0) \equiv (f(1) \equiv [q]\{f(0) \equiv f(q)\})\}$. (D5; T48)

T49 is equiform to the proposition (8), and thus Lemma 3 is true.

LEMMA 4. *If the implication* $\alpha \supset \beta$ *and its antecedent* α *are theorems of* S_2, *then the proposition* β *is also a theorem of that system.*

Proof. The propositions:

T50. $\alpha \supset \beta$;

T51. α

are theorems of S_2 *ex hypothesi*. Hence the following propositions:

T52. $(\alpha \supset \beta) \equiv ([f]\{\alpha \equiv (f(\alpha) \equiv f(\beta))\} \equiv \alpha)$; (D2)

T53. $[f]\{\alpha \equiv (f(\alpha) \equiv f(\beta))\} \equiv \alpha$; (T52; T50)

T54. $[f]\{\alpha \equiv (f(\alpha) \equiv f(\beta))\}$; (T53; T15; T51)

T55. $[f]\{f(\alpha) \equiv f(\beta)\}$; (T54; T51)

T56. $as\,(\alpha) \equiv as\,(\beta)$; (T55)

T57. $\alpha \equiv as\,(\alpha)$; (T15; D1, Section 6)

T58. $as\,(\alpha)$; (T57; T51)

T59. $as\,(\beta)$; (T56; T58)

T60. β (D1, Section 6; T59)

are also theorems of that system. Since T60 is a theorem of S_2, Lemma 4 is true.

LEMMA 5. A1, Section 9, *is a theorem of* S_2.
Proof. The following propositions are theorems of S_2:

T61. $[p, q, r]\{(p \cdot q \equiv p) \equiv (p \supset q)\}$; (T22; D2; D1)

T62. $1 \supset 1$; (T61; T29)

T63. $0 \supset 0$; (T61; T31)

T64. $0 \supset 1$; (T61; T39)

T65. $[p]\{0 \supset p\}$; (T49; T63; T64)

T66. $[p]\{p \supset 1\}$; (T49; T64; T62)

T67. $[p]\{p \supset p\}$. (T49; T63; T62)

The proofs of T65, T66 and T67 have been given in abbreviated form. When giving complete proofs, certain auxiliary definitions and theorems are needed. This completion, however, does not involve any difficulties. In the proofs which are to follow we shall also omit auxiliary definitions and auxiliary theorems connected with them.

T68. $[p]\{1 \equiv (p \supset 1)\}$. (T17; T23; T66)

T69. $[p]\{1 \supset (p \supset 1)\}$. (A2; T68; T62)

T70. $[p, q]\{q \supset (p \supset q)\}$. (T49; T65; T69)

T71. $[p, q]\{q \supset (p \supset p)\}$. (T70; T67; Lemma 4)

T72. $[f]\{f(1) \supset (f(0) \supset f(0))\}$. (T71)

T73. $[f]\{f(1) \supset (f(0) \supset f(1))\}$. (T70)

T74. $[f, q]\{f(1) \supset (f(0) \supset f(q))\}$. (T49; T72; T73)

T75. $1 \equiv [p]\{p \supset p\}$. (T17; T23; T67)

T76. $[f, q]\{f([p]\{p \supset p\}) \supset (f(0) \supset f(q))\}$. (A2; T75; T74)

T77. $[f, q]\{f([p]\{p \supset p\}) \supset (f([q]\{q\}) \supset f(q))\}$. (A2; D5, Section 6; T76)

T77 is equiform to A1, Section 9, and thus Lemma 5 has been proved.

LEMMA 6. *Let α and β stand for any propositional expressions, and let $f_1, \ldots, f_n, g$ stand for the free variables, of any semantic categories, appearing in α and β. We also assume that the variables $f_1, \ldots, f_m, g$ (where $m \leq n$) are all the free variables of the expression β, and that the variable g does not appear in α as a free variable.*

If the proposition

$$[f_1, \ldots, f_n, g]\{\alpha \supset \beta\}$$

is a theorem of S_2, then the proposition

$$[f_1, \ldots, f_n]\{\alpha \supset [g]\{\beta\}\}$$

is also a theorem of that system.

Proof. The following propositions are theorems of S_2:

T78. $(1 \cdot 0 \equiv 1) \equiv 0$; (T21; T35; T23)

T79. $(1 \supset 0) \equiv 0$; (T20; T61; T78)

T80. $(1 \supset 1) \equiv 1$; (T15; T68)

T81. $[p]\{(1 \supset p) \equiv p\}$; (T49; T79; T80)

T82. $[f_1, \ldots, f_m]\{0 \supset [g]\{\beta\}\}$; (T65)

T83. $[f_1, \ldots, f_m, g]\{0 \supset \beta\}$; (T65)

T84. $[f_1, \ldots, f_m]\{(0 \supset [g]\{\beta\}) \equiv [g]\{0 \supset \beta\}\}$; (T17; T82; T83)

T85. $[f_1, \ldots, f_m]\{(1 \supset [g]\{\beta\}) \equiv [g]\{\beta\}\}$; (T81)

T86. $[f_1, \ldots, f_m]\{[g]\{1 \supset \beta\} \equiv [g]\{\beta\}\}$; (T81)

T87. $[f_1, \ldots, f_m]\{(1 \supset [g]\{\beta\}) \equiv [g]\{1 \supset \beta\}\}$; (A1; T85; T86)

T88. $[f_1, \ldots, f_m]\{(p \supset [g]\{\beta\}) \equiv [g]\{p \supset \beta\}\}$; (T49; T84; T87)

T89. $[f_1, \ldots, f_m]\{[g]\{p \supset \beta\} \equiv (p \supset [g]\{\beta\})\}$; (T15; T88)

T90. $[f_1, \ldots, f_n]\{[g]\{\alpha \supset \beta\} \equiv (\alpha \supset [g]\{\beta\})\}$; (T89)

T91. $[f_1, \ldots, f_n, g]\{\alpha \supset \beta\} \equiv [f_1, \ldots, f_n]\{\alpha \supset [g]\{\beta\}\}$. (T90)

Since T91 is a theorem of S_2, Lemma 6 is true.

In view of Lemma 6 the rule of adding the general quantifier to the consequent of an implication whose antecedent does not contain a free variable equiform to the variable bound by that quantifier may be derived from the rules and axioms of S_2.[47]

LEMMA 7. A1, Section 5, *is a theorem of* S_2.

Proof. It can easily be noticed that to prove this lemma it suffices to show that all the zero-one tests of A1, Section 5, are theorems of S_2.

T92. $(0 \supset 0) \supset ((0 \supset 0) \supset (0 \supset 0))$. (T70)

T93. $(1 \supset 0) \supset ((0 \supset 0) \supset (1 \supset 0))$. (T70)

T94. $1 \equiv ((1 \supset 0) \supset (0 \supset 0))$. (T17; T23; T71)

T95. $(0 \supset 1) \supset ((1 \supset 0) \supset (0 \supset 0))$. (A2; T94; T66)

T96. $(0 \supset 0) \supset ((0 \supset 1) \supset (0 \supset 1))$. (T71)

T97. $(1 \supset 1) \supset ((1 \supset 0) \supset (1 \supset 0))$. (T71)

T98. $1 \equiv ((0 \supset 1) \supset (1 \supset 1))$. (T17; T23; T71)

T99. $(1 \supset 0) \supset ((0 \supset 1) \supset (1 \supset 1))$. (A2; T98; T66)

T100. $(0 \supset 1) \supset ((1 \supset 1) \supset (0 \supset 1))$. (T70)

T101. $(1 \supset 1) \supset ((1 \supset 1) \supset (1 \supset 1))$. (T70)

T92, T93, T95–97, T99–101 are all the zero-one tests of A1, Section 5. Consequently, these tests are theorems of S_2. From this and from the fact that T49 is also a theorem of that system, it follows that Lemma 7 is true.

LEMMA 8. *Let β and γ be any propositional expressions, and let $f_1, \ldots, f_n$, g be all the free variables of β and γ. We further assume that $f_1, \ldots, f_m$, g (where $m \leq n$) are all the free variables of β, but we do not make any assumptions as to the semantic categories of these variables.*

If the proposition

$$[f_1, \ldots, f_n, g]\{\beta \supset \gamma\}$$

is a theorem of S_2, then the proposition

$$(9)\ [f_1, \ldots, f_n, g]\{[g]\{\beta\} \supset \gamma\}$$

is also a theorem of that system.

Proof. The following propositions are theorems of S_2:

T102. $[f_1, \ldots, f_m, g]\{[g]\{\beta\} \supset \beta\}$; (T88; T67)

T103. $[f_1, \ldots, f_n, g]\{([g]\{\beta\} \supset \beta) \supset ((\beta \supset \gamma) \supset ([g]\{\beta\} \supset \gamma))\}$; (Lemma 7)

T104. $[f_1, \ldots, f_m, g]\{1 \equiv ([g]\{\beta\} \supset \beta)\}$; (T17; T23; T102)

T105. $[f_1, \ldots, f_n, g]\{1 \supset ((\beta \supset \gamma) \supset ([g]\{\beta\} \supset \gamma))\}$; (A2; T104; T103)

T106. $1 \supset [f_1, \ldots, f_n, g]\{(\beta \supset \gamma) \supset ([g]\{\beta\} \supset \gamma)\}$; (T105; Lemma 6)

T107. $[f_1, \ldots, f_n, g]\{(\beta \supset \gamma) \supset ([g]\{\beta\} \supset \gamma)\}$. (T106; T23; Lemma 4)

According to the assumption of Lemma 8 the proposition

T108. $[f_1, \ldots, f_n, g]\{\beta \supset \gamma\}$

is a theorem of S_2.

Further we have:

T109. $[f_1, \ldots, f_n, g]\{1 \equiv (\beta \supset \gamma)\}$. (T17; T23; T108)

By applying to T109 and T107 the reasoning which was formerly used with reference to T104 and T103 we conclude that the proposition (9) is a theorem of S_2. Thus Lemma 8 has been proved.

According to this lemma, the rule of adding the general quantifier to the antecedent of an implication may be derived from the rules and axioms of S_2.

LEMMA 9. *If the propositions*

(10) $[f_1, \ldots, f_n]\{\alpha \supset \beta\}$

and

(11) $[f_1, \ldots, f_n]\{\beta \supset \alpha\}$

are added to S_1 by applying the rule of definition which is valid in that system, then these propositions are theorems of S_2.

Proof. We shall prove further theorems of S_2.

T110. $(0 \equiv 0) \supset (0 \supset 0)$. (T71)

T111. $(1 \equiv 1) \supset (1 \supset 1)$. (T71)

T112. $(0 \equiv 1) \supset (0 \supset 1)$. (A2; T26; T65)

T113. $(1 \equiv 0) \supset (1 \supset 0)$. (A2; T25; T65)

T114. $[p, q]\{(p \equiv q) \supset (p \supset q)\}$. (T49; T110; T113)

From the fact that the propositions (10) and (11) are added to S_1 on the strength of the rule of definition valid in that system it results that the proposition

$$[f_1, \ldots, f_n]\{\alpha \equiv \beta\}$$

is a theorem of S_2, added to the latter system on the strength of the analogous rule, valid in S_2.

From this, from Lemmata 8, 6 and 4, and from the fact that T114 and T15 are theorems of S_2 it follows that the propositions (10) and (11) are also theorems of S_2.

Thus Lemma 9 has been proved.

LEMMA 10. *If φ is a variable functor of one argument and of at least the second semantic order, then the proposition*

$$(12)\ [f, g, \varphi]\{(f \equiv g) \supset (\varphi(f) \equiv \varphi(g))\}$$

is a theorem of S_2.

Proof. It results from the rule of extensionality which is vaild in S_2 that the following proposition

$$[f, g]\{(f \equiv g) \equiv [\varphi]\{\varphi(f) \equiv \varphi(g)\}\}$$

is a theorem of that system.

From this, from Lemma 4 and in view of the fact that T114 and T88[48] are theorems of S_2 it follows that the proposition (12) is also a theorem of that system.

Thus Lemma 10 has been proved.

THEOREM 2. *Every theorem of S_1 is a theorem of S_2.*

Proof. It follows from Lemma 5 that the axiom of S_1 (A1, Section 9) is a theorem of S_2. It follows from Lemmata 4, 6, 7, 9 and 10 that the rules of S_1 may be derived from the rules and axioms of S_2.

Consequently, Theorem 2 is true.

It may be shown in a similar way that the axioms of S_2 are theorems of S_1, and that the rules of S_2 may be derived from the rules and the axiom of S_1. Consequently, these systems are equivalent.*

Let us also note that from the completeness of system S_1 (Theorem 2, Section 10), from Theorem 2, and from the obvious fact that S_2 is non-contradictory it results immediately that every theorem of S_2 is a theorem of S_1.

THEOREM 3. *S_2 is a complete system.*

Proof. The theorem results from Theorem 2 and from Theorem 2, Section 10.

Thus all the three systems discussed in this paper are complete. These systems are also equivalent.

* [Ed. Note: This is shown in Tanaka (1969); cf. the Protothetic Bibliography.]

NOTES

1 The term "protothetics" means approximatively: "basic statements".
2 With the exception of Section 13 (proof of completeness of protothetics).
3 As stated in the Foreword, this paper is based on the notes taken from Leśniewski's lectures.* We shall quote them as "notes".
4 Since no semantic category is assigned to quantifiers, Leśniewski did not [recreate] the theory of types, cf. Mostowski (1948), pp. 213ff.
5 It is in this way that semantic categories are discussed in Leśniewski (1929), pp. 13ff.
6 Since no semantic category is assigned to quantifiers Leśniewski did not include the general quantifier among the primitive terms of protothetics, although it is not defined in any of the systems of protothetics. The problem whether the general quantifier is a primitive term or not seems to be a terminological one and hence not essential.

The terminology used further in this paper will follow that of Leśniewski. Thus we shall say that the system S_2 has the symbol of equivalence as its sole primitive term although the general quantifier appears in theorems of that system and its definition is not added to it.

Let us also note that as a matter of principle, Leśniewski did not introduce into his systems the existential quantifier. His reasons are, unfortunately, unknown to the author of this paper. It is true that in numerous theorems of Leśniewski's ontology the existential quantifier appears, but it can be replaced there by a system of symbols, consisting of the sign of negation and of the general quantifier. Thus, for example, from the expression $[\exists a]\{f(a)\}$ we would obtain the expression $\sim([a]\{\sim(f(a))\})$. The notes include no theorem of protothetics containing either the existential quantifier or a system of symbols that could be replaced by that quantifier.**
7 An analogous notion is defined in Tarski (1936), p. 69.
8 Elementary protothetics has been defined here by the description of its language. Systems of elementary protothetics may differ in primitive terms, rules of inference and axioms.
9 All the theorems in Leśniewski (1938b) are written in this symbolism.
10 All the theorems in Whitehead–Russell (1925) are written in this symbolism.
11 That is on the strength of such a substitution which permits to replace expressions of the type: $f(x)$ by any expressions belonging to the category of propositions containing the free variable x and having a complex structure.
12 The primitive rules of inference of the systems S and S_1 concerning quantifiers will be discussed in detail in the next section.
13 The primitive rules of inference of S_2 will be discussed in Section 14.
14 Cf. note 6.
15 The system given here is obviously modelled on that given in Łukasiewicz (1929), pp. 154ff.
16 It can easily be noticed that this rule is equivalent to the rule of omitting the general quantifier in the consequent of an implication. Let us also remember that the existential quantifier does not appear in the theorems of protothetics (cf. note 6), so that no rules for using that quantifier are needed in the systems of protothetics.
17 These theorems are discussed both in the notes and in Tajtelbaum-Tarski (1923), pp. 73ff. Let us also remember that elementary protothetics was defined in Section 2. Theorems to be discussed here are theorems of every system of elementary protothetics, regardless of its primitive terms and axioms.
18 The terms defined in D1–5 can also be defined by means of two implications simple and converse: e.g., D1 can be replaced by the following definition:

(α) $[p]\{as(p) \supset p\}$;

---

* [Ed. Note: See Leśniewski (1988) in the Protothetic Bibliography, pp. 3–28.]

** [Ed. Note: Concerning the problem of the absence of existential quantifiers in Protothetic, see Lejewski (1956) in the Protothetic Bibliography.]

(β) $[p]\{p \supset as\,(p)\}$.

The definientia of D2–4 can also be replaced by expressions in which apart from symbols of implication no other constants of the propositional calculus appear. These expressions are as follows:

$$p \supset p, \qquad (p \supset p) \supset [p]\{p\}, \qquad [p]\{p \supset p\}.$$

For the discussion in this and the two following sections the form of the definitions of the terms: *as*, *vr*, *fl*, 1 and 0 – is not essential.

[19] The terms: *law of development* and *theorem on the limit of a function* are taken from the algebra of logic, cf. Couturat (1914).

[20] The proofs of these equivalences are given both in the notes and in Tajtelbaum-Tarski (1923), pp. 77ff.

[21] Let us note that in suppositional proofs we are not permitted to derive new formulae from formulae occurring earlier in the proof by applying substitution, since in this way we might obtain in our system propositions that would be obviously false. For example, we could obtain the proposition $[p, q]\{p \supset q\}$ by substituting the variable q for the variable p which is an assumption of the proof.

[22] In the proof we also distribute general quantifiers.

[23] These propositions are discussed in the notes and in Tajtelbaum-Tarski (1923), pp. 88–89.

[24] Like D1 in the preceding section, D1 and D2 are auxiliary definitions. As in many other auxiliary definitions of Leśniewski's systems, the definienda of these definitions are built of expressions belonging to other semantic categories than the expressions which appear in the definiens. In D1 and D2 variable functors of two propositional arguments appear to the right of the symbol of equivalence, whereas the functors which are being defined are functors of one propositional argument and are dependent on parameters. Moreover, it can easily be seen that these definitions are "creative".

[25] The notion of replacement was discussed in Section 7.

[26] This theorem was proved by Tarski in 1922, but the proof has not been published so far; it is mentioned in the notes.

In giving here this theorem the author avails himself of Professor Tarski's kind permission to include in this paper any results of his research which refer to protothetics and which have so far not been published.

[27] In its structure this theorern clearly resembles the law of verification (d, Section 6) and has the same intuitive meaning.

[28] Cf. note 18.

[29] The system E has been defined in the preceding section.

[30] The supposition that these propositions do not result from the theorems of elementary protothetics, is formulated in Tajtelbaum-Tarski (1923), p. 80.

[31] Mention is made in the notes that the term "verifier" was introduced by Kotarbiński.

[32] The definitions of these terms were given in Section 6 (D4 and D5). Cf. note 18.

[33] The expression $\varphi(f_1, g_1) \cdot \varphi(f_1, g_2) \cdot \ldots \cdot \varphi(f_m, g_n)$ is an abbreviation of a conjunction the components of which are all expressions of the form: $\varphi(f_i, g_j)$, where the index i is a natural number $\leq m$, and the index j is a natural number $\leq n$. In this conjunction each component appears only once.

[34] The theorem that the rule of verification may be replaced by the rule of extensionality has been proved by Tarski. The proof has not been published.

Rules of the same type as the rule of extensionality are called *axiomatic rules*.

[35] The expression $p_1 \cdot q_1 \vee p_1 \cdot q_2 \vee \ldots \vee p_m \cdot q_n$ is an abbreviation of an alternative all the components of which have the form $p_i \cdot q_j$, where the index i is a natural number $\leq m$, and the index j is a natural number $\leq n$. In this alternative each component appears only once.

[36] Cf. note 33.

[37] The notion of equivalence of two propositions, as defined at the beginning of Section 7, is

a special case of this notion.

[38] This rule may be derived from the primitive rules and the axiom of S.

[39] We assume here that no two constant functors belonging to different semantic categories are equiform. This assumption, which does not comply with the rules of symbolism that have been used in this paper so far, will enable us to simplify the reasoning contained in this section without impairing its generality.

[40] Cf. note 39.

[41] These definitions are given in Tajtelbaum-Tarski (1923), pp. 70 and 73.

[42] Proofs of these theorems are given in Tajtelbaum-Tarski (1923), pp. 71–74.

[43] This axiom is quoted in Leśniewski (1938a), p. 31.

[44] As in the case of systems which have implication as their primitive term, we do not introduce the rules of S_2 for using the existential quantifier (cf. note 6).*

[45] The respective proofs for most of these theorems are given in Rasiowa (1947), pp. 23ff.

[46] We have also made use of the rule of distributing the general quantifier preceding an equivalence.

[47] From the fact that T91 is a theorem of S_2 it follows that the rule of omitting the general quantifier which precedes the consequent of an implication may be derived from the rules and the axioms of S_2. The rules of adding and omitting the general quantifier in the consequent of an implication suffice to build a complete system of protothetics, having implication as its primitive term. It is, however, more convenient to base systems of protothetics on the rules of adding the general quantifier to the antecedent and to the consequent of an implication. The systems discussed in previous sections were built in this way. Let us also remember that the rules of adding the general quantifier to the antecedent and to the consequent of an implication were discussed in detail in Section 5. T91 was proved by Tarski, but the proof has not been published so far.**

[48] Let us note that T88 is a theorem of S_2 regardless of the structure of the expression.

REFERENCES

This paper is based on the notes from Leśniewski's lectures: Principles of Protothetics. Section 1 takes also into consideration notes from his other lectures, in the first place those on ontology. The notes from Leśniewski's lectures are referred to as "notes".‡

Couturat, Louis (1914). *L'algèbre de la logique*, Paris.

Leśniewski, Stanisław (1929). 'Grundzüge eines neuen Systems der Grundlagen der Mathematik', *Fundamenta Mathematicae* XIV, 1–81.

Leśniewski, Stanisław (1938a). 'Einleitende Bemerkungen zur Fortsetzung meiner Mitteilung u. d. T. "Grundzüge eines neuen Systems der Grundlagen der Mathematik"', Warszawa.

Leśniewski, Stanisław (1938b). *Grundzüge eines neuen Systems der Grundlagen der Mathematik*, §12, Warszawa.

Leśniewski (1938a, b) appeared as reprints of the first volume of Collectanea Logica, which was not published as a whole.

Łukasiewicz, Jan (1921). 'Two-Valued Logic' [in Polish], *Przegląd Filozoficzny* XXIII, 189–205.

Łukasiewicz, Jan (1929). *Elements of Mathematical Logic* [in Polish], authorised notes from lectures [at the University of Warsaw], prepared by M. Presburger, Warszawa.

* [Ed. Note: For a clarification of this definition of the rule for the distribution of a general quantifier, see Lejewski (1956) in the Protothetic Bibliography.]

** [Ed. Note: The proof of an analogue of T91 is given in Leśniewski (1929), pp. 40–41.]

‡ [Ed. Note: See Leśniewski (1988) in the Protothetic Bibliography.]

Mostowski, Andrzej (1948). *Mathematical Logic* [in Polish], Warszawa–Wrocław.
Rasiowa, Helena (1947). 'Axiomatisation d'un système partiel de la théorie de la déduction', *Comptes rendus des Séances de la Société des Sciences et des Lettres de Varsovie, Classe III* 40, 22–37.
Tajtelbaum-Tarski, Alfred (1923). 'On the Primitive Term of Logistic' [in Polish], *Przegląd Filozoficzny* XXVI, 68–89. [Cf. paper III in this volume – Ed.]
Tarski, Alfred (1936). 'Der Wahrheitsbegriff in den formalisierten Sprachen', *Studia Philosophica* I, 261–405 (translation from the Polish original published in 1933).
Whitehead, Alfred N. and Russell, Bertrand (1925). *Principia Mathematica*, Vol. I, Cambridge.

BOLESŁAW SOBOCIŃSKI

ON THE SINGLE AXIOMS OF PROTOTHETIC *

In this paper I should like to present the results of my unpublished investigations[1] concerning axiom-systems of protothetic. Strictly speaking, only the system of protothetic called $\mathfrak{S}_5$ will be considered here. It seems to me that this investigation may interest students of propositional calculus and the related subjects, since the deductions which will be used, sometimes unexpected and rather difficult, not only explain to some degree the structure of protothetic, but can also throw light upon several problems concerning various systems of propositional calculus. Because, generally, protothetic is still a little known theory, at the beginning I have to give several, possibly short, explanations concerning it. Without them the subject of this paper and the proofs presented below would hardly be understandable for the reader. Thus, in the first section a short description of protothetic and the necessary information about the rules of procedure of the system $\mathfrak{S}_5$ will be given. There will also be added some history of the researches concerning the single axioms of protothetic and related problems. Especially, I shall discuss briefly the metatheorems L (of Leśniewski) and the stronger S (mine). In the second section I shall present a combined proof: (1) that my axiom A_n can serve as a single (and probably the shortest) axiom of the system $\mathfrak{S}_5$ of protothetic, and (2) that the above mentioned metatheorem S is sufficient to check the completeness of any axiom system of protothetic. In the third and the last section it will be shown, in the shortest possible way, how the classical propositional calculus and the quantification theory for protothetical formulas can be obtained in the field of the system $\mathfrak{S}_5$.**

Instead of the authentic symbolism of Leśniewski[2] introduced by him mostly in order to formulate the rules of procedure in the most precise way, I shall use here a more convenient Peano–Russelian symbolism modified in a manner which will satisfy the requirements of protothetic. Anyone who is familiar with [elementary] logic will understand these modifications without difficulty. Only to avoid possible misunderstandings I would like to note:

---

* Bibliographical Note. This paper originally appeared in three parts in the *Notre Dame Journal of Formal Logic* I (1960), 52–73; II (1961), 111–126 and 129–148, by permission of the *Notre Dame Journal of Formal Logic*.

** [Ed. Note: The third section was never published by Sobociński; its intended contents is discussed by Rickey in paper II of this volume.]

(1) that the parentheses of the form "[" and "]" will be used here exclusively as the left and right scopes of the general quantifier, and (2) that if in a formula a quantifier is immediately preceded and followed by identical collections of dots, then the first collection is always considered as a stronger parenthesis than the second one. Concerning the terminology I would like to remark that the term "thesis" will mean throughout this paper a meaningful (i.e., well-formed) formula which is either an axiom or a proved theorem of the system under consideration.

It should be noted that in all systems of Leśniewski: (1) No thesis has free variables; (2) Vacuous quantifiers do not exist; (3) No formula has a form "$[x_1][x_2]\ldots[x_n].\phi(x_1x_2\ldots x_n)$", but always in such a case "$[x_1x_2\ldots x_n].\phi(x_1x_2\ldots x_n)$", where the quantifier "$[x_1x_2\ldots x_n]$" is considered as one unit; (4) If several free variables occur in a formula, they are not ordered in the quantifier. Thus, e.g. the expressions "$[xy].\phi(xy)$" and "$[yx].\phi(xy)$" are equiform formulas. But, obviously, "$[xy].\phi(xy)$" and "$[xy].\phi(yx)$" are different ones. A general quantifier preceding a meaningful formula will be called its main quantifier. For the reasons which will not be discussed here, in protothetic we do not have the existential quantifier.[3] Instead, any formula of the form "$[\exists x].\phi(x)$" has to be expressed by "$\sim([x].\sim(\phi(x)))$".

In the authentic symbolism of Leśniewski any functor precedes its arguments which must be enclosed by one or several pairs of symmetric parentheses. Thus, when using the said symbolism we do not have "$p \equiv q$" but "$\equiv(pq)$". In this paper I shall retain this custom only in the specific protothetical situations. Functors whose arguments are enclosed by several pairs of parentheses are called multi-link functors.[4] For such functors each pair of parentheses must differ from the others. E.g., if "ϕ" is such a functor, then ϕ with its arguments can have the form "$\phi\{p\}\langle q\rangle(r)$". The forms of parentheses which can occur in protothetic are determined by the prescriptions included in the formulation of the rules of procedure of this system. Because this subject is not related strictly to the topic of this paper I shall not further explain this point, remarking only that in any place where the parentheses (which are not dots) are used, it will be done in conformity with the requirements of the said rules.

1

1.1

A system of bi-valued logic of propositions, called protothetic, was constructed by Leśniewski in 1923. In the papers Leśniewski (1929, 1938a, 1938b) he described this theory, outlined a history (up to 1937) of its development and modifications, formulated in a very strict way the rules of procedure of its final formalization, called the system $\mathfrak{S}_5$, and showed how it is possible to deduce the propositional calculus from the original axiom-system (found by himself) of $\mathfrak{S}_5$. Anyone who would like to become more familiar with this theory should study thoroughly the papers mentioned above. Besides Leśniewski's own writings a rather popular characterization of protothetic is given by Słupecki (1953). Unfortunately, in preparing his paper the author preferred to use the notes of former students of Leśniewski rather than the original papers. For this reason, in spite of several valuable remarks and proofs which can be found in Słupecki's paper, many important questions discussed by Leśniewski either are presented inadequately or are even omitted completely (e.g., the whole problem of the, so-called, computable protothetical systems). The brief descriptions of protothetic are given in Prior (1955) and in Church (1956). Grzegorczyk (1955) discusses some problems connected with this theory; it seems that several of his remarks are too hastily formulated. The important results from the field of protothetic obtained by Tarski are published in Tarski (1923a, 1923b, 1924, 1956) and some contributions of Sobociński in (1939, 1949a, 1954).[5]

Here, I have no intention to give a full description of protothetic or to discuss problems unrelated to the topic of this paper. The sole purpose of the explanations *in medias res* which will be presented below is to introduce [protothetic to] readers unfamiliar with Leśniewski's system and his methods of deduction and to make the subject comprehensible to them.

As far as I know, Russell was the first who in (1906) studied the possibility of introducing quantifiers binding the propositional variables into the calculus of propositions. But he did not develop this idea and made no use of it in Whitehead–Russell (1925). Then the propositional calculus with quantifiers was constructed by Łukasiewicz (cf. Łukasiewicz, 1929; Łukasiewicz–Tarski, 1930; Tarski–Łukasiewicz, 1956), who not only formulated the adequate rules of procedure for such a system, but remarked also that it is a stronger theory than the classical calculus of propositions. But Łukasiewicz limited his investigations to the system based on implication as the sole primitive term, and up to now the systems with quantifiers are not elaborated as fully as the ordinary propositional calculi. In 1921, Leśniewski, having his two other theories (mereology and ontology[6]) ready, wanted to formalize his system of the foundations of mathematics by basing it on the strongest possible logic of propositions.[7] To this end, he began to investigate the possibility of constructing a system of propositional logic which could be extended by the addition of not only quantifiers but also variable proposition-forming functors. I.e., in the simplest case he added to the propositional calculus variables which assume the values of four constant functors: negation (N), verum (Vr), falsum (Fl) and assertium (As). As it is well known, in the bi-valued logic we have these and only these constant functors of one propositional argument. This is shown by the following matrix (where 0 represents a false and 1 a true proposition):

M1

p	Np	$\mathrm{Vr}p$	$\mathrm{Fl}p$	$\mathrm{As}p$
0	1	1	0	0
1	0	1	0	1

These considerations and some results obtained by Tarski (1956) enabled Leśniewski to construct a system which is actually a subsystem of the proper protothetic and may be called the restricted system of it.[8] This theory for which Leśniewski formulated suitable rules of procedure can be based, e.g., on the following axiomatic assumptions:

(a) An axiom-system of the complete implicational calculus of propositions, and

(b) An additional axiom Z1:

$$\text{Z1}\ [pf] \therefore f([u].u). \supset : f([u].u. \equiv .[u].u). \supset .f(p)$$

The thesis Z1, called the principle of bivalence for propositions,[9] has, obviously, the following meaning: Assume that an expression "$f(q)$" is a meaningful propositional function, in which "q" is a propositional variable; moreover, that "f" is not an abbreviation of a formula in which "q" occurs but a real variable for which any meaningful formula having q as a free variable can be substituted. Then Z1 says that if "f" is satisfied by "$[u].u$" (i.e., by 0) and by "$[u].u. \equiv .[u].u$" (i.e., by 1), then it is satisfied by any proposition.

As it has been shown by Tarski,[10] the axiom Z1 in this axiom-system can be replaced by the thesis:

$$\text{Z2}\ [pqf] : p \equiv q. \supset .f(p) \supset f(q)$$

which, evidently, is nothing else than the law of extensionality for propositions.

It is obvious that the restricted protothetic is not the strongest possible system of bi-valued logic of propositions, since it can be strengthened by the addition of other variable functors, e.g. of two or more propositional arguments or the variable proposition-forming functors at least one argument of which belongs to a higher type. In other words, the system can be evidently enriched in an analogous way to that which is known in the field of the functional calculi. Therefore, protothetic is a system of logic of propositions in which, besides the propositional variables, we have variable proposition-forming functors of any type that is meaningful according to the theory of semantical categories for protothetic.[11] Moreover, the quantification theory and the laws of extensionality (in the systems $\mathfrak{S}_2$–$\mathfrak{S}_5$) for variables of any category are assumed in this theory.

I shall entirely omit here the history of the transformation of the restricted protothetic into the full system or the gradual modification of the rules of procedure which Leśniewski undertook for this end.[12] Also no explanations will be given concerning the reasons which induced him to adopt this or that position in the final formalization of protothetic.

In 1922, Tarski showed[13] that in the propositional logic having quantifiers and variable functors it is possible to define conjunction by means of equivalence, e.g. in the following way:

$$\text{Z3}\ [pq]\therefore p.q. \equiv: [f] : f(p) \equiv f(q). \equiv p$$

This fact guarantees that any constant functor belonging to propositional calculus can be defined in terms of equivalence alone, since, clearly, negation can be obtained as follows:

$$\text{Z4}\ [p]\therefore \sim p. \equiv: p \equiv .[q].q$$

For several reasons Leśniewski preferred to profit from this result and adopted equivalence as the sole primitive term of protothetic, although for this purpose we can take some other functors, such as implication, Scheffer's functions, etc. The successive formalizations of protothetic based mostly on this primitive term were named the systems $\mathfrak{S}_1$ to $\mathfrak{S}_5$, and the last of them was accepted as final. Besides this, Leśniewski constructed several systems of the so-called computable protothetic. These systems, whose foundations differ sharply from those used in the construction of ordinary protothetic, will not be discussed here. But it has to be noted that the construction of the computable systems of protothetic allowed Leśniewski to prove that $\mathfrak{S}_5$ is a consistent and complete system.[14] The completeness of protothetic requires that for any meaningful formula in which all variables are bound either this formula or its negation is a thesis of $\mathfrak{S}_5$.

1.2

In order to understand a formalization of a theory belonging to Leśniewski's system we have to note that this system differs from others in that it gives no specification regarding the semantical categories (logical types) to which the expressions of the system may belong. Instead of such specification the rules of procedure establish how a new semantical category can be introduced into the system. Every expression occurring in a meaningful formula of a theory, with the exception of brackets and quantifiers, belongs to a defined semantical category. The semantical categories to which the expressions occurring in an axiom-system belong are called the primitive categories of the system. According to the requirements of Leśniewski in a well-constructed axiom-system of any theory the number of different primitive semantical categories should be as small as possible.[15] This demand was fulfilled in the axiom-system of $\mathfrak{S}_1$ constructed by Leśniewski in 1922 which contains the following three theses:

A1 $[pqr] \therefore p \equiv r. \equiv .q \equiv p :\equiv .r \equiv q$

A2 $[pqr] \therefore p \equiv .q \equiv r :\equiv: p \equiv q. \equiv r$

A3 $[gp] :\cdot: [f] :: g(pp). \equiv \therefore [r] : f(rr). \equiv .g(pp) :\equiv$
$: [r] : f(rr). \equiv .g(p \equiv .[q].q, p) ::\equiv .[q].g(qp)$[16]

It can be easily verified that in A1–A3 we have the sole primitive term, viz. equivalence. And, that in these theses there occur only expressions belonging to two semantical categories, viz. to the category of propositions (e.g., "p", "q", etc.) and to the category of proposition-forming functors of two propositional arguments, namely the variables "g" and "f" and the constant "$\equiv$". Hence, these two categories are primitive in the system $\mathfrak{S}_1$.[17]

As proved by Leśniewski,[18] the axioms A1 and A2 together with the customary rules of substitution and detachment (adjusted only to the system without free variables) constitute a complete axiom-system of the bi-valued equivalential calculus of propositions. He called this theory the system $\mathfrak{S}$. The axiom A3 comprises:

(a) The principle of bivalence expressed in terms of equivalence and variable functors of two arguments.
(b) Some forms of the law of extensionality for propositions which together with A1 and A2 enable one to obtain a complete propositional calculus.[19]

It is rather easy to show that on the basis of the rules of $\mathfrak{S}_1$, this axiom-system is inferentially equivalent to the restricted protothetic, given in Section 1.1. In order to obtain a complete system of protothetic Leśniewski did not have to change this axiom-system, but he had only to reformulate and to reinforce the rules of procedure of $\mathfrak{S}_1$. It will become clear, if we remark that the restricted protothetic may be obtained by the addition of the principle of bivalence for propositions (i.e., the thesis Z1) to the complete propositional calculus. The addition to $\mathfrak{S}_1$ of the analogous theses, guaranteeing bivalence for expressions belonging to other semantical categories, carries similar results, giving a stronger and stronger system of protothetic. Theoretically, in the complete system any possible semantical category should be represented; this can be assured only by a suitable rule of procedure which under some conditions allows one to add theses expressing the principle of bivalence for semantical categories other than the category of propositions. The system $\mathfrak{S}_1$ does not have such a rule, but the various formulations of it appear in the following systems. In the complete system of protothetic $\mathfrak{S}_5$ the rules of procedure were given their final formulation.[20] Below, I shall describe briefly only such points concerning the rules of $\mathfrak{S}_5$ as are indispensible for the purposes of this paper. No explanation will be given why this or that formulation was adopted by Leśniewski. But, I should like to remark here that Leśniewski, who tried to have as strong as possible a system of logic (except for existential assumptions), formulated his rules of procedure in an exceptionally constructive manner. This formulation permits one to check their operation by a simple inductive reasoning at any stage of the development of a theory.

Strictly speaking, $\mathfrak{S}_5$ has only one rule of procedure divided into five points. This rule is formulated in such a way that it becomes automatically adjusted to the last thesis belonging to the system. This means that any expression occurring in the axiom-system or in the already proved theses is meaningful with respect to this rule, when we are using it in order to prove a new thesis. The rule says that a formula A, which is meaningful with respect to the last thesis of the system, can be added to it as a new proved thesis when and only when A results from one and only one of the following five conditions:

(α) A is a well-formed definition.

(β) There is in the system such a thesis B that A can be obtained from B by a distribution of its main quantifier.

(γ) There are in the system theses B and C such that A can be obtained by the detachment of C from B.

(δ) There is in the system such a thesis B that A can be obtained from B by substitution.

(ϵ) A is a thesis of extensionality with respect to the expressions belonging to a semantical category which is not a category of propositions, but is already meaningful in the system.[21]

In protothetic, as in all Leśniewski's systems, a definition is not considered as a pure abbreviation, but as a real thesis of the theory. Therefore, there are no prescriptions here concerning the use of definitions, but only the conditions of their construction. A well-formed definition added to the system becomes automatically a proved thesis. Speaking superficially, a definition has always the following form:

$$[a_1 a_2 \ldots a_n].\phi \equiv \psi$$

where the equivalence is the main functor, ϕ a definiens and ψ a definiendum. The variables $a_1, a_2, \ldots, a_n$ from the main quantifier of the definition must belong to semantical categories which are already meaningful in the system, and they must occur in both ϕ and ψ. But there can be cases when a definition does not possess the main quantifier, e.g. "$[u].u. \equiv 0$".[22] A definiens must be a meaningful formula in which the variables bound by quantifiers belonging to it can also occur. The constants occurring in ϕ have to be either the primitive functor or previously defined terms. Any variable occurring in a definiendum has to belong to the main quantifier of a definition and can occur only once in ψ. No quantifier occurs in a definiendum which is either a new constant without arguments or a new constant followed by its variable arguments. The first case occurs when a definition does not have its main quantifier. In the second case the variables are enclosed by one or several pairs of suitable parentheses.[23] The first form does not require further explanation. The second alternative gives the multi-link functors. In order to make these notions more familiar, I shall present here a concrete example. Let us assume that in our system an ordinary conjunction is already defined and that "$\wedge$" is its symbol.[24] Then, obviously, we can define a conjunction of more propositional arguments, e.g.:

$$Z5\ [pqrs] : \wedge(p \wedge (q \wedge (rs))) . \equiv . \wedge_1 (pqrs)$$

where a new constant "$\wedge_1$" belongs to a new semantical category, viz. proposition-forming functors of four propositional arguments. But, besides the definition Z5, the rule allows one to construct also the following definitions:

$$Z6\ [pqrs] : \wedge(p \wedge (q \wedge (rs))) . \equiv . \wedge_2 \langle pqr \rangle (s)$$

$$Z7\ [pqrs] : \wedge(p \wedge (q \wedge (rs))) . \equiv . \wedge_3 \langle pq \rangle (rs)$$

$$Z8\ [pqrs] : \wedge(p \wedge (q \wedge (rs))) . \equiv . \wedge_4 \langle\!\langle p \rangle\!\rangle \langle qs \rangle (r)$$

$$Z9\ [pqrs] : \wedge(p \wedge (q \wedge (rs))) . \equiv . \wedge_5 \langle\!\langle\!\langle p \rangle\!\rangle\!\rangle \langle\!\langle s \rangle\!\rangle \langle r \rangle (q)$$

and other possible combinations.

It should be remarked that the constants $\wedge_2$–$\wedge_5$ defined above belong to different semantical categories. For instance, the category of "$\wedge_2$" is a functor of three propositional arguments which forms a proposition-forming functor of one propositional argument. But the whole expression "$\wedge_2 \langle pqr \rangle$" belongs to the category of proposition-forming functors of one propositional argument. The multi-link functors were introduced by Leśniewski[25] in order to have the fullest possible theory of logical types, i.e., his theory of semantical categories. Moreover, this kind of functors play an important role in the formulation of substitution in his system, as we shall see below. Ending these very superficial remarks about the definitions in protothetic, I should note that if a newly defined constant belongs to a semantical category which was not yet represented in the system, then by this very fact this new category becomes meaningful in the theory, and not only the defined constant, but also variables belonging to this category can be used in substitutions.

The point concerning distribution of quantifiers says that having in the system a thesis B of the form:

$$B \quad [a_1 a_2 \ldots a_n] : \phi . \equiv . \psi$$

i.e., a thesis whose main functor is equivalence and the variables bound by its main quantifier occur either in ϕ or in ψ or in both, we can distribute all or some of these variables between the arguments of the [equivalence]. Hence, if there are n distinct variables in the main quantifier of B (which, evidently, can belong to various semantical categories), then we have $2^n - 1$ possible combinations obtainable from B:

$$A_1\ [a_2 \ldots a_n] : [a_1].\phi. \equiv .[a_1].\psi$$

$$A_2\ [a_1 a_3 \ldots a_n] : [a_2].\phi. \equiv .[a_2].\psi$$

. . . .

$$A_3\ [a_3 \ldots a_n] : [a_1 a_2].\phi. \equiv .[a_1 a_2].\psi$$

. . . .

. . . .

$$A_{2^n-1}\ [a_1 a_2 \ldots a_n].\phi. \equiv .[a_1 a_2 \ldots a_n].\psi$$

and every such formula can be added to the system as a new thesis. Applying this operation, we always have to remember that in protothetic there are no vacuous quantifiers, each variable can occur in the same quantifier only once, and any sequence of quantifiers is combined into one. The rule comprises no other points dealing with quantifiers. In Section 3 a proof will be sketched that the remaining operations with quantifiers are provable in protothetic as theses.

The operation of detachment is formulated as follows: If in the system there are theses B and C having the forms:

$$B\quad \phi \equiv \psi$$

$$C\quad \phi$$

i.e., B is a thesis which has no main quantifier and equivalence is its main functor, and C is a thesis equiform with the first argument of [the equivalence] B, then we can add to the system a new proved thesis A:

$$A\quad \psi$$

which is to be equiform with the second argument of [the equivalence] B.

It should, therefore, be noted, that in protothetic we do not have detachment under quantifiers.

Roughly speaking, in protothetic the restrictions concerning substitution are analogous to those which exist in the functional calculi. However, there is one important difference, viz. in Leśniewski's system only meaningful expressions can be substituted for a variable occurring in the main quantifier of a thesis. This means that: (1) If there is a need to substitute for such a variable a formula which contains expressions belonging to non-primitive semantical categories, then this can be done only if these semantical categories have been previously introduced into the system by way of suitable definitions. I.e., only such constants can be used for substitution which are either primitive or previously defined, and also, that in the case of a substitution in which a variable belonging to a non-primitive category is involved, an arbitrary constant of the same category must be previously defined in the system. (2) Since the so-called incomplete formulas do not belong to any semantical category and, therefore, have no meaning in the system, a substitution of such formulas for the variable functors is not permitted. Instead of performing such forbidden operations in protothetic in order to get the same results as in the other systems, suitable multi-link functors must always be used. The following example will explain the matter briefly. Suppose that having the thesis:

Z10 $[f] : f(f([u].u)). \equiv .[q].f(q)$[26]

in which, evidently, "f" is a variable proposition-forming functor of one argument, we would like to obtain by substitution the thesis:

Z11 $[p] :: p \equiv : p \equiv .[u].u.\therefore \equiv .[q].p \equiv q$

We cannot make this by substituting the incomplete expression "$p \equiv$" for the variable "f", because "$p \equiv$" has no semantical sense as a whole, but first of all we have to add to the system a definition such as:

Z12 $[pq] : p \equiv q. \equiv . \equiv_1 \{p\}(q)$

The expression "$\equiv_1 \{p\}$" occurring in the definiendum of Z12 belongs to the same category as "f" and, therefore, we can substitute it for this variable in Z10. Thus in this way we obtain the thesis:

Z13 $[p] : \equiv_1 \{p\}(\equiv_1 \{p\}([u].u)). \equiv .[q]. \equiv_1 \{p\}(q)$

Now, if there are suitable theses at our disposal of the propositional calculus, and an operation of distribution of quantifiers, then from these assumptions and from Z13 and Z12 we can easily obtain Z11, dismissing from Z13 the defined auxiliary term.

The following final remark concerning substitution must be added. The rule does not allow one to change directly the variables which occur in an interior quantifier of a thesis. But, as will be shown in Section 3, this can always be achieved by appropriate deductions, provided that a new variable will be equiform with no variable occurring in the quantifiers of the thesis under consideration.

As it was mentioned above, in order to obtain a complete system of protothetic from $\mathfrak{S}_1$, the latter system must be strengthened so that for any semantical category, either primitive or introduced by definition of a suitable constant, an appropriate principle of bivalence is provable. Since in protothetic there can theoretically be an unlimited number of various semantical categories, this cannot be accomplished by a construction of an axiom-system having a finite number of theses.[27] On the other hand, due to the fact that equivalence is the sole primitive term of $\mathfrak{S}_1$ the principle of bivalence for propositions formulated in this or that way must be included in any axiom-system of it. Otherwise, we are unable to obtain a full system of $\mathfrak{S}_1$, or even of the complete propositional calculus, unless we would like to change the rules of procedure of protothetic in an unnatural manner.[28] Hence, in order to obtain from $\mathfrak{S}_1$ a complete system of protothetic, Leśniewski at first strengthened $\mathfrak{S}_1$ by the addition of a new rule allowing one to add to the theory a thesis expressing bivalence for any given semantical category, which is already meaningful in the system, other than the category of propositions.

However, a strict formulation of such a rule appeared to be extremely complicated and difficult. For this reason, using a remark of Tarski, Leśniewski later changed it entirely.[29] Tarski had shown that in the field of the complete propositional calculus the principle of bivalence for any semantical category is inferentially equivalent to the law of extensionality for expressions of the same category. Therefore, having the complete propositional calculus included in $\mathfrak{S}_1$, Leśniewski was able to replace the rule concerning the principle of bivalence by the rule of extensionality for variable functors. This modification did not change the range of protothetic, but merely allowed one to formulate the rule [of procedure] in a rather simple way.[30]

In the rule [of procedure] of $\mathfrak{S}_5$ the point concerning the theses of extensionality for variable functors is formulated as follows. Having equivalence as the primitive term, the law of extensionality for propositions can be expressed by the thesis:

Z14 $[pq]\therefore p \equiv q. \equiv: [f] : f(p). \equiv .f(q)$[31]

As it was mentioned, this thesis cannot be added to the system by the rule, but has to be assumed or provable in $\mathfrak{S}_5$ without direct reference to the point ϵ. The structure of theses which this point permits to add is similar [to that of Z14]. Let us suppose that the variables f and g are variable functors (ordinary or multi-link[32]) belonging to the same semantical category α and that the formula "$f \overset{\alpha}{\equiv} g$" expresses the equivalence between these two variables. Then, a suitable law of extensionality can be expressed by the formula:

T1 $[fg]\therefore f \overset{\alpha}{\equiv} g. \equiv: [\phi].\phi\langle f\rangle. \equiv .\phi\langle g\rangle$

where "ϕ" belongs to the category of proposition-forming functors of one argument which belongs to the same category as "f" or "g". Since, obviously, the formula "$f \overset{\alpha}{\equiv} g$" can be defined by means of quantifiers and equivalence in the following way:

$$[a_1 a_2 \ldots a_n] : f(a_1 a_2 \ldots a_n) . \equiv . g(a_1 a_2 \ldots a_n)$$ [33]

for the ordinary variable functors and:

$$[a_1 a_2 \ldots a_n] : f\{a_1\}(a_2 \ldots a_n) . \equiv . g\{a_1\}(a_2 \ldots a_n)$$

etc. for the multi-link functors, the theses of extensionality which the rule allows one to add to the system have the forms always analogous to T1, viz.:

$$\text{T2}\ [fg] \therefore [a_1 a_2 \ldots a_n] : f(a_1 a_2 \ldots a_n) . \equiv . g(a_1 a_2 \ldots a_n) : \equiv : [\phi] : \phi\langle f \rangle . \equiv . \phi\langle g \rangle$$

or

$$\text{T3}\ [fg] \therefore [a_1 a_2 \ldots a_n] : f\{a_1\}(a_2 \ldots a_n) . \equiv . g\{a_1\}(a_2 \ldots a_n) : \equiv : [\phi] : \phi\langle f \rangle . \equiv . \phi\langle g \rangle$$

a.s.o. for any possible similar combinations.

A thesis of extensionality having one of these forms can be added to the system as its new thesis only under the condition that any variable occurring in its main and interior quantifiers belongs to a semantical category which is already meaningful in the system. Hence, if there is a need to add such a thesis and not all the semantical categories involved are yet introduced into the system, then arbitrary constants belonging to these categories must be previously defined.

The rule only says what theses of extensionality may be added to the system, but gives no prescriptions concerning their applications. The situation is entirely analogous to that which we discussed in the case of definitions. A well-formed formula of extensionality added to the system automatically becomes a proved thesis which later can be transformed in any way permitted by the rule of procedure of $\mathfrak{S}_5$. Thus, normally, in protothetic each application of extensional reasoning requires the use of a special definition. E.g., suppose that we have at our disposal propositional calculus, Z14, and theses of the form:[34]

(a) $\alpha \equiv \beta$

(b) $\phi(\alpha)$

and suppose that "$\phi(\alpha)$" is such that from (a) and (b) the formula "$\phi(\beta)$" cannot be proved without Z14. Then, in order to obtain it, we have to make the following steps:[35]

(c) $[f]: f(\alpha). \equiv .f(\beta)$ [From Z14 and a]

(d) $[p]: \phi(p). \equiv \mathfrak{A}(p)$ [A definition]

(e) $\mathfrak{A}(\alpha)$ [d; b]

(f) $\mathfrak{A}(\beta)$ [c; e]

(g) $[p]: \mathfrak{A}(p). \equiv .\phi(p)$ [From d and propositional calculus]

$\phi(\beta)$ [g; f]

The rule of $\mathfrak{S}_5$, outlined above, may appear very heavy and inconvenient in its practical applications. However, its structure is determined by the conditions which, according to Leśniewski's requirement, any well-formalized deductive theory belonging to his system must fulfill. And, if these conditions are taken into consideration, a formulation of the rule appears entirely natural. Moreover, as we shall see in Section 2, in practice a direct application of the prescriptions required by the rule is necessary only for the first few steps, when we are beginning to make the deductions from this or that axiom-system of $\mathfrak{S}_5$. Later, it is easy to prove several metarules of procedure which allow one to obtain the really significant theorems in a short and secure way.[35]

1.3

Having the complete system of protothetic properly constructed and the principles of its rule of procedure established, Leśniewski began to make investigations concerning possible simplifications of the axiom-system of $\mathfrak{S}_5$. These researches appeared to be difficult, often requiring long and complicated deductions. Outlining here the history of these endeavors,[36] I shall give no hints how it is possible to prove that this or that thesis (or a set of several theses) is a single axiom (or an axiom-system) of protothetic, since a simple inspection of the proofs presented in the next section allows one to reconstruct the deductions without substantial difficulties. Also it is entirely unnecessary to give here the proofs that any thesis, presented below, is a true formula of protothetic, because anyone can check this easily using the bivalued logical matrix.[37] The following methatheorem L, stated by Leśniewski, plays an important role in the investigations concerning the axiom-system of protothetic.[38]

METATHEOREM L: An axiom-system of protothetic having the rules of procedure inferentially equivalent to the rule of $\mathfrak{S}_5$ constitutes a complete system, if in its field the following three conditions are satisfied:

(a) The system $\mathfrak{S}$ (i.e., the complete equivalential calculus of propositions) is provable.

(b) The following four laws of the logical product for conjunction:[39]

K1 $1 \equiv 1.1$

K2 $0 \equiv 0.0$

K3 $0 \equiv 0.1$

K4 $0 \equiv 1.0$

(c) The principle of bivalence for propositions is provable as a metarule saying that if "$\phi(p)$" is a meaningful formula in the system and the formulas "$\phi(0)$" and "$\phi(1)$" are already proved, then the formula "$[p].\phi(p)$" is also a thesis of this system.[40]

In fact, it is easy to be convinced that this metatheorem is correct. Having the rule of $\mathfrak{S}_5$, the above mentioned conditions, and using a simple definition:

Z15 $[pq] : p \equiv p.q. \equiv .p \supset q$

we can obtain at once the complete propositional calculus. Hence, we have obviously the logical laws of simplification and identity from which, by substitution and detachment, we get:

Z16 $[f] \therefore f(0). \supset: f(1). \supset .f(0)$

Z17 $[f] \therefore f(0). \supset: f(1). \supset .f(1)$

In virtue of the condition c, the thesis Z1 can be proved from Z16 and Z17 without any difficulty. Thus, the restricted system of protothetic is here provable and, since we have the rule of $\mathfrak{S}_5$, so is the complete system. Therefore, in order to show that an investigated axiom-system of protothetic is complete, it suffices only to prove that this system satisfies the conditions of Leśniewski's metatheorem. This allows one to omit long and uninteresting deductions in many proofs.

The first axiom-system of $\mathfrak{S}_5$ contains, as we know, the following theses:

A1 $[pqr] \therefore p \equiv q. \equiv .q \equiv p :\equiv .r \equiv q$

A2 $[pqr] \therefore p \equiv .q \equiv r :\equiv: p \equiv q. \equiv r$

A3 $[gp] :\cdot: [f] :: g(pp). \equiv \therefore [r] : f(rr). \equiv .g(pp) :\equiv: [r]$
$: f(rr). \equiv .g(p \equiv .[q].q,p) ::\equiv .[q].g(qp)$

Written in the authentic symbolism of Leśniewski, these axioms are expressed using 136 signs counting the variables, symbols of quantifiers and parentheses. In 1923 Leśniewski remarked that in this axiom-system A1 can be substituted by the thesis:[41]

A1* $\lfloor pqr \rfloor \therefore p \equiv q. \equiv: r \equiv q. \equiv .p \equiv r$

which did not change the total length of the axioms, but showed the possibility of abbreviating the first single axiom of protothetic. This thesis, found in 1923 and named A_a, contains 290 signs (in the authentic symbolism):

$$A_a\ [fp] :: f(\lfloor pq \rfloor : p \equiv q. \equiv .q \equiv p, p). \equiv \therefore f([hs] :\cdot: h(\lfloor pqr \rfloor \therefore p \equiv r. \equiv .q \equiv p :\equiv .r \equiv q, s). \equiv:: h([kt] :\cdot: k(\lfloor pqr \rfloor \therefore p \equiv .q \equiv r :\equiv: p \equiv q. \equiv r, t). \equiv:: k([gp] :\cdot: [f] :: g(pp). \equiv \therefore [r] : f(rr). \equiv .g(pp) :\equiv: [r] : f(rr). \equiv .g(p \equiv .[q].q, p) ::\equiv .[q].g(qp), t). \equiv \therefore \lfloor pqr \rfloor \therefore p \equiv .q \equiv r :\equiv: p \equiv q. \equiv r, s) . \equiv \therefore \lfloor pqr \rfloor \therefore p \equiv r. \equiv .q \equiv p :\equiv .r \equiv q, p). \equiv: \lfloor pq \rfloor : p \equiv q. \equiv .q \equiv p$$

In the same year, using A1* instead of A1, Leśniewski replaced A_a by a shorter single axiom (232 signs):

$$A_b\ [hs] :\cdot: h(\lfloor pqr \rfloor \therefore p \equiv q. \equiv: r \equiv q. \equiv .p \equiv r, s). \equiv:: h([kt] :\cdot: k(\lfloor pqr \rfloor \therefore p \equiv .q \equiv r :\equiv: p \equiv q. \equiv r, t). \equiv:: k(\lfloor gp \rfloor :\cdot: [f] :: g(pp). \equiv \therefore [r] : f(rr). \equiv .g(pp) :\equiv: [r] : f(rr). \equiv .g(p \equiv . [q].q, p) ::\equiv .[q].g(qp), t). \equiv \therefore \lfloor pqr \rfloor \therefore p \equiv .q \equiv r :\equiv: p \equiv q. \equiv r, s). \equiv \therefore \lfloor pqr \rfloor \therefore p \equiv q. \equiv: r \equiv q. \equiv .p \equiv r$$

which in 1923 was replaced by the shorter (156 signs):

$$A_c\ [fpqr] :: f(p \equiv p, q). \equiv \therefore f([g] \vdots\cdot\vdots\ g(pp). \equiv \ \vdots\vdots\ g(r \equiv .p \equiv r, p) . \equiv:\cdot: [h] :\cdot: [k] :: [s] : k(ss). \equiv .h(pp) :\equiv \therefore h(pp). \equiv: [s] : k(ss) . \equiv .h(p \equiv .[t].t, p) :\equiv .[t].h(tp), q). \equiv \therefore p \equiv q. \equiv: r \equiv q. \equiv .p \equiv r$$

In 1926 Wajsberg found the first two single axioms for the equivalential calculus of propositions[42] which enabled Leśniewski to find a shorter axiom (124 signs):

$$A_d\ [fhpqrx] \vdots\vdots\ f([k] :: [s] : k(ss). \equiv .h(pp) :\equiv \therefore h(pp). \equiv: [s] : k(ss). \equiv .h(p \equiv: [t].t, p), q). \equiv:\cdot: f([t].h(tp), q) :: p \equiv .q \equiv r :\equiv \therefore r \equiv x. \equiv x :\equiv .p \equiv q$$

In the same year, Wajsberg proved that, on the basis of the rule of $\mathfrak{S}_5$, from the protothetical thesis:

$$\text{W}^{**}\ [pqrst]\therefore p \equiv q. \equiv: [g]: g(r \equiv s. \equiv t, q). \equiv .g(s \equiv t. \equiv r, p)$$

the system $\mathfrak{S}$ can be deduced, which enabled him to construct the axiom (120 signs):

$$A_e\ [fhpqrst] :\cdot: f([g]\therefore h(pp). \equiv: g(q\, h(pp)) \equiv .g(q\, h(p \equiv . [t].t, p)), r) \equiv:: f([t].h(tp), r) \equiv \therefore p \equiv q. \equiv: [g]: g(r \equiv s. \equiv t, q). \equiv .g(s \equiv t. \equiv r, p)$$

which in turn was replaced by a shorter axiom of Leśniewski (116 signs):

$$A_f\ [fhpqrs] :: f([t].h(tp), q). \equiv:\cdot: f([k] :: h(pp). \equiv \therefore [s].k(ss). \equiv \therefore [s]: k(ss). \equiv .h(p \equiv .[t].t, p), q). \equiv:: p \equiv q. \equiv r :\equiv s.\therefore \equiv \therefore s \equiv: p \equiv .q \equiv r$$

Finally, in 1926, Wajsberg found the axiom (106 signs):

$$A_g\ [fp] ::: [s].f(sp). \equiv :\cdot: [g] :\cdot: f(pp). \equiv :: [t].g(tt). \equiv:\cdot: [qrt] :\cdot: g(t \equiv t. \equiv t, t). \equiv:: f(p \equiv .[s].s, p) \equiv \therefore p \equiv .q \equiv r :\equiv: r \equiv .q \equiv p$$

and Leśniewski constructed a single axiom, composed of only 82 signs:

$$A_h\ [fpqrst] :\cdot: p \equiv q. \equiv:: [g] :: f(pf(p[u].u)). \equiv \therefore [u].f(qu). \equiv: g(r \equiv s. \equiv t, q). \equiv .g(s \equiv t. \equiv r, p)$$

This axiom, which Leśniewski used as a base for his formalization of the rule of procedure of $\mathfrak{S}_5$,[43] could not be substituted by a shorter axiom during the next eleven years in spite of the endeavors made by Leśniewski and several of his students.[44] Even a result of Łukasiewicz, who in 1933 proved that each of the following theses:

$$\text{Ł1}\ [pqr]\therefore p \equiv q. \equiv: r \equiv q. \equiv .p \equiv r$$

$$\text{Ł2}\ [pqr]\therefore p \equiv q. \equiv: p \equiv r. \equiv .r \equiv q$$

$$\text{Ł3}\ [pqr]\therefore p \equiv q. \equiv: r \equiv p. \equiv .q \equiv r$$ [45]

is a single shortest axiom of the equivalential calculus of propositions, did not help to solve this problem. It only enabled Leśniewski to replace A_n by the axiom:

$$A_i\ [fpqrst] \therefore\because p \equiv q. \equiv:: [g] :: f(pf(p[u].u)). \equiv \therefore [u].f(qu)$$
$$. \equiv: g(r \equiv t. \equiv .s \equiv r, q). \equiv .g(s \equiv t, p)$$

which has the same length as A_h (82 signs), but from which the necessary deductions can be obtained in a shorter way.

Only in 1937 was I able to advance this stubborn problem. Namely, analyzing the previous results and the contents of metatheorem L, I remarked that the following set of assumptions S:

(a) System $\mathfrak{S}$ (e.g., Łukasiewicz's axiom Ł1)

(b) The protothetical theses:

$$\text{S1}\ [fp] :: f(p). \equiv \therefore f(p \equiv .[u].u). \equiv: [q] : f(p). \equiv .f(q)$$

$$\text{S2}\ [pq] \therefore p \equiv q. \equiv: [f] : f(p). \equiv .f(q)$$

$$\text{S3}\ [pq] :: p \equiv q. \equiv \therefore [f] \therefore f(p). \equiv .f(q) :\equiv .p \equiv q$$

constitutes a complete axiom-system of $\mathfrak{S}_5$.[46] In fact, these assumptions satisfy metatheorem L. Namely: the point a evidently gives the condition $\mathfrak{a}$. Thesis S1, which is nothing else but the principle of bivalence for propositions expressed in terms of equivalence,[47] fulfills, obviously, the condition $\mathfrak{c}$. And in virtue of the system $\mathfrak{S}$ and the theses S2 and S3, the condition $\mathfrak{b}$ can be obtained in a very easy way, if the thesis:

$$\text{Z18}\ [pq] :: [f] \therefore f(p). \equiv .f(q) :\equiv q \therefore \equiv .p.q$$

is adopted as a definition of conjunction.[48]

In the same year 1937, I proved that the set of assumptions S is not mutually independent, hence either the thesis S2 or the thesis S3 is a consequence of the remaining axioms. In other words, I showed that in the condition $\mathfrak{b}$ of Leśniewski's metatheorem either the thesis K4 or the thesis K3 is superfluous, because, in virtue of the remaining conditions, K4 is a consequence of {K1; K2; K3} and K3 of {K1; K2; K4}, provided that an adopted definition of conjunction is analogous to Z18.

Applying this result to a thesis which I knew previously, viz.:

$$A_j\ [pq] \because p \equiv q. \equiv:\because [fst] :\because f(pf(p[u].u)). \equiv:: [r].f(qr). \equiv \therefore$$
$$s \equiv t. \equiv q :\equiv: t \equiv .s \equiv p$$

I obtained at once a new single axiom of protothetic composed of only 72 signs[49] and immediately after that Leśniewski improved this result, showing that the thesis

$$A_k \ [pq] \vdots\vdots p \equiv q. \equiv\because [f] \because f(pf(p[u].u)). \equiv:: [rs] :: f(qr). \equiv \therefore s \equiv r. \equiv q :\equiv: r \equiv .s \equiv p$$

consisting of 71 signs can also serve this purpose.

In 1938, using the point ϵ of the rule, i.e., the point concerning the laws of higher extensionalities,[50] I proved that in the axiom-system S both theses S1 and S2 are superfluous. By this I showed that the condition $\mathfrak{b}$ of Leśniewski's metatheorem can be dropped entirely, as the theses K1, K2, K3, and K4 are provable from the remaining conditions of **L**. This result not only elucidated a role which the rule of extensionality plays in the inner structure of protothetic, but enabled me to remark that the thesis (66 signs long):

$$A_l \ [fpq] \because f(pf(p[u].u)). \equiv:: [r] :: f(p \equiv q. \equiv q, r). \equiv \therefore p \equiv q. \equiv: r \equiv q. \equiv .p \equiv r$$

can be adopted as a single axiom of protothetic.[51]

In the same year, starting from my aforesaid results and using my incidental remark that probably the condition $\mathfrak{a}$ could be weakened too, Leśniewski established the thesis (62 signs):

$$A_m \ [pq] \vdots\vdots p \equiv q. \equiv\because [f] \because f(qf(q[u].u)). \equiv:: [r] :: f(pr). \equiv \therefore r \equiv: q \equiv .r \equiv p$$

as a single axiom.[52]

Finally, in 1945 I proved that the thesis:

$$A_n \ [pq] :: p \equiv q. \equiv \therefore [f] \therefore f(pf(p[u].u)). \equiv: [r] : f(qr). \equiv .q \equiv p$$

is a single axiom of the system $\mathfrak{S}_5$ of protothetic. This thesis, whose length is 54 when expressed in the authentic symbolism of protothetic, seems to be also the shortest single axiom of the system $\mathfrak{S}_5$, although I have not been able to prove this yet.

In addition, I remarked that each of the following theses (54 signs):

$$A_o \ [pq] :: p \equiv q. \equiv \therefore [f] \therefore f(qf(q[u].u)). \equiv: [r] : f(pr). \equiv .q \equiv p$$

$$A_p \ [pq] :: p \equiv q. \equiv \therefore [f] \therefore f(pf(q[u].u)). \equiv: [r] : f(qr). \equiv .q \equiv p$$

A_q $[pq] :: p \equiv q. \equiv .\therefore [f].\therefore f(qf(p[u].u)). \equiv: [r] : f(pr). \equiv .q \equiv p$

can also serve as a single axiom of protothetic.[53]

These axioms differ in some respects from the axioms A_j–A_m. In order to establish that each of these theses can be taken as a single axiom of $\mathfrak{S}_5$, it suffices to be acquainted with my aforenamed results and to apply this or that modification of the methods of deduction previously used for the same purpose in regard to the theses A_a–A_i. The proof of the sufficiency of A_n requires further deductions, rather long and elaborate, which will be presented in Section 2.

Recollecting these problems in 1952, I found that the condition $\mathfrak{a}$ of Leśniewski's metatheorem L can be replaced by a weaker assertion, viz., that the system $\mathfrak{S}$ can be substituted by a small fragment of it.

Hence, summarizing my researches from this field, I was able to formulate the following metatheorem S.

METATHEOREM S: An axiom-system of protothetic having the rules of procedure inferentially equivalent to the rule of $\mathfrak{S}_5$ constitutes a complete system, if in its field the following conditions are satisfied:

I. The following two theses are provable:

F1 $[u].u. \equiv .[u].u$

F2 $[pq].\therefore p \equiv: q \equiv p. \equiv q$

II. The principle of bivalence for propositions is provable as a metarule saying that if "$\phi(p)$" is a meaningful formula in the system and the formulas "$\phi([u].u)$" and "$\phi([u].u. \equiv .[u].u)$" are already proved, then the formula "$[p].\phi(p)$" is also a thesis of this system.

A question remains open whether the thesis F1 is indispensible or whether it can be proved from condition II and thesis F2.

2*

In this section I propose to present the proofs relevant to the preceding discussion of the single axioms of the system $\mathfrak{S}_5$ of protothetic. First of all I will prove: (1) that thesis A_n mentioned in Section 1.3, and also each of the theses A_o, A_p and A_q, can serve as a single axiom of $\mathfrak{S}_5$,[54] and (2) that Leśniewski's metatheorem L, discussed in Section 1.3,[55] can be replaced by my metatheorem S,[56] whose conditions constitute a relatively small fragment of the original prerequisites set out by Leśniewski. In addition, a number of

* Bibliographical Note. This part of the paper originally appeared in the *Notre Dame Journal of Formal Logic* II (1961), 111–126.

questions closely connected with the topics mentioned above will also be discussed.

It is clear that if metatheorem S is true, then in order to establish that this or that protothetical thesis in which equivalence occurs as the only constant term can be adopted as a single axiom of $\mathfrak{S}_5$, one must be able to prove that, relatively to the rule of procedure of $\mathfrak{S}_5$, the thesis under consideration satisfies the requirements of S. It is evident too, that the truth of metatheorem S depends exclusively on whether or not it is possible, by applying the said rule, to deduce metatheorem L from the assumptions of S. Thus, from the methodological point of view, it would appear that thesis A_n should be discussed after metatheorem S has been established. There are, however, serious reasons for adopting a reversed order of presentation. For it so happens that the proof concerning A_n, though long and complicated, is much more elementary than the deductions required for the proof of S. Hence, it seems to me that in the beginning it is better to show that A_n (and also A_o, A_p and A_q) implies the conditions of S. This will enable anyone who is not familiar with the methods of deduction in protothetic to understand subsequent proofs better. Moreover, the proof of the main theorem, due to which metatheorem S holds and due to which almost all the results discussed here hold, is patterned on the idea underlying my 1937 investigations of the mutual relations among theses K1, K2, K3 and K4,[57] although, it must be said, these investigations do not involve deductions as complicated as those which establish metatheorem S. Although the results obtained by me in 1937 have been, in a sense, superseded by metarule S, I will present them here because they will make the proof of the metarule more understandable.

The detailed plan of this section is as follows: In Section 2.1, I shall prove that thesis A_n implies the conditions $\mathfrak{a}$ and $\mathfrak{c}$ of metatheorem L. Hence, *a fortiori* it satisfies metatheorem S which (assuming the correctness of S) proves automatically that A_n can serve as a single axiom of $\mathfrak{S}_5$. In Section 2.2 a proof will be outlined that the same holds for A_o, A_p and A_q. In Section 2.3, I shall show that, given the first four points of the rule of $\mathfrak{S}_5$, the condition $\mathfrak{c}$ of L and a small fragment of the theory $\mathfrak{S}$, viz. theses F1 and F2,[58] imply the whole of $\mathfrak{S}$, i.e., the condition $\mathfrak{a}$ of Leśniewski's metatheorem L. Incidentally, several new metatheorems concerning the completeness of $\mathfrak{S}_5$ will be established in Sections 2.2 and 2.3. In Section 2.4 a proof will be presented that without the application of the point ϵ (concerning higher extensionalities) of the rule of procedure either thesis K4 is a consequence of K1, K2 and K3 or K3 follows from K1, K2 and K4, provided that we have the theory $\mathfrak{S}$, the condition $\mathfrak{c}$ of L, and the extensional form of the definition of conjunction.[59] In addition, the axiom-system S, mentioned in Section 1.3,[60] will be discussed in this section. Finally, in Section 2.5, it will be demonstrated with the help of the point ϵ of the rule of $\mathfrak{S}_5$ that the condition $\mathfrak{b}$ of metatheorem L results from the condition $\mathfrak{c}$ and the theory $\mathfrak{S}$, which means that it follows from the condition $\mathfrak{c}$ and theses F1 and F2 in virtue of deductions shown in Section 2.3.

Thus, Sections 2.3 and 2.5 will contain the proof of metatheorem $\mathfrak{S}$, Section 2.4 will form an explanatory and historical introduction to Section 2.5, and Sections 2.1 and 2.2 will contain deductions concerning theses A_n–A_o.

The rule of procedure of $\mathfrak{S}_5$ allows us to add a new thesis to the system on condition that this thesis results from one and only one of the points which constitute the rule.[61] Thus, e.g., if a thesis A is a consequence of two theses B and C (previously proved) in virtue of suitable substitutions in B and C, the distribution of the quantifier in B, and the subsequent detachment, then, in fact, four new theses have to be inscribed in the roll of the system. I shall follow this prescription only in the first few steps in Section 2.1 in order to show how it works. Later, I shall use combined proof lines in which the application of the points β, γ and δ of the rule will be indicated jointly. The applications of the points α and ϵ will always be pointed out separately as the rule requires. Moreover, in the course of deductions, several metarules of procedure will be established and put to use. In this way about a thousand uninteresting steps will be omitted without affecting the rigorousness of the deductions.

Concerning the proof lines, mentioned above, it should be noted that:

(1) Each such line will end with parentheses of the form "[" and "]"; (2) An inscription, e.g., "$[B, p \mid q; C, r \mid s]$" indicates that in B, which is a thesis of the system, a meaningful formula q is substituted for the variable p; similarly in C, which again is a thesis of the system, s is substituted for r; then the required distributions of the quantifiers are made in B and C, and, finally, the operation of detachment is performed; (3) Whenever a distribution of the quantifier is made as a separate step in the course of deductions, it will be indicated by the letter β (e.g., when this point of the rule is applied to a result of detachment); (4) The proof line of theses obtained in virtue of points α and ϵ have the form "$[\alpha]$" and "$[\epsilon]$"; (5) Symbols: *SI*, *SII*, etc., occurring in the proof lines indicate the application of derivative metarules of procedure, whose validity has already been proved.

Also, it should be noted that the symbols "0" and "1" are not defined constants, but typographical abbreviations of "$[u].u$" and "$[u].u. \equiv .[u].u$", respectively.[62] The use of these symbols allows us to present theses and proofs in a shorter, clearer, and more understandable manner.

2.1

As a single axiom of protothetic we assume:

$$A_n \;\; [pq] :: p \equiv q. \equiv .\therefore [f] .\therefore f(pf(p[u].u)). \equiv: [r] : f(qr). \equiv .q \equiv p$$

We then adjust to it the rule of procedure of the system $\mathfrak{S}_5$.[63] In virtue of this rule, so adjusted, we deduce from A_n the following theses:

D1 $[p].p \equiv \mathrm{As}\,(p)$ [In virtue of the point α of the rule]

A1 $[p] :: p \equiv \mathrm{As}\,(p). \equiv .\therefore [f] .\therefore f(pf(p[u].u)). \equiv: [r] : f(\mathrm{As}\,(p)r). \equiv .$
$\mathrm{As}\,(p) \equiv p$ (A_n; in virtue of the point δ of the rule]

A2 $[p].p \equiv \mathrm{As}\,(p). \equiv .\therefore [pf] .\therefore f(pf(p[u].u)). \equiv: [r] : f(\mathrm{As}\,(p)r). \equiv .$
$\mathrm{As}\,(p) \equiv p$ (A1; in virtue of the point β of the rule]

A3 $[pf] .\therefore f(pf(p[u].u)). \equiv: [r] : f(\mathrm{As}\,(p)r). \equiv .\mathrm{As}\,(p) \equiv p$
(A2; D1; in virtue of the point γ of the rule]

D2 $[pq] .\therefore p \equiv q. \equiv .\mathrm{As}\,(p \equiv q) :\equiv \mathrm{Vr}\,(pq)$ [By α]

The proof of D1–D2 shows clearly how the points α, β, γ and δ of the rule are to be used if we wish to add new theses to the system. In what follows, abridged proofs will be given which involve steps made in virtue of a joint application of β, γ and δ. Care will be taken that this does not lead to misunderstandings. It should be noted that D1 is a definition of "assertium" of one propositional argument and D2 is a definition of "verum" of two propositional arguments. Obviously, the defined constant functor "As" belongs to the semantical category of proposition-forming functors of one propositional argument. D1 introduces this new category into the system. On the other hand, no new category is introduced by D2, since the defined constant "Vr" belongs to the same category as the variable "f" or the constant "$\equiv$" in A_n or A3, which are proposition-forming functors of two propositional arguments.

A4 $[pq].\mathrm{Vr}\,(pq)$ [By β, δ, γ: D2; D1, $p \mid p \equiv q$]

A5 $[pr] : \mathrm{Vr}\,(\mathrm{As}\,(p)r). \equiv .\mathrm{As}\,(p) \equiv p$ [A3, $f \mid \mathrm{Vr}$; A4, $q \mid \mathrm{Vr}\,(p[u].u)$]

A6 $[p].\mathrm{As}\,(p) \equiv p$ [A5; A4, $p \mid \mathrm{As}\,(p), q \mid r$]

Now, we are able to prove the following two metarules of procedure:

SI. *If in the field of our system we have a thesis of the following structure*

$$[a, b, c, \ldots] : \Phi. \equiv .\Psi$$

(where either the main quantifier does not exist or $a, b, c, \ldots$ are the variables which belong to this quantifier and occur as free in the formula "$\Phi. \equiv .\Psi$"), *then we can always add the following thesis to the system*:

$$[a, b, c, \ldots] : \Psi. \equiv .\Phi$$

The proof is evident for anyone who understands how A6 is obtained from D1.

Dem.:

(𝔞) $[a, b, c, \ldots] : \Phi . \equiv . \Psi$ [The assumption]

(𝔟) $[a', b', c', \ldots] : \Phi . \equiv . \Psi$ [From the point 𝔞; by means of substitution we change the variables occurring in the main quantifier of 𝔞 so that none of them remains equiform with any variable occurring in any of the quantifiers in A_n.]

(𝔠) $[a', b', c', \ldots, f] :: f(\Phi f(\Phi[u].u)) . \equiv \therefore [r] \therefore f(\Psi r) . \equiv : \Psi . \equiv . \Phi$

$[A_n, p \mid \Phi, q \mid \Psi;$ 𝔟$]$

(𝔡) $[a', b', c', \ldots, r] \therefore \mathrm{Vr}\,(\Psi r) . \equiv : \Psi . \equiv . \Phi$

[𝔠, $f \mid \mathrm{Vr}$; A4, $p \mid \Phi, q \mid \mathrm{Vr}\,(\Phi[u].u)$]

(𝔢) $[a', b', c', \ldots] : \Psi . \equiv . \Phi$ [𝔡; A4, $p \mid \Psi, q \mid r$]

(𝔣) $[a, b, c, \ldots] : \Psi . \equiv . \Phi$ [𝔢; by means of substitution we return to the same variables which occur in the main quantifier of 𝔞.]

Q.E.D.

SII. *If in the field of the system we have two theses of the following structures*:

$$[a, b, c, \ldots] : \Phi . \equiv . \Psi$$

and

$$[a, b, c, \ldots] . \Psi$$

(where $a, b, c, \ldots$ are variables which belong to the main quantifiers of these theses and occur either in "Φ" or in "Ψ" or in both), *then we can always add the following thesis to the system*:

$$[a, b, c, \ldots] . \Phi$$

The proof of SII is obvious in virtue of SI and the points β and γ of the rule. As a scheme of reasoning, SI says that although we have not yet obtained the thesis:

$[pq] : p \equiv q . \equiv . q \equiv p$[64]

nevertheless, we can always transform any thesis of the form "$[a, b, c, \ldots] : \Phi. \equiv .\Psi$" into a thesis of the form "$[a, b, c, \ldots] : \Psi. \equiv .\Phi$". Evidently, SII is a sort of detachment rule working from right to left.

We proceed by proving:

A7 $[pf] \therefore f(\mathrm{As}\,(p) f(\mathrm{As}\,(p)[u].u)). \equiv: [r] : f(pr). \equiv .p \equiv \mathrm{As}\,(p)$
$[A_n, p \mid \mathrm{As}\,(p), q \mid p; \mathrm{A6}]$

A8 $[pr] : \mathrm{Vr}\,(pr). \equiv .p \equiv \mathrm{As}\,(p)$
$[\mathrm{A7}, f \mid \mathrm{Vr}; \mathrm{A4}, p \mid \mathrm{As}\,(p), q \mid \mathrm{Vr}\,(\mathrm{As}\,(p)[u].u)]$

D3 $[pqr] \therefore \mathrm{Vr}\,(pr). \equiv .p \equiv q :\equiv .\Phi\alpha\!\prec\! p \!\succ\!(qr)$ [α[65]]

A9 $[pr].\Phi\alpha\!\prec\! p \!\succ\!(\mathrm{As}\,(p)r)$ $[\mathrm{D3}, q \mid \mathrm{As}\,(p); \mathrm{A8}]$

A10 $[pr] : \Phi\alpha\!\prec\! p \!\succ\!(pr). \equiv .p \equiv \mathrm{As}\,(p)$
$[\mathrm{A7}, f \mid \Phi\alpha\!\prec\! p \!\succ; \mathrm{A9}, r \mid \Phi\alpha\!\prec\! p \!\succ\!(\mathrm{As}\,(p)[u].u)]$

A11 $[pr].\Phi\alpha\!\prec\! p \!\succ\!(pr)$ [A10; D1; SII[66]]

A12 $[pr] : \mathrm{Vr}\,(pr). \equiv .p \equiv p$ $(\mathrm{D3}, q \mid p; \mathrm{A11}; \mathrm{SII}]$

A13 $[p].p \equiv p$ $[\mathrm{A12}, r \mid q; \mathrm{A4}]$

Now, we are going to prove two metarules of procedure which make clear that although we have not yet got in the system the thesis:

$[pq] \therefore p \equiv q. \equiv: [f] : f(p). \equiv .f(q)$[67]

or any other thesis which, at this stage of the [development of the] system, could allow us to make the extensional deductions directly, we nevertheless can reason in accordance with the law of extensionality for any expression which is a proposition.

SIII. *If two formulas α and β belong to the semantical category of propositions and if in the field of our system we have two theses of the following structures*:

$[a, b, c, \ldots] : \alpha. \equiv .\beta$

and

$[a, b, c, \ldots].\Phi(\alpha)$

(where "Φ" is a constant or a multi-link proposition-forming functor of one propositional argument, and where the variables $a, b, c, \ldots$ may occur as free variables not only in "α" and "β" but also in "Φ"), *then we can always add the following thesis to the system*:

$$[a, b, c, \ldots].\Phi(\beta)$$

Dem.:

($\mathfrak{a}$) $[a, b, c, \ldots] : \alpha. \equiv .\beta$ [The assumption]

($\mathfrak{b}$) $[a, b, c, \ldots].\Phi(\alpha)$ [The assumption]

($\mathfrak{c}$) $[a', b', c', \ldots] : \alpha. \equiv .\beta$ [$\mathfrak{a}$; as in the point $\mathfrak{b}$ of the proof of SI]

($\mathfrak{d}$) $[a', b', c', \ldots].\Phi(\alpha)$ [$\mathfrak{b}$; as in the point $\mathfrak{b}$ of the proof of SI]

($\mathfrak{e}$) $[a', b', c', \ldots, f] :: f(\alpha f(\alpha[u].u)). \equiv .\therefore [r] \therefore f(\beta r). \equiv: \beta. \equiv .\alpha$
[A_n, $p \mid \alpha$, $q \mid \beta$, $\mathfrak{c}$]

($\mathfrak{f}$) $[a', b', c', \ldots, p, q, r] :: \mathrm{Vr}\,(\Phi(p)r). \equiv: \Phi(q). \equiv .\Phi(p)$
$\therefore \equiv .\Psi\alpha\!\!\prec\! q \!\succ\!(pr)$ [In virtue of the point α of the rule. Thus, expression $\mathfrak{f}$ is a scheme of a definition. If any free variables occur in "Φ", "$\Psi\alpha$" is a multi-link functor for these variables. E.g., "$\Psi\alpha$" can have the following form "$\Psi\alpha\!\prec\! a', b', c', \ldots \!\succ$"]

($\mathfrak{g}$) $[a', b', c', \ldots, p, r].\Psi\alpha\!\prec\! p \!\succ\!(pr)$ [$\mathfrak{f}$, $q \mid p$; A12, $p \mid \Phi(p)$]

($\mathfrak{h}$) $[a', b', c', \ldots, r] \therefore \Psi\alpha\!\prec\! \alpha \!\succ\!(\beta r). \equiv: \beta. \equiv .\alpha$
[$\mathfrak{e}$, $f \mid \Psi\alpha\!\prec\! \alpha \!\succ$; $\mathfrak{g}$, $p \mid \alpha$, $r \mid \Psi\alpha\!\prec\! \alpha \!\succ\!(\alpha[u].u)$]

($\mathfrak{i}$) $[a', b', c', \ldots] : \beta. \equiv .\alpha$ [$\mathfrak{c}$; SI]

($\mathfrak{j}$) $[a', b', c', \ldots, r].\Psi\alpha\!\prec\! \alpha \!\succ\!(\beta r)$ [$\mathfrak{h}$; $\mathfrak{i}$; SII]

($\mathfrak{k}$) $[a', b', c', \ldots, r] \therefore \mathrm{Vr}\,(\Phi(\beta)r). \equiv: \Phi(\alpha). \equiv .\Phi(\beta)$
[$\mathfrak{f}$, $p \mid \beta$, $q \mid \alpha$; $\mathfrak{j}$; SII]

($\mathfrak{l}$) $[a', b', c', \ldots] : \Phi(\alpha). \equiv .\Phi(\beta)$ [$\mathfrak{f}$; A4, $p \mid \Phi(\beta)$, $q \mid r$]

($\mathfrak{m}$) $[a', b', c', \ldots].\Phi(\beta)$ [$\mathfrak{l}$; $\mathfrak{d}$]

(𝔫) $[a, b, c, \ldots].\Phi(\beta)$ [𝔪; as in the point 𝔣 of the proof of SI]

Q.E.D.

SIV. *If under the same assumptions as in* SIII *in the field of the system we have two theses of the following structures*:

$$[a, b, c, \ldots] : \alpha. \equiv .\beta$$

and

$$[a, b, c, \ldots].\Phi(\beta)$$

then we can always add the following thesis to the system:

$$[a, b, c, \ldots].\Phi(\alpha)$$

The proof of SIV follows at once from SI and SIII.

It has to be noted that whenever we apply SIII (or SIV) we must introduce two auxiliary definitions into our system in order to perform the required deductions. We need one definition in order to transform the thesis, with respect to which we want to apply extensional reasoning, into a thesis of the form similar to our assumption 𝔟, i.e. into a thesis which is formed by means of a proposition-forming functor of one propositional argument. The other definition is required in order to apply the point 𝔣 in the proof of SIII. It is not difficult to construct such definitions for any given case and I shall omit them when using SIII or SIV. Only for the purpose of illustration will I give the definitions when SIII is used for the first time. Namely:

A15 $[pq] \therefore p \equiv q. \equiv .\mathrm{As}\,(p \equiv q) :\equiv .p \equiv p$ [A12, $r \mid q$; D2; SIII]

In order to give a complete proof of A15, we must introduce the definition:

D𝔄 $[pqr] :: p \equiv q. \equiv .\mathrm{As}\,(p \equiv q) :\equiv r \therefore \equiv .\chi\alpha\!\!\prec\!\! pq \!\!\succ\!\!(r)$ [α]

From D𝔄 and D2 we get:

I $[pq].\chi\alpha\!\!\prec\!\! pq \!\!\succ\!\!(\mathrm{Vr}\,(pq))$ [D𝔄, $r \mid \mathrm{Vr}\,(pq)$; D2]

Thesis I corresponds to our assumption 𝔟 and thesis A12 ($r \mid q$) to 𝔞. Then, we can easily prove the theses corresponding to 𝔠–𝔢. Next, we introduce the second definition corresponding to the point 𝔣:

D𝔅 $[pqrst] :: \mathrm{Vr}\,(\chi\alpha\!\!\prec\!\! st \!\!\succ\!\!(p)r). \equiv : \chi\alpha\!\!\prec\!\! st \!\!\succ\!\!(p). \equiv .\chi\alpha\!\!\prec\!\! st \!\!\succ\!\!(q)$
$\therefore \equiv .\chi\beta\!\!\Prec\!\! st \!\!\Succ\!\!\prec\!\! q \!\!\succ\!\!(pr)$ [α]

and, subsequently, we get the theses corresponding to the points $\mathfrak{g}$–$\mathfrak{n}$, i.e., we finally obtain the thesis:

II $[pq].\chi\alpha\langle pq\rangle(p \equiv p)$

which in virtue of D$\mathfrak{A}$ $(r \mid p \equiv p)$ and SII gives A15.

A16 $[pq]\therefore p \equiv q. \equiv .p \equiv q :\equiv .p \equiv p$ [A15; A6, $p \mid p \equiv q$; SIII]

A17 $[pq]\therefore p \equiv p. \equiv: p \equiv q. \equiv .p \equiv q$ [A16; SI]

D4 $[pq]\therefore p \equiv p. \equiv .q \equiv q :\equiv .\mathrm{Vr}_1(qp)$ [α^{68}]

Now, in order to make the formulas clearer and easier to read, I introduce the following two abbreviations: "0" will be used for "$[u].u$" and "1" for "$[u].u. \equiv .[u].u$". We should always remember that these symbols ("0" and "1") are used here only as typographical abbreviations; they are not introduced into the system by means of definitions although the definitions:[69]

DI $[u].u. \equiv .0$

and

DII $[u].u. \equiv .[u].u :\equiv .1$

could be added to the system in accordance with the rule of procedure. Hence, e.g., the real form of A18, to be proved next, is:

$\mathrm{Vr}_1([u].u, [u].u. \equiv .[u].u)$

A18 $\mathrm{Vr}_1(0\ 1)$

[D4, $p \mid [u].u. \equiv .[u].u$, $q \mid [u].u$; A16, $p \mid [u].u$, $q \mid [u].u$]

A19 $1. \equiv .\mathrm{Vr}_1(0\ 0)$ [A17, $p \mid 0$, $q \mid 0$; D4, $p \mid 0$, $q \mid 0$; SIII]

A20 $\mathrm{Vr}_1(0\ \mathrm{Vr}_1(0\ 0))$ [A18; A19; SIII]

A21 $[r] : \mathrm{Vr}_1(0\ r). \equiv .0 \equiv 0$ [A14, $p \mid 0$, $f \mid \mathrm{Vr}_1$; A20]

A22 $[r].\mathrm{Vr}_1(0\ r)$ [A21; A13, $p \mid 0$; SII]

A23 $[p] : p \equiv p. \equiv .1$ [D4, $q \mid 0$; A22, $r \mid p$; SII]

A24 $[p] : 1. \equiv .p \equiv p$ [A23; SI]

A25 $1 \equiv 1. \equiv .1 \equiv 0 :\equiv: 1 \equiv .1 \equiv 0$

[A13, $p \mid 1 \equiv .1 \equiv 0$; A23, $p \mid 1$; SIV]

A26 $[p] :: 0 \equiv .0 \equiv 1 :\equiv .\cdot.0 \equiv: 0 \equiv .p \equiv p$

[A13, $p \mid 0 \equiv .0 \equiv 1$; A24; SIII]

A27 $[p] :\cdot: 0 \equiv .0 \equiv 1 :\equiv 1.\cdot. \equiv:: 0 \equiv: 0 \equiv .p \equiv p.\cdot. \equiv 1$

[A13, $p \mid 0 \equiv .0 \equiv 1 :\equiv 1$; A26; SIII]

D5 $[pq] :: p \equiv: p \equiv .q \equiv q.\cdot. \equiv .\mathrm{Vr}_2(pq)$ [α]

A28 $[p].\cdot.0 \equiv .0 \equiv 1 :\equiv \mathrm{Vr}_2(0p)$ [A26; D5, $p \mid 0$, $q \mid p$; SIV]

A29 $[r] :: 0 \equiv .0 \equiv 1 :\equiv 1.\cdot. \equiv: \mathrm{Vr}_2(0\,r). \equiv .1$

[A27, $p \mid r$; D5, $p \mid 0$, $q \mid r$; SIII]

A30 $0 \equiv .0 \equiv 1 :\equiv 1.\cdot. \equiv: [r] : \mathrm{Vr}_2(0\,r). \equiv .1$ [A29; β]

The proof of A30 deserves our attention, because A30 follows from A29 by the distribution of the quantifier. Since we do not know any other way of obtaining A30 at this stage and since it is a necessary step to get A32 and, consequently, the important A37, it is clear that the point β of the rule plays an essential role in the deductions presented in this section.

A31 $0 \equiv .0 \equiv 1 :\equiv: [r] : \mathrm{Vr}_2(0\,r). \equiv .1$

[A14, $p \mid 0$, $f \mid \mathrm{Vr}_2$; A28, $p \mid \mathrm{Vr}_2(0\,0)$; SIV]

A32 $0 \equiv .0 \equiv 1 :\equiv .\cdot.0 \equiv .0 \equiv 1 :\equiv 1$ [A31; A30; SIV]

D6 $[pq].\cdot.p \equiv .p \equiv q :\equiv .\Phi\beta(qp)$ [α]

A33 $\Phi\beta(1\,0). \equiv: \Phi\beta(1\,0). \equiv 1$ [A32; D6, $p \mid 0$, $q \mid 1$; SIII]

A34 $\Phi\beta(1\;\Phi\beta(1\,0))$ [D6, $p \mid \Phi\beta(1\,0)$, $q \mid 1$; A33]

A35 $[r] : \Phi\beta(1\,r). \equiv .1 \equiv 1$ [A14, $p \mid 1$, $f \mid \Phi\beta$; A34]

A36 $[r].\Phi\beta(1\,r)$ [A35; A13, $p \mid 1$; SII]

A37 $[p] : p \equiv .p \equiv 1$ [D6, $q \mid 1$; A36, $r \mid p$; SII]

A38 $[p] : p \equiv 1. \equiv p$ [A37; SI]

A39 $[pq] \therefore p \equiv : p \equiv .q \equiv q$ [A37; A24, $p \mid q$; SIII]

A40 $[pq] \therefore p \equiv q. \equiv : p \equiv 1. \equiv q$ [A13, $p \mid p \equiv q$; A37; SIII]

A41 $[pq] \therefore p \equiv q. \equiv : p \equiv 1. \equiv .q \equiv 1$ [A40; A37, $p \mid q$; SIII]

A42 $[p] \therefore p \equiv .p \equiv 1 : \equiv 1$ [A23; A37; SIII]

A43 $[p] \therefore 1 \equiv : p \equiv .p \equiv 1$ [A42; SI]

A44 $[p] \therefore 1 \equiv .1 \equiv 1 : \equiv : p \equiv .p \equiv 1$ [A43, $p \mid 1$; A43; SIII]

Now, although we have not yet got the thesis:

$$[f] :: f([u].u). \equiv \therefore f([u].u. \equiv .[u].u). \equiv : [p] : f([u].u). \equiv .f(p)$$ [70]

or any other thesis which, at this stage of the [development of the] system, could allow us to make the deductions of generalization directly, we are in a position to prove a scheme of reasoning which says that in the field of the system we can always reason in accordance with the principle of bivalence for propositions. Viz., we prove the following metarule of procedure:

SV. *If in the field of the system a formula*

$$[a, b, c, \ldots, p].\Phi(p)$$

is meaningful, i.e., if it is a well-formed formula, in which "p" belongs to the semantical category of propositions, and if the variables $a, b, c, \ldots$ are free in "Φ" which is a simple or a multi-link functor belonging to the category of proposition-forming functors of one propositional argument, *and, finally, if the following two formulas*

$$[a, b, c, \ldots].\Phi([u].u)$$

and

$$[a, b, c, \ldots].\Phi([u].u. \equiv .[u].u)$$

are already proved in the system, then we can always add to it as a new thesis the formula

$$[a, b, c, \ldots, p].\Phi(p)$$

Dem.:

(𝔞) $[a, b, c, \ldots].\Phi(0)$ [The assumption]

(𝔟) $[a, b, c, \ldots].\Phi(1)$ [The assumption]

(𝔠) $[a', b', c', \ldots].\Phi(0)$ [𝔞, as in the point 𝔟 of the proof of SI]

(𝔡) $[a', b', c', \ldots].\Phi(1)$ [𝔟, as in the point 𝔟 of the proof of SI]

(𝔢) $[a', b', c', \ldots, p] : \Phi(0). \equiv .p \equiv p$ [A39, $p \mid \Phi(0)$, $q \mid p$; 𝔠]

(𝔣) $[a', b', c', \ldots, p] : \Phi(1). \equiv .p \equiv p$ [A39, $p \mid \Phi(1)$, $q \mid p$; 𝔡]

(𝔤) $[a', b', c', \ldots, p, q] \therefore \Phi(q). \equiv .p \equiv p :\equiv .\Psi\beta(pq)$

[In virtue of the point α of the rule, as in the point 𝔣 of the proof of SIII]

(𝔥) $[a', b', c', \ldots, p].\Psi\beta(p0)$ [𝔤, $q \mid 0$; 𝔢]

(𝔦) $[a', b', c', \ldots, p].\Psi\beta(p1)$ [𝔤, $q \mid 1$; 𝔣]

(𝔧) $[a', b', c', \ldots, p] : \Psi\beta(p0). \equiv .1$ [A37, $p \mid \Psi\beta(p0)$; 𝔥]

(𝔨) $[a', b', c', \ldots, p].\Psi\beta(p\Psi\beta(p[u].u))$ [𝔦; 𝔧; SIV]

(𝔩) $[a', b', c', \ldots, p, r] : \Psi\beta(pr). \equiv .p \equiv p$ [A14, $f \mid \Psi\beta$; 𝔨]

(𝔪) $[a', b', c', \ldots, p, r].\Psi\beta(pr)$ [𝔩; A13; SII]

(𝔫) $[a', b', c', \ldots, p] : \Phi(p). \equiv .p \equiv p$ [𝔤, $q \mid p$; 𝔪, $r \mid p$; SII]

(𝔬) $[a', b', c', \ldots, p].\Phi(p)$ [𝔫; A13; SII]

(𝔭) $[a, b, c, \ldots, p].\Phi(p)$ [𝔬; as in the point 𝔣 of the proof of SI]

Q.E.D.

The proof of SV shows that A_n satisfies the condition 𝔠 of metatheorems L and S. Just as in the case of applying SIII (or SIV), whenever we use SV, we must introduce into our system two auxiliary definitions. Since the construction of such definitions is not difficult, I shall omit them when using SV. Only for the purpose of illustration will I give the definitions, when SV is used for the first time. Viz.:

A45 $[p] \therefore 1 \equiv p. \equiv: 1 \equiv 0. \equiv .p \equiv 0$

[A17, $p \mid 1$, $q \mid 0$; A39, $p \mid 1 \equiv 0$, $q \mid 0$; SV]

In order to have a complete proof of A45, we must introduce the following definition:

D$\mathfrak{C}$ $[p] :: 1 \equiv p. \equiv: 1 \equiv 0. \equiv .p \equiv 0. \therefore \equiv .\chi\gamma(p)$ $[\alpha^{71}]$

From D$\mathfrak{C}$ and A39 we get:

III $\chi\gamma(0)$ [D$\mathfrak{C}$, $p \mid 0$; A39, $p \mid 1 \equiv 0$, $q \mid 0$]

and from D$\mathfrak{C}$ and A17:

IV $\chi\gamma(1)$ [D$\mathfrak{C}$, $p \mid 1$; A17, $p \mid 1$, $q|0$]

Theses III and IV correspond to our assumptions $\mathfrak{a}$ and $\mathfrak{b}$, respectively. Then, an easy proof gives theses corresponding to the points $\mathfrak{c}$–$\mathfrak{f}$. Next, we introduce the second definition according to the point $\mathfrak{g}$:

D$\mathfrak{D}$ $[pq] \therefore \chi\gamma(q). \equiv .p \equiv p :\equiv .\chi\delta(pq)$ $[\alpha]$

and, subsequently, we get the theses corresponding to the points $\mathfrak{h}$–$\mathfrak{p}$, i.e., we finally obtain

V $[p].\chi\gamma(p)$

which, in virtue of D$\mathfrak{C}$ and SII, gives A45.

Our deductions proceed as follows:

A46 $[p] :: 1 \equiv .1 \equiv p :\equiv \therefore 1 \equiv 0. \equiv: 1 \equiv 0. \equiv p$

[A45, $p \mid 1 \equiv 0$; A44, $p \mid 1 \equiv 0$; SV]

A47 $1 \equiv: 1 \equiv .1 \equiv 0. \therefore \equiv .1 \equiv 0$

[A46, $p \mid 1 \equiv 0$; A39, $p \mid 1 \equiv 0$, $q \mid 1 \equiv 0$; SIV]

A48 $[p] \therefore 1 \equiv p. \equiv .1 \equiv 0 :\equiv: 1 \equiv .p \equiv 0$ [A25; A16, $p \mid 1$, $q \mid 0$; SV]

A49 $1 \equiv \therefore 1 \equiv .1 \equiv 0 :\equiv 0$ [A48, $p \mid 1 \equiv .1 \equiv 0$; A47]

A50 $1 \equiv .1 \equiv 0 :\equiv 0$ [A49; A13, $p \mid 0$]

A51 $[q] \therefore 1 \equiv .1 \equiv q :\equiv q$ [A50; A42, $p \mid 1$; SV]

A52 $[q] \therefore 0 \equiv .0 \equiv q :\equiv q$ [A38, $p \mid 0$; A42, $p \mid 0$; SV]

A53 $[pq] \therefore p \equiv .p \equiv q :\equiv q$ [A52; A51; SV[72]]

A54 $[pq] \therefore q \equiv: p \equiv .p \equiv q$ [A53; SI]

A55 $[pq] \therefore q \equiv 1. \equiv: p \equiv .p \equiv q$ [A54; A37, $p \mid q$; SIII]

A56 $[q] \therefore 0 \equiv q. \equiv: 1 \equiv .q \equiv 0$ [A55, $p \mid 1, q \mid 0$; A37, $p \mid 0 \equiv 0$; SV]

A57 $[pq] \therefore p \equiv q. \equiv: p \equiv 0. \equiv .q \equiv 0$ [A56; A45, $p \mid q$; SV]

A58 $[pqr] \therefore p \equiv q. \equiv: p \equiv r. \equiv .q \equiv r$ [A57; A41; SV]

A59 $[pqr] :: p \equiv q. \equiv \therefore p \equiv .q \equiv r :\equiv r$

[A58, $r \mid q \equiv r$; A53, $p \mid q, q \mid r$; SIII]

A60 $[pq] :: p \equiv \therefore p \equiv .1 \equiv q :\equiv q$ [A59, $q \mid 1, \mid q$; A38; SIII]

A61 $[pq] :: p \equiv .1 \equiv q :\equiv q \therefore \equiv p$ [A60; SI]

A62 $[pq] :\cdot: p \equiv:: p \equiv .1 \equiv q :\equiv 1 \therefore \equiv q$

[A60; A37, $p \mid p \equiv .1 \equiv q$; SIII]

A63 $[pq] \vdots\vdots p \equiv:\cdot: p \equiv .1 \equiv q :\equiv 1 \therefore \equiv 1 ::\equiv q$

[A60; A37, $p \mid p \equiv .1 \equiv q :\equiv 1$; SIII]

A64 $[p] \vdots\cdot\vdots p \equiv .1 \equiv p :\equiv:\cdot: p \equiv .1 \equiv p :\equiv 1 \therefore \equiv 1 ::\equiv p \vdots\vdots \equiv:\cdot: p \equiv .1 \equiv p :\equiv 1 \therefore \equiv 1 ::\equiv p$ [A61, $q \mid p$; A63, $q \mid p$; SIII]

D7 $[pq] :\cdot: 0 \equiv .p \equiv 0 :\equiv .q \equiv 0 \therefore \equiv .q \equiv 0 ::\equiv .\Phi\gamma(pq)$ [α]

A65 $0 \equiv .1 \equiv 0 :\equiv .\Phi\gamma(1\,0) \equiv 0 \therefore \equiv .\Phi\gamma(1\,0) \equiv 0$

[A64, $p \mid 0$; D7, $p \mid 1, q \mid 0$; SIII]

A66 $\Phi\gamma(1\Phi\gamma(1\,0))$ [D7, $p \mid 1, q \mid \Phi\gamma(1\,0)$; A65]

A67 $[r] : \Phi\gamma(1\ r). \equiv .1 \equiv 1$ [A14, $p \mid 1, f \mid \Phi\gamma$; A66]

A68 $[q].\Phi\gamma(1\ q)$ [A67, $r \mid q$; A13, $p \mid 1$; SII]

A69 $[q] :: 0 \equiv .1 \equiv 0 :\equiv .q \equiv 0.\therefore \equiv .q \equiv 0$ [D7, $p \mid 1$; A68; SII]

A70 $0 \equiv .1 \equiv 0 :\equiv .0 \equiv 0$ [A69, $q \mid 0$; A13, $p \mid 0$; SII]

A71 $0 \equiv .1 \equiv 0$ [A70; A13, $p \mid 0$; SII]

A71 constitutes a crucial point in this proof, since it gives:

A72 $[p] : p \equiv .1 \equiv p$ [A71; A37, $p \mid 1$; SV]

A73 $[p] : 1 \equiv p. \equiv .p \equiv 1$ [A37; A72; SIII]

A74 $[p] : p \equiv 1. \equiv .1 \equiv p$ [A73; SI]

A75 $[p] : p \equiv 0. \equiv .0 \equiv p$ [A13, $p \mid 0 \equiv 0$; A73, $p \mid 0$; SV]

A76 $[pq] : p \equiv q. \equiv .q \equiv p$ [A75; A74; SV]

Thus, we have proved that equivalence is symmetrical, and that from now on we can dispense with SI and SII.

A77 $[pqr]\therefore p \equiv q. \equiv: q \equiv r. \equiv .p \equiv r$

[A58; A76, $p \mid p \equiv r, q \mid q \equiv r$; SIII]

A78 $[pqr]\therefore p \equiv q. \equiv: r \equiv q. \equiv .p \equiv r$ [A77; A76, $p \mid q, q \mid r$; SIII]

Since A78 is Łukasiewicz's axiom Ł1 of the classical equivalential propositional calculus and since the rules of procedure of that theory are, obviously, included into the rule of $\mathfrak{S}_5$, I have proved that from A_n the system $\mathfrak{S}$ can be obtained, which satisfies the condition $\mathfrak{a}$ of metatheorem L. Condition $\mathfrak{c}$ of that metatheorem is also satisfied by A_n as can be seen from metarule SV. Since the condition $\mathfrak{b}$ of L is superfluous, as will be shown in Section 2.5, we have a proof that A_n can serve as a single axiom of the system $\mathfrak{S}_5$ of protothetic. It is worth noting that although we have not used the point ϵ of the rule of procedure so far, we have been able to obtain very strong deductive results. On the other hand, I am unable to prove that A_n (or A_o or A_p or A_q) is a single axiom of $\mathfrak{S}_5$ without the application of reasonings which will be discussed in Section 2.5 and in which the point ϵ plays an essential role.

Evidently, F1 and F2 (appearing in metarule S) can be derived from A78 alone by substitution and detachment. F1 has already been proved as it results by substitution from A13 ($p \mid [u].u$) while F2 is easily obtainable as follows:

A79 $[pq]\therefore p \equiv: q \equiv p. \equiv q$ [A54, $p \mid q, q \mid p$; A76, $p \mid q, q \mid q \equiv p$; SIII]

2.2

A simple inspection of the deductions presented in the preceding subsection will convince us that each of the theses:

$$A_o\ [pq] :: p \equiv q. \equiv \therefore [f]\therefore f(qf(q[u].u)). \equiv: [r] : f(pr). \equiv .q \equiv p$$

$$A_p\ [pq] :: p \equiv q. \equiv \therefore [f]\therefore f(pf(q[u].u)). \equiv: [r] : f(qr). \equiv .q \equiv p$$

$$A_q\ [pq] :: p \equiv q. \equiv \therefore [f]\therefore f(qf(p[u].u)). \equiv: [r] : f(pr). \equiv .q \equiv p$$

can serve as a single axiom of the system $\mathfrak{S}_5$. It is evident that the theses A1–A14 and metarules SI and SII can be obtained from each of the discussed theses in exactly the same way as they were obtained from A_n. An entirely analogous proof of SIII exists for A_p. In the case of A_o and A_q a little modification has to be introduced in order to get this metarule. Viz., a scheme of a definition used in the point $\mathfrak{f}$ has to have now the following form:

$$(\mathfrak{f}^*)\ [a', b', c', \ldots, p, q, r] :: \mathrm{Vr}\,(\Phi(q)r). \equiv: \Phi(p). \equiv .\Phi(q)\therefore \equiv .$$
$$\Psi^*\alpha \langle q \rangle (pr)$$

from which and the points $\mathfrak{a}$–$\mathfrak{e}$ the analogues of the points $\mathfrak{g}$–$\mathfrak{n}$ can be obtained without any difficulty. Now, metarules SIV and SV and theses A15–A78 are provable in each of these cases in exactly the same way as in Section 2.1, since no direct application of an axiom is used in order to obtain them. Hence, each of the discussed theses can serve as a single axiom of the system $\mathfrak{S}_5$.

A more penetrating analysis of the demonstrations given in Section 2.1 allows us to establish the following metatheorem S*:

METATHEOREM S*: An axiom-system of protothetic having the rules of procedure inferentially equivalent to the rule of $\mathfrak{S}_5$ yields a complete system, if in its field the following conditions are satisfied:

I. Thesis A14 is provable.
II. Metarules SI and SIII are provable.

In fact in virtue of the rule of $\mathfrak{S}_5$ alone we have D1, D2, A4; SII and SIV follow from SI and SIII. Hence, as an inspection of Section 2.1 shows, one can obtain A12, A13, SV and A15–A78 using only the conditions of S*.

Therefore, having A78 and SV, we get the required conditions of S. It seems to me that without the application of the point ϵ (concerning higher extensionalities) it is impossible to prove A_n (or A_o, A_p, A_q) from the conditions of S*. But it is easy to deduce axiom A_l mentioned in Section 1.3.

Dem.: Obviously, in virtue of S* we have A77 and A78. Due to Łukasiewicz's proof, the following two theses can be obtained from A78.

M1 $[pqrst] :\cdot: s \equiv .t \equiv t :\equiv:: s \equiv .\therefore p \equiv q. \equiv: r \equiv q. \equiv .p \equiv r$

M2 $[pq] .\therefore p \equiv: p \equiv q. \equiv q$

Therefore, we have:

M3 $[fpq] :\cdot: [r] : f(pr). \equiv .p \equiv p :\equiv:: [r] :: f(pr). \equiv .\therefore p \equiv q. \equiv: r \equiv q. \equiv .p \equiv r$ [M1, $s \mid f(pr)$, $t \mid p$; and the application of the point β of the rule with respect to r]

M4 $[fpt] :: [r] : f(pr). \equiv .p \equiv p :\equiv t.\therefore \equiv: f(pf(p[u].u)). \equiv t$ [A77, $p \mid f(pf(p[u].u))$, $q \mid [r] : f(pr). \equiv .p \equiv p, r \mid t$; β; A14, in virtue of S*]

M5 $[fpq] :\cdot: f(pf(p[u].u)). \equiv:: [r] :: f(pr). \equiv .\therefore p \equiv q. \equiv : r \equiv q. \equiv .p \equiv r$

[M4, $t \mid [r] :: f(pr). \equiv .\therefore p \equiv q. \equiv: r \equiv q. \equiv .p \equiv r$; β; M3]

A_l $[fpq] :\cdot: f(pf(p[u].u)). \equiv:: [r] :: f(p \equiv q. \equiv q, r). \equiv .\therefore p \equiv q. \equiv: r \equiv q. \equiv .p \equiv r$ [M5; M2; SIII, in virtue of S*]

Q.E.D.

On the other hand, there is no simple way of obtaining the conditions of S* from A_l alone. Indeed, using elementary deductions we can get A78 from it at once as, in virtue of the rule, we have D1, D2 and A4 at our disposal. Therefore, we have also SI (by A78) and we can prove easily:

N1 $[fpq] : f(pf(p[u].u)). \equiv .[r].f(p \equiv q. \equiv qr)$

But, A14 and SIII cannot be deduced directly from A78 and N1, as the last thesis gives only an extremely narrow possibility of extensional deductions. In order to obtain the discussed thesis A14 and metarule SIII from those assumptions we have to use metatheorem S, i.e. we have to prove first that its condition c (SV) follows from A78 and N1. This can be done only by an application of the point ϵ of the rule and the reasonings which *mutatis mutandis* are similar to those to be presented in Section 2.5. I will return to this problem in Section 3 of the present paper.

To conclude these considerations concerning the structure of A_n, I would like to make the following historical remark. The methods of deductions applied to obtain SI and SV in the preceding subsection, had been established by Leśniewski and applied to axioms A_a–A_m. On the other hand, the proofs by which A13, A37, A71, and metarule SIII are obtained from A_n in Section 2.1 were previously unknown. The "decompositions" of A_a–A_m are much more simple than the decomposition of A_n and do not involve so complicated deductions.

*2.3**

When in 1938 I proved that the condition $\mathfrak{b}$ of metarule L follows from the remaining conditions $\mathfrak{a}$ and $\mathfrak{c}$,[73] I also remarked that probably the condition $\mathfrak{a}$ of L could be weakened too. Viz., I thought that the system $\mathfrak{S}$ could be substituted by a suitable fragment of it, and that this would certainly be done, if we had SIII at our disposal, which, as we remember, enables us to make extensional deductions for expressions belonging to the semantical category of propositions. This suggestion of mine and the result concerning the redundancy of the condition $\mathfrak{b}$ enabled Leśniewski to construct his axiom A_m:[74]

$$A_m \ [pq] :: p \equiv q. \equiv :\cdot: [f] :\cdot: f(qf(q[u].u)). \equiv :: [r] :: f(pr). \equiv .\!\cdot\!. \ r \equiv : q \equiv .r \equiv p$$

In fact, starting from A_m alone and using deductions in some respects similar but less complicated than those presented in Section 2.1, one can easily prove:

$$\text{P1} \ [pq] \therefore p \equiv : q \equiv .p \equiv q$$

A13, and metarules SII, SIII and SV, which amounts to the condition $\mathfrak{c}$. Now, given the rule of $\mathfrak{S}_5$, P1, A13, SII, SIII and SV suffice for the purpose of deriving A78, i.e., the condition $\mathfrak{a}$. Moreover, this can be done in a very elementary way. These deductions of Leśniewski confirmed the correctness of my remark. For, obviously, in the set of assumptions {A13; P1; SII; SIII; SV} we can replace SII by A76, but on the basis of the rule of procedure of $\mathfrak{S}$ the theses A13, A76 and P1 do not constitute a complete system of $\mathfrak{S}$. We can prove this using the following matrix:

---

* Bibliographical Note. Sections 2.3–2.5 originally appeared in the *Notre Dame Journal of Formal Logic* II (1961), 129–148.

M2

≡	1	2	3	4	5	6	7	8	9
* 1	1	2	3	4	5	6	7	8	9
2	2	1	5	7	3	1	4	1	1
3	3	5	1	6	2	4	1	1	1
4	4	7	6	1	1	3	2	1	1
5	5	3	2	1	1	1	1	9	8
6	6	1	4	3	1	1	8	7	1
7	7	4	1	2	1	8	1	6	1
8	8	1	1	1	9	7	6	1	5
9	9	1	1	1	8	1	1	5	1

which is hereditary with respect to the system $\mathfrak{S}$ and its rules of procedure, and in which the designated value is 1. This matrix verifies A13, A76 and P1, but falsifies A78, for $p = 2$, $q = 3$ and $r = 4$. Viz.: $[2 \equiv 3. \equiv: 4 \equiv 3. \equiv .2 \equiv 4] = [5 \equiv .6 \equiv 7] = [5 \equiv 8] = 9$.[75]

An analysis of Leśniewski's deductions connected with axiom A_m allowed me to formulate the following metatheorem:

METATHEOREM L*: An axiom-system of protothetic having the rules of procedure inferentially equivalent to the rule of $\mathfrak{S}_5$ yields a complete system, if in its field the following conditions are satisfied.

I. Metarules SII (or the thesis A76), SIII and SV are provable.

II. Theses A13 and P1 are provable.

My further investigations have shown that L* can be expressed in a more refined manner as metatheorem W or metatheorem W*. In metatheorem W the reference to A76 in the condition I of L* is dropped while the condition II demands that the following theses should be provable:

G1 $1 \equiv 1$

G2 $1 \equiv 0. \equiv 0$

G3 $0 \equiv .0 \equiv 1$

Metatheorem W* differs from metatheorem W in that SII is removed from the condition I of W, whereas the condition II demands that A76 and G2 should be provable. All we have to do in order to show the correctness of these metatheorems is to obtain the theory $\mathfrak{S}$ from their conditions. For the condition $\mathfrak{c}$ of metatheorem L is guaranteed by SV. Since SII is contained in W, we are able to apply detachment in both directions. This is indispensable since the application of SIII and SV involves the construction of suitable definitions and the method of getting rid of the defined constants in the final result. As the function of these definitions was discussed in Section 2.1, I omit them in the proof that follows.

On the basis of the rule $\mathfrak{S}_5$ and assuming the theses G1, G2, G3, and the metarules SII, SIII and SV, we obtain:

G4 $0 \equiv 0$ [α; G2; SII; SIII]

G5 $[p].p \equiv p$ [α; G4; G1; SII; SV]

G6 $1 \equiv: 0 \equiv .0 \equiv 1$ [α; G1; G3; SII; SIII]

G7 $[p]{\therefore}p \equiv: 0 \equiv .0 \equiv p$ [α; G3; G6; SII; SV]

G8 $[p]{\therefore}0 \equiv: 0 \equiv .p \equiv p$ [G7, $p \mid p \equiv p$; G5]

G9 $1 \equiv 0. \equiv .0 \equiv 1$ [α; G2; G3; SII; SIII]

G10 $[p] : p \equiv 0. \equiv .0 \equiv p$ [α; G1; G9; SII; SV]

G11 $0 \equiv .1 \equiv 0$ [G10, $p \mid 1 \equiv 0$; G2]

G12 $0 \equiv: 1 \equiv .1 \equiv 0$ [α; G11; SII; SIII]

G13 $[p]{\therefore}0 \equiv .p \equiv p :\equiv: 1 \equiv .1 \equiv 0$ [α; G12; G8; SII; SIII]

G14 $[p]{\therefore}p \equiv .1 \equiv 1 :\equiv: 1 \equiv .1 \equiv p$
[G13, $p \mid 1$; G5, $p \mid 1 \equiv .1 \equiv 1$; SII; SV]

G15 $1 \equiv: 1 \equiv .1 \equiv 1$ [G14, $p \mid 1 \equiv 1$; G5, $p \mid 1 \equiv 1$]

G16 $1 \equiv .1 \equiv 1$ [G15; G4]

G17 $[p] : p \equiv .p \equiv 1$ [α; G3; G16; SII; SV]

G18 $[p] : p \equiv .1 \equiv p$ [α; G11; G16; SII; SV]

G19 $[p] : p \equiv 1. \equiv .1 \equiv p$ [α; G17; G18; SII; SIII]

G20 $[pq] : p \equiv q. \equiv .q \equiv p$ [α; G10; G19; SII; SV]

We can get G20, i.e. A76, directly from G1, G2 and G3 without the use of G16, but this would not shorten the deductions, as G16 is indispensable, being, in fact, a crucial point in the proof. Having G20 we can dispense with SII.

G21 $[q]\therefore 0 \equiv q. \equiv: 1 \equiv q. \equiv 0$ [α; G20, $p \mid 0$; G18, $p \mid q$; SII; SIII]

G22 $[q]\therefore 0 \equiv q. \equiv: 1 \equiv q. \equiv .0 \equiv 1$ [α; G21; G3; SII; SIII]

G23 $[qr]\therefore 0 \equiv q. \equiv: r \equiv q. \equiv .0 \equiv r$ [α; G17, $p \mid 0 \equiv q$; G22; SII; SV]

G24 $[q]\therefore q \equiv: 0 \equiv q. \equiv 0$ [α; G27, $p \mid q$; G20, $p \mid 0, q \mid 0 \equiv q$; SII; SIII]

G25 $[q]\therefore 1 \equiv q. \equiv: 0 \equiv q. \equiv 0$ [α; G24; G18, $p \mid q$; SII; SIII]

G26 $[q]\therefore 1 \equiv q. \equiv: 0 \equiv q. \equiv .1 \equiv 0$ [α; G25; G11; SII; SIII]

G27 $[p]\therefore p \equiv: p \equiv .1 \equiv 1$ [α; G17; G16; SII; SIII]

G28 $[qr]\therefore 1 \equiv q. \equiv: r \equiv q. \equiv .1 \equiv r$ [α; G26; G27, $p \mid 1 \equiv q$; SII; SV]

A78 $[pqr]\therefore p \equiv q. \equiv: r \equiv q. \equiv .p \equiv r$ [α; G23; G28; SII; SV]

Thus, we have proved that W fulfills the condition $\mathfrak{a}$. It is evident that W*, too, satisfies this condition as SII and G1 follow directly from A76, and G3 can be obtained from A76 and G2 by an application of SIII. A simple inspection of the proof of W shows the correctness of L* too.

When in 1952 I analyzed once more these metatheorems and similar constructions (in several of them neither SII nor SIII occur), I found, finally, that on the basis of the rule of $\mathfrak{S}_5$ and given SV it is possible to obtain the theory $\mathfrak{S}$ (i.e., the condition $\mathfrak{a}$) from the following two theses:

F1 $[u].u. \equiv .[u].u$

F2 $[pq]\therefore p \equiv: q \equiv p. \equiv q$

which constitute a very small fragment of the complete equivalential calculus of propositions. This can be done not only without metarule SIII but also without an application of the point ϵ of the rule of $\mathfrak{S}_5$.

The proof proceeds as follows.

On the basis of the rule of $\mathfrak{S}_5$ and thesis F2 one can establish the following two metarules of procedure:

(1) Metarule SII formulated exactly in the same way as in Section 2.1. Its proof is apparent from thesis F2.

(2) Metarule SVI:

SVI. *If two formulas of the following forms:*

$$[a, b, c, \ldots].\Phi$$

and

$$[r, s, t, \ldots].\Psi$$

(in which the variables $a, b, c, \ldots$ occur and are free in Φ, and in which the variables $r, s, t, \ldots$ occur and are free in Ψ), *are theses of the system, then we can add to the system a new thesis of the following form:*

$$[a, b, c, \ldots, r, s, t, \ldots] : \Phi. \equiv .\Psi$$

Dem.:

($\mathfrak{a}$) $[a, b, c, \ldots].\Phi$	[The assumption]
($\mathfrak{b}$) $[r, s, t, \ldots].\Psi$	[The assumption]
($\mathfrak{c}$) $[a, b, c, \ldots, r, s, t, \ldots] \therefore \Phi. \equiv .\Psi :\equiv: \Phi$	[F2, $p \mid \Psi$, $q \mid \Phi$; $\mathfrak{b}$]
($\mathfrak{d}$) $[a, b, c, \ldots, r, s, t, \ldots] : \Phi. \equiv .\Psi$	[$\mathfrak{c}$; $\mathfrak{a}$; SII]

Q.E.D.

We continue by deducing:

F3 $[q] : q \equiv 1. \equiv q$	[F2, $p \mid 1$; F1]
F4 $1 \equiv 1$	[F3, $q \mid 1$; F1; SII]
F5 $1 \equiv .1 \equiv 1$	[F1; F4; SVI]
F6 $1 \equiv: 0 \equiv .1 \equiv 0$	[F1; F2, $p \mid 0$, $q \mid 0$; SVI]
F7 $1 \equiv 1. \equiv: 0 \equiv .1 \equiv 0$	[F4; F2, $p \mid 0$, $q \mid 0$; SVI]
F8 $0 \equiv .1 \equiv 0 :\equiv .1 \equiv 1$	[F2, $p \mid 0$, $q \mid 0$; F4; SVI]
F9 $[q] \therefore q \equiv .1 \equiv 1 :\equiv q$	[F2, $p \mid 1 \equiv 1$; F4]
F10 $[p].p \equiv p$	[α; F1; F4; SII; SV]

F11 $[p] : p \equiv .1 \equiv p$ $[\alpha; \text{F2}, p \mid 0, q \mid 0; \text{F5}; \text{SII}; \text{SV}]$

F12 $[p] \therefore p \equiv 1. \equiv: 0 \equiv .p \equiv 0$ $[\alpha; \text{F10}, p \mid 0 \equiv 1; \text{F7}; \text{SII}; \text{SV}]$

F13 $0 \equiv: 1 \equiv 1. \equiv 0$ $[\text{F12}, p \mid 1 \equiv 1; \text{F3}, q \mid 1]$

F14 $[p] \therefore p \equiv: 1 \equiv 1. \equiv p$ $[\alpha; \text{F13}; \text{F2}, p \mid 1, q \mid 1; \text{SII}; \text{SV}]$

F15 $[p] \therefore p \equiv .1 \equiv 0 :\equiv: 1 \equiv .p \equiv 0$ $[\alpha; \text{F8}; \text{F10}, p \mid 1 \equiv .1 \equiv 0; \text{SII}; \text{SV}]$

F16 $[pq] \therefore p \equiv .q \equiv 0 :\equiv: q \equiv .p \equiv 0$ $[\alpha; \text{F12}; \text{F15}; \text{SII}; \text{SV}]$

F17 $[q] : 0 \equiv q. \equiv .q \equiv 0$ $[\text{F16}, p \mid q, q \mid 0 \equiv q; \text{F2}, p \mid q, q \mid 0]$

F18 $0 \equiv .0 \equiv 1$ $[\text{F17}, q \mid 0 \equiv 1; \text{F3}, q \mid 0; \text{SII}]$

F19 $1 \equiv 1. \equiv: 0 \equiv .0 \equiv 1$ $[\text{F18}; \text{F4}; \text{SVI}]$

F20 $[p] \therefore 1 \equiv .0 \equiv p :\equiv: p \equiv .0 \equiv 1$

$[\alpha; \text{F19}; \text{F10}, p \mid 1 \equiv .0 \equiv 1; \text{SII}; \text{SV}]$

F21 $1 \equiv 0. \equiv .0 \equiv 1$ $[\text{F20}, p \mid 1 \equiv 0; \text{F6}]$

F22 $[q] : 1 \equiv q. \equiv .q \equiv 1$ $[\alpha; \text{F21}; \text{F10}, p \mid 1 \equiv 1; \text{SII}; \text{SV}]$

F23 $[pq] : p \equiv q. \equiv .q \equiv p$ $[\alpha; \text{F17}; \text{F22}; \text{SII}; \text{SV}]$

F24 $[q] : q. \equiv .q \equiv 1$ $[\text{F23}, p \mid q \equiv 1; \text{F3}]$

F25 $[q] \therefore q \equiv: q \equiv .1 \equiv 1$ $[\text{F23}, p \mid q \equiv .1 \equiv 1; \text{F9}]$

F26 $[pqr] \therefore p \equiv p. \equiv: r \equiv q. \equiv .q \equiv r$ $[\text{F10}; \text{F23}, p \mid r; \text{SVI}]$

F27 $[r] \therefore 1 \equiv 0. \equiv: r \equiv 0. \equiv .1 \equiv r$

$[\alpha; \text{F11}, p \mid 1 \equiv 0; \text{F25}, q \mid 1 \equiv 0; \text{SII}; \text{SV}]$

F28 $[qr] \therefore 1 \equiv q. \equiv: r \equiv q. \equiv .1 \equiv r$ $[\alpha; \text{F27}; \text{F26}, p \mid 1, q \mid 1; \text{SII}; \text{SV}]$

F29 $[r] \therefore 0 \equiv 1. \equiv: r \equiv 1. \equiv .0 \equiv r$

$[\alpha; \text{F24}, q \mid 0 \equiv 1; \text{F14}, p \mid 0 \equiv 1; \text{SII}; \text{SV}]$

F30 $[qr] \therefore 0 \equiv q. \equiv: r \equiv q. \equiv .0 \equiv r$ $[\alpha$; F26, $p \mid 0$, $q \mid 0$; F29; SII; SV]

A78 $[pqr] \therefore p \equiv q. \equiv: r \equiv q. \equiv .p \equiv r$ $[\alpha$; F30; F28; SII; SV]

This concludes the proof, since A78 is Łukasiewicz's single axiom of the theory $\mathfrak{S}$. As it was mentioned in Section 1.3,[76] I have no proof whether F1 is indispensable or whether it can be obtained from F2 and SV.

2.4

My 1952 result, demonstrated in the preceding subsection, allowed me to get the final formulation of metarule **S**. Namely, it has shown that the condition $\mathfrak{a}$ of **L** can be substituted by a much weaker one. The proof that the condition $\mathfrak{b}$ of Leśniewski is superfluous had been obtained in two steps and several years earlier. In 1937 I proved without the help of the point ϵ of the rule of $\mathfrak{S}_5$ that the formulation of $\mathfrak{b}$ given by Leśniewski is too strong. In 1938, with the help of the point ϵ of the rule and using an idea of demonstration analogous to that which was applied in the proof of my 1937 result, I was able to establish that the condition $\mathfrak{b}$ follows from points $\mathfrak{a}$ and $\mathfrak{c}$ of metarule **L**.

In this subsection I shall present my deductions from 1937, since: (1) they not only allowed me to establish single axioms of protothetic shorter than A_h or A_i,[77] but also explained several important connections in the field of protothetic, and (2) their analysis allows one to understand better the deductions which will be presented in the next subsection. To put it very generally, the result to be discussed now can be formulated as follows.

THEOREM **A1**. *If*:

(1) *we have the theory* $\mathfrak{S}$ (i.e., the complete equivalential calculus of propositions), *and*

(2) *we can reason in accordance with the law of extensionality for propositions and in accordance with the principle of bivalence for propositions* (i.e., no matter whether both these laws are given as theses or as metarules), *and*

(3) *we have three laws* (out of possible four) *of logical product for conjunction*, viz.[78]

	K1 $0 \equiv .0.0$		K1 $0 \equiv .0.0$
either	K2 $1 \equiv .1.1$	*or*	K2 $1 \equiv .1.1$
	K3 $0 \equiv .0.1$		K4 $0 \equiv .1.0$

(where conjunction is not a primitive term but is defined in terms of equivalence by means of a definition which has an "extensional" form[79]), *then, by using only the first four points of the rule of $\mathfrak{S}_5$, we can always obtain either K4 from K1, K2 and K3 or K3 from K1, K2 and K4.*

In order to prove this theorem let us assume only the first four points of the rule of procedure and that we have:

($\mathfrak{A}$) Theory $\mathfrak{S}$.

($\mathfrak{B}$) The possibility to make deductions in accordance with the law of extensionality for propositions.

($\mathfrak{C}$) The possibility to make deductions in accordance with the principle of bivalence for propositions.

($\mathfrak{D}$) The arbitrarily chosen "extensional" definition of conjunction in terms of equivalence, e.g.:

Z18 $[pq] :: [f] .\therefore f(p). \equiv .f(q) :\equiv q .\therefore \equiv .p.q$

We need such a definition only for the purpose of making conjunction meaningful in the field of the rule of $\mathfrak{S}_5$.

($\mathfrak{E}$) The laws K1 and K2.

Obviously, in order to prove Theorem **A1** we have to consider two cases separately: the first, when K3 is assumed and, the second, when K4 is chosen.

It is evident that in both cases we are able to obtain theses A13, A37, A54, A76 (by $\mathfrak{A}$) and, in virtue of points $\mathfrak{A}$, $\mathfrak{C}$ and $\mathfrak{E}$, the theses:

K5 $[p] : p \equiv .p.p$ [K1; K2; $\mathfrak{C}$]

and

K6 1.1 [K2; A13, $p \mid 0$]

Hence, having A76 we can establish that metarule SII holds for our assumptions. Moreover, in virtue of the point α of the rule of procedure of $\mathfrak{S}_5$, we add the definitions D1 and:

D8 $[p] : p \equiv p. \equiv .\mathrm{Vr}(p)$ [α^{80}]

Now, if we assume K3, then we can make the following deductions:

B1 $[p] : 0 \equiv .0.p$ [α; K1; K3; $\mathfrak{C}$]

B2 $1.(1.(0 \equiv 0))$ [K6; K5, $p \mid 1$; $\mathfrak{B}$]

B3 $0 \equiv .0.(1.0) :\equiv .1.(1.(0 \equiv 0))$

[A54, $p \mid 0 \equiv .0.(1.0)$, $q \mid 1.(1.(0 \equiv 0))$; B1, $p \mid 1.0$; B2]

B4 $1.(1.0). \equiv: 1 \equiv .1.(1.0)$ [A54, $p \mid 1$, $q \mid 1.(1.0)$; A13, $p \mid 0$]

B5 $1 \equiv .1.(1.0) :\equiv .1.(1.(0 \equiv 1))$

[α; A76, $p \mid 1.(1.0)$, $q \mid 1 \equiv .1.(1.0)$; B4; A37, $p \mid 0$; $\mathfrak{B}$]

B6 $[p] \therefore p \equiv .p.(1.0) :\equiv .1.(1.(0 \equiv p))$ [α; B3; B5; $\mathfrak{C}$]

B7 $1.(1.(0 \equiv .1.0))$ [B6, $p \mid 1.0$; K5, $p \mid 1.0$]

B8 $[f] \therefore f(1). \equiv .f(1.(0 \equiv .1.0)) :\equiv .1.(0 \equiv .1.0)$

[Z18, $p \mid 1$, $q \mid 1.(0 \equiv .1.0)$; B7; SII]

B9 $[p].\mathrm{Vr}\,(p)$ [D8; A13]

B10 $[pq] : \mathrm{Vr}\,(p). \equiv .\mathrm{Vr}\,(q)$ [A54, $p \mid \mathrm{Vr}\,(p)$, $q \mid \mathrm{Vr}\,(q)$; B9; B9, $p \mid q$]

B11 $1.(0 \equiv .1.0)$ [B8, $f \mid \mathrm{Vr}$; B10, $p \mid 1$, $q \mid 1.(0 \equiv .1.0)$]

B12 $[f] \therefore f(1). \equiv .f(0 \equiv .1.0) :\equiv: 0 \equiv .1.0$

[Z18, $p \mid 1$, $q \mid 0 \equiv .1.0$; B11; SII]

K4 $0 \equiv .1.0$ [B12, $f \mid \mathrm{Vr}$; B10, $p \mid 1$, $q \mid 0 \equiv .1.0$]

Thus, our theorem holds for the first case. Now, let us assume K4. We can deduce K3 as follows:

C1 $[p] : 0 \equiv .p.0$ [α; K1; K4; $\mathfrak{C}$]

C2 $((0 \equiv 0).1).1$ [α; K6; K5, $p \mid 1$; $\mathfrak{B}$]

C3 $0 \equiv .(0.1).0 :\equiv .((0 \equiv 0).1).1$

[A54, $p \mid 0 \equiv .(0.1).0$, $q \mid ((0 \equiv 0).1).1$; C1, $p \mid 0.1$; C2]

C4 $(0.1).1. \equiv: 1 \equiv .(0.1).1$ [A54, $p \mid 1$, $q \mid (0.1).1$; A13, $p \mid 0$]

C5 $1 \equiv .(0.1).1 :\equiv .((0 \equiv 1).1).1$

[A76, $p \mid (0.1).1$, $q \mid 1 \equiv .(0.1).1$; C4; A37, $p \mid 0$; $\mathfrak{B}$]

C6 $[p] \therefore p \equiv .(0.1).p :\equiv .((0 \equiv p).1).1$ [α; C3; C5; $\mathfrak{C}$]

C7 $((0 \equiv .0.1).1).1$ [C6, $p \mid 0.1$; K5, $p \mid 0.1$]

C8 $[f] \therefore f((0 \equiv .0.1).1). \equiv .f(1) :\equiv 1$

[Z18, $p \mid (0 \equiv .0.1).1$, $q \mid 1$; C7; SII]

C9 As (1) [D1, $p \mid 1$; A13, $p \mid 0$]

C10 As $((0 \equiv .0.1).1)$ [C8, $f \mid$ As; A13, $p \mid 0$; SII; C9; SII]

C11 $(0 \equiv .0.1).1$ [D1, $p \mid (0 \equiv .0.1).1$; C10; SII]

C12 $[f] \therefore f(0 \equiv .0.1). \equiv .f(1) :\equiv 1$ [Z18, $p \mid 0 \equiv .0.1$, $q \mid 1$; C11; SII]

C13 As $(0 \equiv .0.1)$ [C12, $f \mid$ As; A13, $p \mid 0$; SII; C9; SII]

K3 $0 \equiv .0.1$ [D1, $p \mid 0 \equiv .0.1$; C13; SII]

Hence, the theorem holds also for the second case and, therefore, the proof is completed. I do not know, however, whether a modification of Theorem **A1** holds, viz. whether K2 can be obtained from K1, K3, K4 or whether K1 follows from K2, K3, K4, provided of course that the other conditions of **A1** are unchanged and that the definition of conjunction is adopted in such a way that K1 or K2 cannot be obtained from it and the theory $\mathfrak{S}$.[81]

Now, consider the set of assumptions S:

(a) System $\mathfrak{S}$.

S1 $[fp] :: f(p). \equiv \therefore f(p \equiv .[u].u). \equiv: [q] : f(p). \equiv .f(q)$

S2 $[pq] \therefore p \equiv q. \equiv: [f] : f(p). \equiv .f(q)$

S3 $[pq] :: p \equiv q. \equiv \therefore [f] \therefore f(p). \equiv .f(q) :\equiv .p \equiv q$

which, as we know, constitutes a complete axiom-system of $\mathfrak{S}_5$.[82] Theorem **A1** shows at once that in this set of assumptions either S2 or S3 is a consequence of the remaining axioms. In fact, it is evident that the set $\{\mathfrak{S}, \mathrm{S1}, \mathrm{S2}\}$ satisfies the first two points of **A1**. Hence, if we assume $\mathfrak{S}$, S1, S2 and adopt Z18 as a definition of conjunction, then we have, in virtue of $\mathfrak{S}$, the theses A39, A76, A77 and

S4 $[pq] \therefore p \equiv : q \equiv q . \equiv p$

and we can proceed as follows:

S5 $[p] :: p \equiv \therefore [f] \therefore f(p) . \equiv . f(p) : \equiv p$ [S4, $q \mid f(p)$; point β of the rule]

S6 $[p] : p \equiv . p . p$

[A77, $q \mid [f] : f(p) . \equiv . f(p) : \equiv p, r \mid p . p$; S5; Z18, $q \mid p$]

Thus, we have obtained K1 and K2, since they follow from S6 by substitution.

S7 $[pqr] :: [f] : f(p) . \equiv . f(q) : \equiv \therefore [f] \therefore f(p) . \equiv . f(q) : \equiv . r \equiv r$

[A39, $p \mid [f] : f(p) . \equiv . f(q), q \mid r; \beta$]

S8 $[pqr] :: p \equiv q . \equiv \therefore [f] \therefore f(p) . \equiv . f(q) : \equiv . r \equiv r$

[A77, $p \mid p \equiv q, q \mid [f] : f(p) . \equiv . f(q)$,
$r \mid [f] : f(p) . \equiv . f(q) : \equiv . r \equiv r$; S2; S7]

S9 $0 \equiv 1 . \equiv . 0 . 1$

[A77, $p \mid 0, q \mid 0 \equiv 1, r \mid 0.1$; S8, $p \mid 0, q \mid 1, r \mid 0$; Z18, $p \mid 0, q \mid 1$]

K3 $0 \equiv . 0 . 1$ [A77, $p \mid 0, q \mid 0 \equiv 1, r \mid 0.1$; A39, $p \mid 0, q \mid 1$; S9]

Since we have the law of extensionality and the principle of bivalence for propositions, and since we have K1, K2 and K3, an application of Theorem **A1** gives K4. Therefore:

S10 $[pq] :: p . q . \equiv \therefore [f] \therefore f(p) . \equiv . f(q) : \equiv q$

[A76, $p \mid p . q, q \mid [f] \therefore f(p) . \equiv . f(q) : \equiv q$; Z18]

S11 $0 \equiv \therefore [f] \therefore f(1) . \equiv . f(0) : \equiv 0$

[A77, $p \mid 0, q \mid 1.0, r \mid [f] \therefore f(1) . \equiv . f(0) : \equiv 0$; K4; S10, $p \mid 1, q \mid 0$]

Having S11, $\mathfrak{S}$, S2 (i.e., the possibility to reason in accordance with the law of extensionality for propositions) and using A39, A76 and S4, we can establish without any difficulty:

S12 $0 \equiv 1 . \equiv \therefore [f] \therefore f(0) . \equiv . f(1) : \equiv . 0 \equiv 1$ [α; S11; A39; A76; S2]

S13 $1 \equiv 0. \equiv .\therefore [f] \therefore f(1). \equiv .f(0) :\equiv .1 \equiv 0$ [α; S11; A39; A76; S2]

In virtue of $\mathfrak{S}$ and S1, from S8 ($p \mid 0, q \mid 0, r \mid 0$), S8 ($p \mid 1, q \mid 1, r \mid 1$), S12 and S13 one can obtain S3.

On the other hand, if $\mathfrak{S}$, S1 and S3 are assumed and if definition Z18 is adopted, then, in virtue of $\mathfrak{S}$, we have A39, A76, A77, S4 and, therefore, also S5, S6 (i.e., K1 and K2), S7 and S10. Moreover, due to A76 and S3, we can establish metarule SIII[83] as follows:

Dem.:

($\mathfrak{a}$) $\alpha \equiv \beta$ [The assumption]

($\mathfrak{b}$) $\Phi(\alpha)$ [The assumption]

($\mathfrak{c}$) $[f] \therefore f(\alpha). \equiv .f(\beta) :\equiv .\alpha \equiv \beta$ [S3, $p \mid \alpha$, $q \mid \beta$; $\mathfrak{a}$]

($\mathfrak{d}$) $\Phi(\beta)$ [A76, $p \mid \Phi(\alpha). \equiv .\Phi(\beta)$, $q \mid \alpha \equiv \beta$; $\mathfrak{c}$, $f \mid \phi$; $\mathfrak{a}$; $\mathfrak{b}$]

Q.E.D.

This shows that the conditions of the point 2 of Theorem **A1** are satisfied by $\mathfrak{S}$, S1 and S3. Furthermore:

S14 $[pq] \therefore q \equiv q. \equiv p :\equiv p$ [A76, $q \mid q \equiv q. \equiv p$; S4]

The application of metarule SIII to S14 ($p \mid 0, q \mid 0$) and S3 ($p \mid 1, q \mid 0$) gives at once:

S15 $0 \equiv .\therefore [f] \therefore f(1). \equiv .f(0) :\equiv 0$

Hence:

K4 $0 \equiv .1.0$

[A77, $p \mid 0$, $q \mid [f] \therefore f(1). \equiv .f(0) :\equiv 0$, $r \mid 1.0$; S15; Z18, $p \mid 1, q \mid 0$]

and, therefore, in virtue of $\mathfrak{S}$, S1, SIII and Theorem **A1**, we have K3. We further have:

S16 $0 \equiv .\therefore [f] \therefore f(0). \equiv .f(1) :\equiv 1$

[A77, $p \mid 0$, $q \mid 0.1$, $r \mid [f] \therefore f(0). \equiv .f(1) :\equiv 1$; K3; Z18, $p \mid 0, q \mid 1$]

S17 $[pqr] :: [f] \therefore f(p). \equiv .f(q) :\equiv .r \equiv r \therefore \equiv: [f] : f(p). \equiv .f(q)$

$$[A76, p \mid [f] : f(p). \equiv .f(q), q \mid [f] \therefore f(p). \equiv .f(q) :\equiv .r \equiv r; S7]$$

S18 $0 \equiv: [f] : f(0). \equiv .f(1)$

$$[A76, p \mid [f] \therefore f(0). \equiv .f(1) :\equiv 1, r \mid [f] : f(0). \equiv .f(1); S16; S17, p \mid 0, q \mid 1, r \mid 0]$$

Now, the elementary applications of A39, S4, A76 and A77 to S18 allow us to establish:

S19 $0 \equiv 1. \equiv: [f] : f(0). \equiv .f(1)$

and

S20 $1 \equiv 0. \equiv: [f] : f(1). \equiv .f(0)$

Besides, $\mathfrak{S}$ gives:

S21 $[pq] : p \equiv p. \equiv .q \equiv q$

From S21 and the points β and δ of the rule we get at once:

S22 $[p] \therefore p \equiv p. \equiv: [f] : f(p). \equiv .f(p)$

Therefore, without any difficulty we can deduce S2 from S22 ($p \mid 0$), S22 ($p \mid 1$), S19 and S20, in virtue of $\mathfrak{S}$ and S1.

Thus, the proof that S3 follows from $\mathfrak{S}$, S1 and S2, and that S2 can be obtained from $\mathfrak{S}$, S1 and S3 is completed.

A simple inspection of the above proof shows that Theorem **A1** can be applied *mutatis mutandis* to metatheorem **L** of Leśniewski, provided that a definition of conjunction which makes K1–K4 meaningful will be akin in its structure to Z18, i.e., it will be the "extensional" kind of definition of conjunction in terms of equivalence.

2.5

In 1938 I proved a theorem much stronger than **A1**. Viz.:

THEOREM **A2**. *On the basis of the rule of* $\mathfrak{S}_5$, *the system* $\mathfrak{S}$, *and the possibility of reasoning in accordance with the principle of bivalence for propositions* (i.e., no matter whether this law is given as a thesis, e.g. S1, or as a metarule, e.g. SV) *yield K1, K2, K3 and K4.*

Evidently, a direct consequence of **A2** is that $\mathfrak{S}$ and S1 imply S2 and S3, and also, that the point $\mathfrak{b}$ of metarule **L** is entirely superfluous. The proof of **A2** is rather long and complicated, and since in some respect its idea is patterned after the proof of **A1**, I shall stress carefully the analogous points. In the proof that follows, the theses which are consequences of $\mathfrak{S}$ alone will be indicated by J1, J2, J3, etc., but the definitions will continue to be enumerated following the order established in Section 2.1.

Proof. We assume the system $\mathfrak{S}$ and metarule SV. Therefore, in virtue of $\mathfrak{S}$ we, obviously, have metarule SII and the following theses:

J1 $[pqr] :: p \equiv .q \equiv q :\equiv r.\therefore \equiv .p \equiv r$

J2 $[pqrs] :: p \equiv r. \equiv \therefore q \equiv s. \equiv : p \equiv q. \equiv .r \equiv s$

J3 $[p].p \equiv p$

J4 $[pq]\therefore p \equiv : q \equiv .q \equiv p$

J5 $[pqr] :: p \equiv r. \equiv \therefore p \equiv : q \equiv q. \equiv r$

J6 $[pqr]\therefore p \equiv q. \equiv : q \equiv r. \equiv .p \equiv r$

J7 $[pqr] :: p \equiv .q \equiv q :\equiv r.\therefore \equiv .p \equiv r$

J8 $[pq]\therefore p \equiv p. \equiv q :\equiv q$

J9 $[pqr]\therefore p \equiv q. \equiv : r \equiv p. \equiv .r \equiv q$

Now, we introduce the following definition:

D9 $[pq]\therefore p \equiv .q \equiv q :\equiv .\mathfrak{G}\langle p\rangle(q)$ $[\alpha^{84}]$

H1 $[pq] : p \equiv .\mathfrak{G}\langle p\rangle(q)$ $[J1, r \mid \mathfrak{G}\langle p\rangle(q); D9]$

H2 $[pqt]\therefore p \equiv q. \equiv : \mathfrak{G}\langle p\rangle(t). \equiv .\mathfrak{G}\langle q\rangle(t)$
$[J2, r \mid \mathfrak{G}\langle p\rangle(t), s \mid \mathfrak{G}\langle q\rangle(t); H1, q \mid t; H1, p \mid q, q \mid t]$

H3 $[pq]\therefore p \equiv q. \equiv : [t] : \mathfrak{G}\langle p\rangle(t). \equiv .\mathfrak{G}\langle q\rangle(t)$ $[H2; \beta]$

H4 $[t].\mathfrak{G}\langle 1\rangle(t)$ $[H1, p \mid 0 \equiv 0, q \mid t; J3, p \mid 0]$

H5 $[q]\therefore q \equiv: q \equiv .[t].\mathfrak{G}\{1\}(t)$ [J4, $p \mid [t].\mathfrak{G}\{1\}(t)$; H4]

H6 $[p] : p \equiv .[t].\mathfrak{G}\{p\}(t)$ H1, $q \mid t; \beta$]

H7 $[pq]\therefore p \equiv: q \equiv q. \equiv .[t].\mathfrak{G}\{p\}(t)$ [J5, $r \mid [t].\mathfrak{G}\{p\}(t)$; H6]

For purposes which will become clear later, I have to be able to use propositional variables as parts of expressions belonging to a semantical category higher than the category of propositions. The multi-link definition [D9] serves this end. It is evident that the logical value of "$\mathfrak{G}\{p\}(q)$" is always the same as the value of "p". On the other hand, "$\mathfrak{G}\{p\}$" belongs to the semantical category of proposition-forming functors of one propositional argument. And, as we shall see, expressions of this category play an essential role in our proof.

D1 $[p].p = \mathrm{As}\,(p)$ [α]

D10 $[f]\therefore[p] : f(p). \equiv .f(p) :\equiv \mathrm{Vr}\,\langle f\rangle$ [α]

The sole purpose of D1 is to introduce into the system expressions belonging to the semantical category of proposition-forming functors of one propositional argument. Without D1 we cannot add to the system D10, where the variable "f" belongs to this category. The constant "Vr", defined in D10, belongs to the semantical category of proposition-forming functors of one argument which belongs to the semantical category of proposition-forming functors of one propositional argument. Obviously, this constant is a kind of "Verum" and for this reason the symbol "Vr" is used. This leads to no misunderstanding, since with this "Vr" parentheses of the form "⟨" and "⟩" are associated. Hence, we are able to distinguish "Vr $\langle f\rangle$" from "Vr (p)", defined in D8, and due to this we know that the argument of the first "Vr" belongs to the semantical category different from the semantical category to which the second "Vr" belongs. On the other hand, the use of the same symbol indicates that the terms, defined in D8 and D10, are analogous.[85]

H8 $[f].\mathrm{Vr}\,\langle f\rangle$ [D10; J3, $p \mid f(p)$]

H9 $[fg] : \mathrm{Vr}\,\langle f\rangle. \equiv .\mathrm{Vr}\,\langle g\rangle$ [J4, $p \mid \mathrm{Vr}\,\langle g\rangle$, $q \mid \mathrm{Vr}\,\langle f\rangle$; H8, $f \mid g$; H8]

H10 $[p] :: p \equiv \therefore[\Psi]\therefore \Psi\langle\mathfrak{G}\{p\}\rangle. \equiv .\Psi\langle\mathfrak{G}\{p\}\rangle :\equiv .[t].\mathfrak{G}\{p\}(t)$
[H7, $q \mid \Psi\langle\mathfrak{G}\{p\}\rangle$; β]

Although H10 follows directly from H7 by substitution and by the distribution of quantifiers, it cannot be obtained without D10 (or a similar definition), since the variable "Ψ" involved in the substitution belongs to the semantical category which is not primitive. The constant "Vr", defined in D10, belongs to the needed category and, therefore, when making substitution in H7 we can use variables of that category. The formula "$\Psi\langle \mathfrak{G}\{p\} \rangle$" is meaningful, since the multi-link functor "$\mathfrak{G}\{p\}$" belongs to the same category as "f" in D10. Concerning H10, let us remark that in some respect this thesis is analogous to S6.

$$\text{E1 } [fg]\therefore [t] : f(t). \equiv .g(t) :\equiv: [\Psi] : \Psi\langle f \rangle. \equiv .\Psi\langle g \rangle \qquad [\epsilon]$$

Thesis E1 is the law of extensionality for expressions belonging to the semantical category of proposition-forming functors of one propositional argument. We can add E1 to the system in virtue of the point ϵ of the rule of procedure, since every expression which occurs in E1 belongs to a semantical category already introduced into the system. In fact, besides the primitive categories, the variables "f", "g" and "Ψ" are meaningful due to definitions D1 and D10. Thus, the point of the rule of procedure of $\mathfrak{S}_5$ concerning higher extensionalities is used in the proof of **A2**.

$$\text{H11 } [pq]\therefore p \equiv q. \equiv: [\Psi] : \Psi\langle \mathfrak{G}\{p\} \rangle. \equiv .\Psi\langle \mathfrak{G}\{q\} \rangle$$
$$[\text{J6}, p \mid p \equiv q, q \mid [t] : \mathfrak{G}\{p\}(t). \equiv .\mathfrak{G}\{q\}(t),$$
$$r \mid [\Psi] : \Psi\langle \mathfrak{G}\{p\} \rangle. \equiv .\Psi\langle \mathfrak{G}\{q\} \rangle; \text{H3}; \text{E1}, f \mid \mathfrak{G}\{p\}, g \mid \mathfrak{G}\{q\}]$$

The assumptions of Theorem **A2** give neither the law of extensionality for propositions nor metarule SIII. However, due to E1 and D9, we obtained H11 which allows us to reason as follows: if the formulas: (1) $\alpha \equiv \beta$, and (2) $\Phi\langle \mathfrak{G}\{\alpha\} \rangle$ have been proved, then the formula $\Phi\langle \mathfrak{G}\{\beta\} \rangle$ is provable. Hence, H11 can be considered as a kind of *Ersatz* of the law of extensionality for propositions. Indeed, it is a very weak form of this law, but, as we shall see, it suffices for our purposes.

$$\text{H12 } [\Psi] : \Psi\langle \mathfrak{G}\{0\} \rangle. \equiv .\Psi\langle \mathfrak{G}\{0 \equiv 1\} \rangle$$
$$[\text{H11}, p \mid 0, q \mid 0 \equiv 1; \text{J4}, p \mid 0, q \mid 0]$$

$$\text{H13 } 0 \equiv: [\Psi] : \Psi\langle \mathfrak{G}\{0\} \rangle. \equiv .\Psi\langle \mathfrak{G}\{1\} \rangle$$
$$[\text{J7}, p \mid 0, q \mid 0, r \mid [\Psi] : \Psi\langle \mathfrak{G}\{0\} \rangle. \equiv .\Psi\langle \mathfrak{G}\{1\} \rangle; \text{H11}, p \mid 0, q \mid 0 \equiv 0]$$

In some respect H13 is analogous to S18.

H14 $[\Psi] : \Psi\langle\mathfrak{S}\{0\}\rangle . \equiv . \Psi\langle\mathfrak{S}\{1\}\rangle :\equiv .\therefore [\Psi] .\therefore \Psi\langle\mathfrak{S}\{0\}\rangle . \equiv .$
$\Psi\langle\mathfrak{S}\{1\}\rangle :\equiv . [t] . \mathfrak{S}\{1\}(t)$ $[\mathrm{H5}, q \mid \Psi\langle\mathfrak{S}\{0\}\rangle . \equiv . \Psi\langle\mathfrak{S}\{1\}\rangle; \beta]$

H15 $0 \equiv .\therefore [\Psi] .\therefore \Psi\langle\mathfrak{S}\{0\}\rangle . \equiv . \Psi\langle\mathfrak{S}\{1\}\rangle :\equiv . [t] . \mathfrak{S}\{1\}(t)$
$[\mathrm{J6}, p \mid 0, q \mid [\Psi] : \Psi\langle\mathfrak{S}\{0\}\rangle . \equiv . \Psi\langle\mathfrak{S}\{1\}\rangle, r \mid [\Psi] .\therefore$
$\Psi\langle\mathfrak{S}\{0\}\rangle . \equiv . \Psi\langle\mathfrak{S}\{1\}\rangle :\equiv . [t] . \mathfrak{S}\{1\}(t); \mathrm{H13}; \mathrm{H14}]$

Confront H15 with S16.

D11 $[fg] :: [\Psi] .\therefore \Psi\langle f\rangle . \equiv . \Psi\langle g\rangle :\equiv . [t] . g(t) .\therefore \equiv . \mathfrak{N}\langle fg\rangle$ $[\alpha]$

D11 is a definition of conjunction of two proposition-forming functors of one propositional argument. As far as its form is regarded, it is analogous to definition Z18. Obviously, the constants "$\mathfrak{N}$" and "Vr" in D10 belong to different semantical categories, but we can associate with "$\mathfrak{N}$" the same kind of parentheses, as are given in D10, since in D11 these parentheses enclose two variables, each of which belongs to the same category as "f" in D10.

Now, having $\mathfrak{S}$, metarule SV, H11, H15 and D11 we will be able to obtain formula H31 which, in some respect, is analogous to B5. Consequently, H31 implies H38 which is indispensable for our purpose.

H16 $[p] : p \equiv . \mathfrak{N}\langle\mathfrak{S}\{p\}\mathfrak{S}\{p\}\rangle$
$[\mathrm{J6}, q \mid [\Psi] .\therefore \Psi\langle\mathfrak{S}\{p\}\rangle . \equiv . \Psi\langle\mathfrak{S}\{p\}\rangle :\equiv . [t] . \mathfrak{S}\{p\}(t),$
$r \mid \mathfrak{N}\langle\mathfrak{S}\{p\}\mathfrak{S}\{p\}\rangle; \mathrm{H10}; \mathrm{D11}, f \mid \mathfrak{S}\{p\}, g \mid \mathfrak{S}\{p\}]$

Cf. K5, K1 and K2.

H17 $\mathfrak{N}\langle\mathfrak{S}\{1\}\mathfrak{S}\{1\}\rangle$ $[\mathrm{H16}, p \mid 0 \equiv 0; \mathrm{J13}, p \mid 0]$

Cf. K6.

H18 $[p\Psi] : \Psi\langle\mathfrak{S}\{1\}\rangle . \equiv . \Psi\langle\mathfrak{S}\{\mathfrak{N}\langle\mathfrak{S}\{p\}\mathfrak{S}\{p\}\rangle\}\rangle$
$[\mathrm{H11}, q \mid \mathfrak{N}\langle\mathfrak{S}\{p\}\mathfrak{S}\{p\}\rangle; \mathrm{H16}]$

H19 $0 . \equiv . \mathfrak{N}\langle\mathfrak{S}\{0\}\mathfrak{S}\{1\}\rangle$
$[\mathrm{J6}, p \mid 0, q \mid [\Psi] .\therefore \Psi\langle\mathfrak{S}\{0\}\rangle . \equiv . \Psi\langle\mathfrak{S}\{1\}\rangle :\equiv . [t] . \mathfrak{S}\{1\}(t),$
$r \mid \mathfrak{N}\langle\mathfrak{S}\{0\}\mathfrak{S}\{1\}\rangle; \mathrm{H15}; \mathrm{D11}, f \mid \mathfrak{S}\{0\}, g \mid \mathfrak{S}\{1\}]$

Cf. K3.

H20 $[p] : 0. \equiv .\mathfrak{N}\langle\mathfrak{G}\{0\}\mathfrak{G}\{p\}\rangle$ [H16, $p \mid 0$; H19; SV]

In some respect H20 is analogous to B1. It is clear that an application of SV involves an auxiliary definition and metarule SII. Since by now these steps have become rather trivial, I have omitted them.

D12 $[f] : \mathfrak{N}\langle\mathfrak{G}\{1\}f\rangle. \equiv .\Theta\alpha\langle f\rangle$ [α]

D12 is an auxiliary definition needed for extensional deductions.

H21 $\Theta\alpha\langle\mathfrak{G}\{1\}\rangle$ [D12, $f \mid \mathfrak{G}\{1\}$; H17]

H22 $\Theta\alpha\langle\mathfrak{G}\{\mathfrak{N}\langle\mathfrak{G}\{1\}\mathfrak{G}\{1\}\rangle\}\rangle$ [H18, $p \mid 1$, $\Psi \mid \Theta\alpha$; H21]

H23 $\mathfrak{N}\langle\mathfrak{G}\{1\}\mathfrak{G}\{\mathfrak{N}\langle\mathfrak{G}\{1\}\mathfrak{G}\{1\}\rangle\}\rangle$

[D12, $f \mid \mathfrak{G}\{\mathfrak{N}\langle\mathfrak{G}\{1\}\mathfrak{G}\{1\}\rangle\}$; H22; SII]

Cf. B2.

H24 $[p] \therefore 0. \equiv .\mathfrak{N}\langle\mathfrak{G}\{0\}\mathfrak{G}\{p\}\rangle :\equiv .\mathfrak{N}\langle\mathfrak{G}\{1\}\mathfrak{G}\{\mathfrak{N}\langle\mathfrak{G}\{1\}\mathfrak{G}\{1\}\rangle\}\rangle$

[J4, $p \mid \mathfrak{N}\langle\mathfrak{G}\{1\}\mathfrak{G}\{\mathfrak{N}\langle\mathfrak{G}\{1\}\mathfrak{G}\{1\}\rangle\}\rangle$;

$q \mid 0. \equiv .\mathfrak{N}\langle\mathfrak{G}\{0\}\mathfrak{G}\{p\}\rangle$; H23; H20]

Cf. B3.

D13 $[f] :: 1. \equiv .\mathfrak{N}\langle\mathfrak{G}\{1\}\mathfrak{G}\{\mathfrak{N}\langle\mathfrak{G}\{1\}\mathfrak{G}\{0\}\rangle\}\rangle :\equiv .$
$\mathfrak{N}\langle\mathfrak{G}\{1\}\mathfrak{G}\{\mathfrak{N}\langle\mathfrak{G}\{1\}f\rangle\}\rangle \therefore \equiv .\Theta\beta\langle f\rangle$ [α]

This is an auxiliary definition.

H25 $\Theta\beta\langle\mathfrak{G}\{0\}\rangle$

[D13, $f \mid \mathfrak{G}\{0\}$; J8, $p \mid 0$, $q \mid \mathfrak{N}\langle\mathfrak{G}\{1\}\mathfrak{G}\{\mathfrak{N}\langle\mathfrak{G}\{1\}\mathfrak{G}\{0\}\rangle\}\rangle$]

H26 $\Theta\beta\langle\mathfrak{G}\{0 \equiv 1\}\rangle$ [H12, $\Psi \mid \Theta\beta$; H25]

H27 $1. \equiv .\mathfrak{N}\langle\mathfrak{G}\{1\}\mathfrak{G}\{\mathfrak{N}\langle\mathfrak{G}\{1\}\mathfrak{G}\{0\}\rangle\}\rangle :\equiv .$
$\mathfrak{N}\langle\mathfrak{G}\{1\}\mathfrak{G}\{\mathfrak{N}\langle\mathfrak{G}\{1\}\mathfrak{G}\{0 \equiv 1\}\rangle\}\rangle$

[D13, $f \mid \mathfrak{G}\{0 \equiv 1\}$; H26; SII]

Cf. B5. Now, I am applying SV to H24 and H27. Since these formulas are a little complicated, I will introduce the required auxiliary definition, although the construction of such a definition is a simple matter of routine. Viz.:

D14 $[p] :: p. \equiv .\mathfrak{N}\langle\mathfrak{G}\{p\}\mathfrak{G}\{\mathfrak{N}\langle\mathfrak{G}\{1\}\mathfrak{G}\{0\}\rangle\}\rangle :\equiv .$
$\mathfrak{N}\langle\mathfrak{G}\{1\}\mathfrak{G}\{\mathfrak{N}\langle\mathfrak{G}\{1\}\mathfrak{G}\{0 \equiv p\}\rangle\}\rangle .\therefore \equiv .\Theta\gamma(p)$ [α]

H28 $\Theta\gamma(0)$ [D14, $p \mid 0$; H24, $p \mid \mathfrak{N}\langle\mathfrak{G}\{1\}\mathfrak{G}\{0\}\rangle$]

H29 $\Theta\gamma(1)$ [D14, $p \mid 1$; H27]

H30 $[p].\Theta\gamma(p)$ [SV; H28; H29]

H31 $[p] .\therefore p. \equiv .\mathfrak{N}\langle\mathfrak{G}\{p\}\mathfrak{G}\{\mathfrak{N}\langle\mathfrak{G}\{1\}\mathfrak{G}\{0\}\rangle\}\rangle :\equiv .$
$\mathfrak{N}\langle\mathfrak{G}\{1\}\mathfrak{G}\{\mathfrak{N}\langle\mathfrak{G}\{1\}\mathfrak{G}\{0 \equiv p\}\rangle\}\rangle$ [D14; H30; SII]

Cf. B5. The derivation of H31 constitutes a crucial point in this proof.

H32 $\mathfrak{N}\langle\mathfrak{G}\{1\}\mathfrak{G}\{\mathfrak{N}\langle\mathfrak{G}\{1\}\mathfrak{G}\{0. \equiv .\mathfrak{N}\langle\mathfrak{G}\{1\}\mathfrak{G}\{0\}\rangle\}\rangle\}\rangle$
[H31, $p \mid \mathfrak{N}\langle\mathfrak{G}\{1\}\mathfrak{G}\{0\}\rangle$; H16, $p \mid \mathfrak{N}\langle\mathfrak{G}\{1\}\mathfrak{G}\{0\}\rangle$]

Cf. B6.

H33 $[\Psi] .\therefore \Psi\langle\mathfrak{G}\{1\}\rangle. \equiv .\Psi\langle\mathfrak{G}\{\mathfrak{N}\langle\mathfrak{G}\{1\}\mathfrak{G}\{0. \equiv .$
$\mathfrak{N}\langle\mathfrak{G}\{1\}\mathfrak{G}\{0\}\rangle\}\rangle\}\rangle :\equiv .[t].\mathfrak{G}\{\mathfrak{N}\langle\mathfrak{G}\{1\}\mathfrak{G}\{0. \equiv .$
$\mathfrak{N}\langle\mathfrak{G}\{1\}\mathfrak{G}\{0\}\rangle\}\rangle\}(t)$
[D11, $f \mid \mathfrak{G}\{1\}$, $g \mid \mathfrak{G}\{\mathfrak{N}\langle\mathfrak{G}\{1\}\mathfrak{G}\{0. \equiv .\mathfrak{N}\langle\mathfrak{G}\{1\}\mathfrak{G}\{0\}\rangle\}\rangle\}$;
H32; SII]

H34 $[t].\mathfrak{G}\{\mathfrak{N}\langle\mathfrak{G}\{1\}\mathfrak{G}\{0. \equiv .\mathfrak{N}\langle\mathfrak{G}\{1\}\mathfrak{G}\{0\}\rangle\}\rangle\}(t)$
[H33, $\Psi \mid$ Vr; H9, $f \mid \mathfrak{G}\{1\}$,
$g \mid \mathfrak{G}\{\mathfrak{N}\langle\mathfrak{G}\{1\}\mathfrak{G}\{0. \equiv .\mathfrak{N}\langle\mathfrak{G}\{1\}\mathfrak{G}\{0\}\rangle\}\rangle\}$]

H35 $\mathfrak{N}\langle\mathfrak{G}\{1\}\mathfrak{G}\{0. \equiv .\mathfrak{N}\langle\mathfrak{G}\{1\}\mathfrak{G}\{0\}\rangle\}\rangle$
[H6, $p \mid \mathfrak{N}\langle\mathfrak{G}\{1\}\mathfrak{G}\{0. \equiv .\mathfrak{N}\langle\mathfrak{G}\{1\}\mathfrak{G}\{0\}\rangle\}\rangle$; H34; SII]

H36 $[\Psi] .\therefore \Psi\langle\mathfrak{G}\{1\}\rangle. \equiv .\Psi\langle\mathfrak{G}\{0. \equiv .\mathfrak{N}\langle\mathfrak{G}\{1\}\mathfrak{G}\{0\}\rangle\}\rangle :\equiv .$
$[t].\mathfrak{G}\{0. \equiv .\mathfrak{N}\langle\mathfrak{G}\{1\}\mathfrak{G}\{0\}\rangle\}(t)$
[D11, $f \mid \mathfrak{G}\{1\}$, $g \mid \mathfrak{G}\{0. \equiv .\mathfrak{N}\langle\mathfrak{G}\{1\}\mathfrak{G}\{0\}\rangle\}$; H35; SII]

H37 $[t].\mathfrak{G}\{0. \equiv .\mathfrak{R}\langle\mathfrak{G}\{1\}\mathfrak{G}\{0\}\rangle\}(t)$

$[H36, \Psi \mid \mathrm{Vr}; H9, f \mid \mathfrak{G}\{1\}, g \mid \mathfrak{G}\{0. \equiv .\mathfrak{R}\langle\mathfrak{G}\{1\}\mathfrak{G}\{0\}\rangle\}]$

H38 $0. \equiv .\mathfrak{R}\langle\mathfrak{G}\{1\}\mathfrak{G}\{0\}\rangle$ $[H6, p \mid 0. \equiv .\mathfrak{R}\langle\mathfrak{G}\{1\}\mathfrak{G}\{0\}\rangle; H37; SII]$

In some respect H38 is analogous to K4. The derivation of H38 actually concludes the proof of Theorem **A2**, since we can define conjunction of propositions in the following way:

D15 $[pq] : \mathfrak{R}\langle\mathfrak{G}\{p\}\mathfrak{G}\{q\}\rangle. \equiv .p.q$ $[\alpha]$

which gives at once:

B1 $[p] : p \equiv .p.p$ $[J6, q \mid \mathfrak{R}\langle\mathfrak{G}\{p\}\mathfrak{G}\{p\}\rangle, r \mid p.p; H16; D15, q \mid p]$

B1 includes K1 and K2.

K3 $0 \equiv .0.1$

$[J6, p \mid 0, q \mid \mathfrak{R}\langle\mathfrak{G}\{0\}\mathfrak{G}\{1\}\rangle, r \mid 0.1; H19; D15, p \mid 0, q \mid 1]$

K4 $0 \equiv .1.0$

$[J6, p \mid 0, q \mid \mathfrak{R}\langle\mathfrak{G}\{1\}\mathfrak{G}\{0\}\rangle, r \mid 1.0; H38; D15, p \mid 1, q \mid 0]$

Thus, Theorem **A2** has been proved, since all the four laws of logical product for conjunction have been obtained. It remains to note that definition D15 shows that conjunction of propositions is a particular case of a more general term, viz. the functor "$\mathfrak{R}$" which is defined in D11. Obviously, from D15 this functor can be dismissed easily:

H39 $[pq] :: [\Psi] \therefore \Psi\langle\mathfrak{G}\{p\}\rangle. \equiv .\Psi\langle\mathfrak{G}\{q\}\rangle :\equiv .[t].\mathfrak{G}\{q\}(t) \therefore \equiv .p.q$

$[J6, p \mid [\Psi] \therefore \Psi\langle\mathfrak{G}\{p\}\rangle. \equiv .\Psi\langle\mathfrak{G}\{q\}\rangle :\equiv .[t].\mathfrak{G}\{q\}(t),$

$q \mid \mathfrak{R}\langle\mathfrak{G}\{p\}\mathfrak{G}\{q\}\rangle, r \mid p.q; D11, f \mid \mathfrak{G}\{p\}, g \mid \mathfrak{G}\{q\}; D15]$

and we can even obtain a simpler formula, as follows:

H40 $[pq] :: [\Psi] \therefore \Psi\langle\mathfrak{G}\{p\}\rangle. \equiv .\Psi\langle\mathfrak{G}\{q\}\rangle :\equiv .q \therefore \equiv \therefore [\Psi] \therefore$

$\Psi\langle\mathfrak{G}\{p\}\rangle. \equiv .\Psi\langle\mathfrak{G}\{q\}\rangle :\equiv .[t].\mathfrak{G}\{q\}(t)$

$[J9, p \mid q, q \mid [t].\mathfrak{G}\{q\}(t), r \mid \Psi\langle\mathfrak{G}\{p\}\rangle. \equiv .\Psi\langle\mathfrak{G}\{q\}\rangle; H6, p \mid q; \beta]$

H41 $[pq] :: [\Psi] .\therefore \Psi\langle\mathfrak{S}\{p\}\rangle . \equiv . \Psi\langle\mathfrak{S}\{q\}\rangle :\equiv .q .\therefore \equiv .p.q$

$$[J6, p \mid [\Psi] : \Psi\langle\mathfrak{S}\{p\}\rangle . \equiv . \Psi\langle\mathfrak{S}\{q\}\rangle :\equiv .q, q \mid [\Psi] .\therefore \Psi\langle\mathfrak{S}\{p\}\rangle . \equiv . \Psi\langle\mathfrak{S}\{q\}\rangle :\equiv .[t].\mathfrak{S}\{q\}(t), r \mid p.q; \text{H40; H39}]$$

It is obvious that D15, H39 and H41 are inferentially equivalent. These theses are analogous to Z18. The proof that under the assumptions of **A2**, D15 really defines the same constant as Z18, will be outlined in Section 3 of this paper. Also, it should be noted that instead of D11 we can adopt the following definition:

D11* $[fg] :: [\Psi] .\therefore \Psi\langle f\rangle . \equiv . \Psi\langle g\rangle :\equiv .[t].f(t) .\therefore \equiv .\mathfrak{R}\langle fg\rangle$

In this case the proof of Theorem **A2** has to be patterned on the proof of the second part of Theorem **A1**.

Since Theorem **A2** implies that condition $\mathfrak{b}$ of Leśniewski's metatheorem **L** is superfluous, and the proof given in Section 2.3 of this paper shows that the system $\mathfrak{S}$ follows from F1, F2 and SV, my metarule **S** is proved. Hence, each of the axioms A_n, A_o, A_p and A_q can serve as a single axiom of the system $\mathfrak{S}_5$, because each of them, according to Sections 2.1–2.2 of this paper, satisfies the conditions of **S**.

NOTES

[1] These results were obtained in the years 1937, 1938, 1945 and 1952. Some of them were included in my Polish paper "O aksjomatykach prototetyki" (On the Axiom-Systems of Protothetic) which was to be published in the first volume of *Collectanea Logica*. During the siege of Warsaw in September 1939, this almost ready volume perished when a printing office was burnt. Cf. the introduction to Sobociński (1949a). In 1953 I published a short resume in Polish concerning these researches, cf. Sobociński (1954).

[2] A description of this symbolism is given in Leśniewski (1938a), cf. pp. 21–23. Cf. also Leśniewski (1938b).

[3] Leśniewski maintained that any rule of procedure for his system must be formulated as fully as possible. Hence, a possible rule concerning quantifiers other than the general ones would have to be formulated in such a way that it would be possible to introduce into the system not only the existential quantifiers but also other kinds of quantifiers possible in the bi-valued logic. E.g., the quantifier saying that a formula holds for at least two different instances of a variable bound by it, etc. On the other hand, in Leśniewski's system the formalization of the rules of procedure is strictly constructive. This fact prevented a possibility to formulate such a general rule concerning quantifiers. Therefore, not wishing to have only partially formulated rule [of procedure of his system,] he dropped the existential quantifier from his system. However, we can always use it "unofficially" as a pure abbreviation. Cf. Leśniewski (1929), pp. 59–78, and Sobociński (1934). Also cf. Lejewski (1956).

[4] The meaning of multi-link functors will be explained later. In Leśniewski's system the form of parentheses is determined by a constant functor to which this kind of parentheses was prescribed for the first time. Consequently, the same form of parentheses must be used for any other functor belonging to the same semantical category as the first one. Cf. Leśniewski

(1929), pp. 59–78.

[5] Cf. also Sobociński (1949b), p. 257.

[6] Cf. Leśniewski (1929, 1938a), Sobociński (1934) and Sobociński (1949b), pp. 254–257.

[7] We can base ontology and mereology on the ordinary classical propositional calculus, but in such a case their rules of procedure must be strengthened by the addition of a rule concerning the operations with the quantifiers.

[8] A description of this system is given in Słupecki (1953), pp. 55–79. The restricted system of protothetic cannot be confounded with the extended propositional calculus constructed by Łukasiewicz, cf. Łukasiewicz (1951), Meredith (1951). In that system Łukasiewicz accepted much stronger rules of procedure than we have in protothetic. Besides, in the extended propositional calculus there are free variables.

[9] Concerning this principle, cf. remarks of Łukasiewicz in (1921) and (1952). Cf. also Sobociński (1956a), pp. 22–27.

[10] Cf. Tarski (1956) and Słupecki (1953).

[11] In Leśniewski's system this theory is adopted instead of a theory of logical types. It applies also to expressions belonging to protothetic. A short description of this theory is given in Słupecki (1953), pp. 45–47. Cf. a remark of Lejewski in (1956).

[12] The history of these modifications is given in Leśniewski (1929), pp. 30–59.

[13] Cf. Tarski (1956) and Leśniewski (1929), pp. 9–13. In Sobociński (1939, 1949a) I presented the definitions of conjunction in terms of equivalence; their forms differ from the definitions of Tarski, although they are based essentially on his idea. E.g:

$$[pq]\therefore p.q. \equiv: [f] : f(pq). \equiv .f(qp \equiv p)$$

[14] The computable systems of protothetic are discussed in Leśniewski (1938a), pp. 35–43. Leśniewski's proof was never published, but it can be easily reconstructed. The proof of completeness of protothetic, given in Słupecki (1953), pp. 90–97, is entirely different.

[15] Cf. Sobociński (1956b), pp. 61–62.

[16] Cf. Leśniewski (1929), pp. 30–35 and Leśniewski (1938a), pp. 5–6.

[17] The quantifiers and parentheses do not have semantical meaning in Leśniewski's system. But in protothetic any meaningful propositional formula with bound or free variables belongs to the category of propositions, e.g., A3 or any meaningful propositional part of it.

[18] Cf. Leśniewski (1929), pp. 15–30 and Leśniewski (1938a), pp. 17–21. Cf. also Łukasiewicz (1939) and some remarks in Rasiowa (1947).

[19] A complete proof of it is given in Leśniewski (1938b).

[20] An exceptionally precise formulation of them is given in Leśniewski (1929), pp. 59–78.

[21] The formal properties of protothetic are such that either an axiom-system of $\mathfrak{S}_5$ includes the law of extensionality concerning propositions or this law can be obtained from a suitable axiom-system by an application of the point ϵ of the discussed rule. Hence, this point allows to add to the system only the "higher" theses of extensionality. See Section 2 of this paper, the proof of metarule S.

[22] A definition without the main quantifier is called an absolute protothetical definition; the definition with the main quantifier is called a relative protothetical definition. The forms in some respect similar to the absolute protothetical definitions can be found in Whitehead–Russell (1925), p. XII.

[23] The forms of these parentheses are determined by the semantical category of a functor to which they belong. In the case of a multi-link functor, each pair of its parentheses has to have a form different from the other pairs.

[24] Ordinarily in this paper I use a dot as a symbol of conjunction, as in Whitehead–Russell (1925). However, a dot is very inconvenient when the parentheses are used. Although in Z5 "$\wedge$" and "$\wedge_1$" belong to different semantical categories, we can use the same form of parentheses for both functors because the arguments of "$\wedge$" and "$\wedge_1$" belong to the same semantical category and the number of these arguments enclosed by parentheses indicates the category of a functor.

[25] An idea of multi-link functors is akin to Schönfinkel's theory of superposition of functions (cf. Schönfinkel, 1924). Leśniewski introduced these functors to his system independently of Schönfinkel. He was rather influenced by some formulas in Whitehead–Russell (1925), e.g.. *31.14. Cf. Leśniewski(1938a), p. 44; Leśniewski (1929), p. 66, and Sobociński (1934), p. 159.

[26] Concerning this thesis cf. Tarski (1956), p. 17 and Leśniewski (1938a), pp. 30–55. In Meredith (1951) it is shown that this formula can serve as a single axiom of the extended propositional calculus of Łukasiewicz. Meredith's proof cannot be carried out in protothetic since in this theory the rule of substitution adopted by Łukasiewicz is not valid.

[27] Similarly as in the functional calculus of omega degree.

[28] This fact is connected with the properties of the possible definitions of conjunction by means of equivalence.

[29] Cf. Leśniewski (1929), pp. 41–44.

[30] Cf. T.E.IL in Leśniewski (1929), pp. 74–75.

[31] In Z14, obviously, there occur expressions belonging to three different semantical categories. Namely, "f" belongs to the category of proposition-forming functors of one propositional argument.

[32] Discussing this point, Słupecki omitted the case of multi-link functors, cf. Słupecki (1953), pp. 89 and 100. Cf. also Lejewski (1956).

[33] After Słupecki (1953), p. 81, a functor of a type "$\overset{\alpha}{\equiv}$" can be called an interfunctorial equivalence.

[34] In order to simplify the situation I omitted in the formulas (a) and (b) the possible quantifiers. If they are present, we have to define in the point (d) a multi-link functor.

[35] Several such metarules will be established in Section 2.

[36] In Leśniewski (1938a), pp. 23–35, the history of the successive simplifications of the axiom system of $\mathfrak{S}_5$ is discussed up to 1936. Here I repeat the essential points concerning this subject from that paper as only a few copies of that publication are preserved. The results obtained after 1936 are not discussed in Leśniewski (1938a).

[37] Cf., e.g., Słupecki (1953), pp. 79–84, where a method of verification of protothetical functors by the bi-valued matrix is given.

[38] In Leśniewski's writings this metarule is not stated explicitly but he used it constantly. E.g., cf. Leśniewski (1938a), p. 34.

[39] Here "0" and "1" are not defined constants, but the abbreviations of "$[u].u$" and "$[u].u. \equiv .[u].u$", respectively.

[40] Obviously, in "ϕ" other variables can occur.

[41] Cf. Leśniewski (1938a), p. 24. Only in 1933 has Łukasiewicz proved that A1* is a single axiom of the equivalential propositional calculus. Cf. Łukasiewicz (1939, 1948).

[42] Cf. Leśniewski (1938a), p. 27 and Wajsberg (1937).

[43] Cf. Leśniewski (1929), pp. 59–78, and Leśniewki (1938a), pp. 14–16.

[44] In connection with this problem, Leśniewski, preparing (1938a) in 1937, made the following remark (p. 23): "Das von mir oben in dem Resüme des §11 meiner Mitteilung angegebene einzige Axiom der nach Direktiven des Systems $\mathfrak{S}_5$ gebauten Protothetik stellte sich leider bisher "siegreich" seit schon elf Jahren von anderen Forschern und von mir unternommenen Versuchen entgegen, um es wenigstens ein einziges Wort zu verkürzen. Die Sachen haben sich hier jedoch so gestaltet, dass ich beinahe bis zum heutigen Tag wiederholt von immer neuen diese Aufgabe betreffenden Lösungsideen hörte. Alles dies bewegt mich, dass ich ohne zu warten, bis die mit der Entstehung des eben erwähnten Axioms und mit seinen kleinen weiteren Umgestaltungen verbundenen Probleme eine angemessene Besprechung an einer passenden Stelle im Rahmen der Forsetzung meiner Mitteilung finden, hier im Interesse des Lesers, welcher sich mit selbständigen Forschungen in diesem Gebiet beschäftigen möchte, eine ganz allgemeine Übersicht der wichtigsten theoretischen Positionen angebe, die zum Entwicklungsbild des einzigen Axioms der Protothetik beitragen und verschiedene auf diesem Boden angewandte nützliche konstruktive Kunstgriffe illustrieren."

[45] Cf. Łukasiewicz (1939), Sobociński (1949a), p. 10 and Leśniewski (1938a), p. 31.
[46] In Słupecki (1953), p. 99 and Grzegorczyk (1955) the authors do not mention my name when discussing this axiom-system. Cf. Sobociński (1956b), p. 60, note 17, and Sobociński (1954).
[47] Thesis S1 was established by Leśniewski.
[48] Obviously, Z15 is inferentially equivalent to Z3.
[49] In the authentic symbolism of Leśniewski.
[50] In 1928, applying the rule of extensionality, I simplified the axiom-system of Leśniewski's ontology, cf. Sobociński (1934). However, the proof given there has nothing in common with the deductions used in protothetic.
[51] The thesis A_l is not organic, while the axioms A_h–A_k are. Concerning the notion of "organic formula", cf. Sobociński (1956b), p. 60.
[52] It was the last investigation made by Leśniewski before his fatal illness. From the notes he had left I prepared a paper concerning this result with the intention to publish it in *Collectanea Logica*. Due to World War II this was not accomplished. Cf. Sobociński (1949a), the introduction.
[53] It was not mentioned in Sobociński (1954).
[54] Cf. Section 1.3, pp. 170–171.
[55] Cf. Section 1.3, pp. 165–166.
[56] Cf. Section 1.3, p. 171.
[57] Cf. Section 1.3, pp. 169–170.
[58] Concerning this rule cf. Section 1.2, p. 159. Also, cf. Leśniewski (1929), pp. 59–78.
[59] It is well known, cf., e.g., Tarski (1956), pp. 1–8 and Słupecki (1953), pp. 97–100, that Tarski established that it is possible to define conjunction in terms of equivalence, provided that one is allowed to use quantifiers and variable functors. I.e., such definitions can be established in a system of the calculus of propositions in which variables of higher semantical categories are allowed, and in which quantifiers are used with appropriate rules to bind variables of any category. It is also known that definitions suggested by Tarski are of two different types.

The following propositions can serve as examples of the first type of definitions:

1. $[pq] :\cdot: [f] :: p \equiv .\therefore [r] : p \equiv .f(r) :\equiv: [r] : q \equiv .f(r) ::\equiv .p.q$

2. $[pq] :\cdot: [f] :: q \equiv .\therefore [r] : p \equiv .f(r) :\equiv: [r] : q \equiv .f(r) ::\equiv .p.q$

In this type of definitions the arguments of definiendum ("p" and "q" in the examples given above) do not occur as arguments of the variable functors in the definiens (we have "$f(r)$" in the definiens but not "$f(p)$" and "$f(q)$").

The other type of definitions can be exemplified with the aid of the following propositions:

3. $[pq] :: [f] .\therefore p \equiv: f(p). \equiv .f(q) .\therefore \equiv .p.q$

4. $[pq] :: [f] .\therefore q \equiv: f(p). \equiv .f(q) .\therefore \equiv .p.q$

In 3 and 4 the arguments of the definiendum (i.e., "p" and "q" in 3 and 4) occur as arguments of the variable functor in the definiens (in the formulas 3 and 4 we have "$f(p)$" and "$f(q)$").

Only in the field of a theory of propositions which is enriched by the addition of the law of extensionality for propositions we are able to give a proof that both types of these definitions are inferentially equivalent. Cf. Tarski (1923a, 1923b); (1956), pp. 1–8 and Słupecki (1953), pp. 97–100. It is quite obvious that in the field of full systems of protothetic, such as $\mathfrak{S}_1$–$\mathfrak{S}_5$, this equivalence holds.

Definitions of both types admit modifications. Thus, for instance, we have Tarski's thesis:

5. $[pq] :\cdot: [f] :: p \equiv .\because [r].f(r). \equiv: [r] : q \equiv .f(r) ::\equiv .p.q$

cf. Leśniewski (1938a), p. 27, which belongs to the first type. On the other hand, the formulas established by Leśniewski:

6. $[pq] :: [f].\because p \equiv: f(q). \equiv .f(1).\because \equiv .p.q$

7. $[pq] :: [f].\because q \equiv: f(p). \equiv .f(1).\because \equiv .p.q$

cf. Leśniewski (1938a), p. 24, and the definitions suggested by myself:

8. $[pq].\because [f] : f(pq). \equiv .f(11) :\equiv .p.q$

9. $[pq].\because [f] : f(p1). \equiv .f(1q) :\equiv .p.q$

10. $[pq].\because [f] : f(1p). \equiv .f(q1) :\equiv .p.q$

cf. Sobociński (1939, 1949a), belong to the second class. In Section 2.5 of this paper I shall use a definition of conjunction in terms of equivalence, which is of the second type but much more complicated than the formulas 3 or 4. The results discussed in Section 2.4 hold for such definitions of the second kind which possess strictly "extensional" form, as the theses 3 and 4 or their suitable modifications. Namely, in order to obtain the deductions presented in Section 2.4, we can adopt an arbitrary definition of conjunction by means of equivalence, but this definition must satisfy the following additional condition: Both arguments of the definiendum must occur as the n-th arguments (for n: $1, 2, 3, \ldots$) of the variable functor in the definiens. This additional condition is satisfied by the formulas 3 and 4, but not by the theses 6–10.

The meaning of the symbol "1", which is used in the formulas 6–10, is explained at the end of this introduction.

[60] Cf. Section 1.3, p. 169.

[61] Cf. Section 1.2, p. 159, Leśniewski (1938b) and (1929), p. 76. Leśniewski observed this requirement consequently in Leśniewski (1938b).

[62] Cf. note 39.

[63] The formulation of the rule of procedure of $\mathfrak{S}_5$ requires to adjust this rule to each particular axiom–system of $\mathfrak{S}_5$. There exists a method which we can use to make such adjustment automatically. Cf. Leśniewski (1929), pp. 59–76, especially p. 63, T.E.I–T.E.IV.

[64] This theorem will be proved later, viz. as the thesis A76.

[65] In this and the subsequent definitions I am using the symbols "$\Phi\alpha$", "$\Phi\beta$", ..., "$\Psi\alpha$", "$\Psi\beta$", etc. when the defined terms possess no important logical meanings, i.e., when the suitable definitions have purely auxiliary character.

Obviously, the expression "$\Phi\alpha\langle p\rangle$" occurring in the definiendum of D3 belongs to the same semantical category as "f" in A7, and, therefore, we can substitute it for this variable in order to obtain A10.

[66] Evidently, instead of SII we can use SI, but in such a case the deductions would be longer.

[67] Cf. formula Z14, p. 163.

[68] In this and the subsequent definitions I am using the symbols "Vr_1", "Vr_2", etc., in order to indicate that the defined terms are really the same as "Vr", introduced by D2. At this stage we are not yet able to show that these terms are the same. This will be possible at a later stage, when the system is sufficiently developed. Cf. the use of such symbols in Leśniewski (1938b), e.g. D9, p. 126 and D10, p. 130.

[69] The formulas DI and DII belong to the type of definitions which in Leśniewski's system are called the absolute protothetical definitions (in protothetic). Cf. note 22.

[70] In connection with this thesis cf. thesis SI. Obviously, if we have SI at our disposal, then the rule of $\mathfrak{S}_5$ allows us to prove that these theses are inferentially equivalent.

[71] Obviously, the symbols "0" and "1" which occur in the definiens of this definition are only typographical abbreviations. Cf. the introduction to Section 2 of this paper.

[72] The application of SV in this case requires, obviously, an auxiliary definition which must be multi-link, viz.:

$$[pq] :: p \equiv .p \equiv q :\equiv q.\because \equiv .\chi\delta\langle q\rangle(p)$$

Similarly, in the proof of A57, etc.

[73] Cf. this paper, pp. 170–171, Sobociński (1954), and the introduction to Sobociński (1949a), p. 17.

[74] Cf. Sobociński (1954) and (1949a), pp. 9 and 18.

[75] The analysis of matrix **M2** allows us to establish easily some interesting properties of the bi-valued equivalential propositional calculus, i.e., of the system $\mathfrak{S}$. Namely, using **M2** and Leśniewski's theorem which will be given below, we can prove that:

1. *A single axiom of the system $\mathfrak{S}$ must be at least nine letters long* (in Łukasiewicz's symbolism).

and

2. *In any complete axiom-system of $\mathfrak{S}$ at least one axiom must contain three different variables.*

The first point confirms a proof given by Łukasiewicz that each of his theses Ł1 (A78), Ł2 and Ł3 is the shortest axiom of $\mathfrak{S}$. See Section 1.3, pp. 168–169. Cf. Łukasiewicz (1939, 1948), Leśniewski (1938a), p. 31, Sobociński (1954) and (1949a), p. 10. The second point says that a known theorem of Wajsberg (which was later obtained independently by Diamond and McKinsey) holds not only for the complete bi-valued propositional calculus, but also for the system $\mathfrak{S}$. Cf. Wajsberg (1931), Diamond–McKinsey (1947), Łukasiewicz–Tarski (1930) and Tarski–Łukasiewicz (1956), p. 47. A simple consequence of this point and Leśniewski's theorem mentioned below is that:

3. *The smallest number n such that the system $\mathfrak{S}$ can be based on a set of axioms each of which contains at most n variables (counted with multiplicity) is equal to 6.*

This fact is rather interesting, since an analogous number for the complete classical propositional calculus is equal to 5, cf. Sobociński (1955). It has to be noted that the related results were published by Jaśkowski for the bi-valued propositional calculus or any fragment of it, if in these systems: (1) substitution and detachment (with respect to implication) are the only rules of procedure; and (2) the thesis $CCpqCCqrCpr$ holds. Cf. Jaśkowski (1948) and Sobociński (1955), p. 114.

A proof of points 1–3 requires the use of a known theorem of Leśniewski which says that:

Any well-formed equivalential formula is a thesis of $\mathfrak{S}$ if and only if each variable occurs in it an even number of times.

Besides Leśniewski's treatise Leśniewski (1929), §3, pp. 15–30, there are several papers in which this theorem and its modifications are discussed, cf. Rasiowa (1947, 1955), Götlind (1946, 1951), Mihailescu (1937a, 1937b, 1938), Rose (1951), Stone (1937).

Now, we can outline the proof of 1–3 as follows.

A simple inspection of **M2** shows that this matrix is symmetrical and that in its field the values of arbitrary meaningful equivalential formulas which have the following forms (in Łukasiewicz's symbolism): $E\alpha E\alpha\beta$ and $E1\gamma$ are equal to the values of β and γ, respectively. Moreover, **M2** verifies the following 16 theses:

1. *Epp*; 2. *EpEpEqq*; 3. *EpEqEpq*; 4. *EpEqEqp*; 5. *EEppEqq*;

6. *EEpqEpq*; 7. *EEpqEqp*; 8. *EpEEqqp*; 9. *EpEEpqq*; 10. *EpEEqpq*;

11. *EEpEqqp*; 12. *EEqEpqp*; 13. *EEqEqpp*; 14. *EEEqqpp*;

15. *EEEpqqp*; 16. *EEEqpqp*

and **M2** falsifies the theses Ł1 (A78), Ł2 and Ł3. This being the case, an elementary proof by induction shows that the points 1–3 follow from these properties of matrix **M2** and from Leśniewski's theorem.

[76] See p. 171.

[77] Viz. A_j and A_k. The axioms A_l, A_m, and A_n–A_q require the method proved in Section 2.5 of this paper.

[78] Cf. note 39.

[79] Cf. note 59.

[80] The term "Vr" defined in D8 is a constant term different from "Vr" defined in D2, since it belongs to the semantical category of proposition forming functors of one propositional argument. In both cases we can use the same symbols and the same form of parentheses prescribed to these symbols in their definitions, since in the first case the parentheses enclose one variable and in the second two variables. Hence, no misunderstanding is possible. Such use of symbols is permitted by the rule of $\mathfrak{S}_5$.

[81] If this modification of Theorem **A1** were valid, it would be possible to prove without the aid of the point ϵ of the rule of $\mathfrak{S}_5$ that each of the theses A_l–A_p is a single axiom of $\mathfrak{S}_5$.

[82] Cf. note 46.

[83] Cf. Section 2.1, pp. 176–177.

[84] The term "$\mathfrak{G}$" defined in D9 is a multi-link form of the functor *Gpq* whose logical matrix is:

M3

	G	0	1
	0	0	0
*	1	1	1

Letter *G* is used as a symbol of this constant in my paper Sobociński (1953), p. 79.

[85] Cf. note 80.

REFERENCES

Church, Alonzo (1956). *Introduction to Mathematical Logic*, Vol. 1. Princeton University Press, Princeton, NJ.

Diamond, A.H. and McKinsey, J.C.C. (1947). 'Algebras and Their Subalgebras', *Bulletin of the American Mathematical Society* 53, 959–962.

Götlind, Erik (1946). 'Om några ekvivalenssatser i tvåvärdelogiken (On Some Equivalence Propositions in Two-Valued Logic)', *Norsk matematisk tidsskrift* 28, 71–75.

Götlind, Eric (1951). 'A Leśniewski–Mihailescu Theorem for m-Valued Propositional Calculi', *Portugaliae Mathematica* 10, 97–102.

Grzegorczyk, Andrzej (1955). 'The System of Leśniewski in Relation to Contemporary Logical Research', *Studia Logica* III, 77–97.

Hiż, Henry (1952). 'On Primitive Terms of Logic', *The Journal of Symbolic Logic* 17, 156–157.

Jaśkowski, Stanisław (1948). 'Trois contributions au calcul des propositions bivalent', *Studia Societatis Scientiarum Torunensis*, section A, I(1), 1–15.

Lejewski, Czesław (1956). Review of Słupecki (1953). *The Journal of Symbolic Logic* 21, 188–191.
Leśniewski, Stanisław (1929). 'Grundzüge eines neuen Systems der Grundlagen der Mathematik', *Fundamenta Mathematicae* XIV, 1–81.
Leśniewski, Stanisław (1938a). 'Einleitende Bemerkungen zur Fortsetzung meiner Mitteilung u.d.T. "Grundzüge eines neuen Systems der Grundlagen der Mathematik"', *Collectanea Logica* I, 1–60.
Leśniewski, Stanisław (1938b). 'Grundzüge eines neuen Systems der Grundlagen der Mathematik, §12', *Collectanea Logica* I, 61–144.
Łukasiewicz, Jan (1921). 'Logika dwuwartościowa (Two-Valued Logic)', *Przegląd Filozoficzny* XXIII, 189–205.
Łukasiewicz, Jan (1929). *Elementy logiki matematycznej* (Elements of Mathematical Logic), Warszawa.
Łukasiewicz, Jan (1939). 'Der Äquivalenzenkalkül', *Collectanea Logica* I, 145–169.
Łukasiewicz, Jan (1948). 'The Shortest Axiom of the Implicational Calculus of Propositions', *Proceedings of the Royal Irish Academy* 52A, 25–33.
Łukasiewicz, Jan (1951). 'On Variable Functors of Propositional Arguments', *Proceedings of the Royal Irish Academy* 54A, 25–35.
Łukasiewicz, Jan (1952). 'On the Intuitionistic Theory of Deduction', *Proceedings of Koninklijke Nederlandse Akademie van Wetenschappen* 14A, 202–212.
Łukasiewicz, Jan and Tarski, Alfred (1930). 'Untersuchungen über den Aussagenkalkül', *Comptes rendus des Séances de la Société des Sciences et des Lettres de Varsovie, Classe III* 23, 30–50. Cf. Tarski–Łukasiewicz (1956).
Meredith, Carew (1951). 'On an Extended System of the Propositional Calculus', *Proceedings of the Royal Irish Academy* 54A, 37–47.
Mihailescu, Eugen Gh. (1937a). 'Recherches sur un sous-système du calcul des propositions', *Annales Scientifiques de l'Université de Jassy* 23, 106–124.
Mihailescu, Eugen Gh. (1937b). 'Recherches sur la négation et l'équivalence dans le calcul des propositions', *Annales Scientifiques de l'Université de Jassy* 23, 369–408.
Mihailescu, Eugen Gh. (1938). 'Recherches sur l'équivalence et la réciprocité dans le calcul des propositions', *Annales Scientifiques de l'Université de Jassy* 24, 116–153.
Mihailescu, Eugen Gh. (1951). 'Cercetări asupra subsistemelor calculului propoziţiilor (Researches on Sub-Systems of the Propositional Calculus)', *Studii şi cercetări matematice. Academia Republicii Populare Romane. Institut de Matematica* 2, 1–44.
Prior, Arthur N. (1955). *Formal Logic*, Clarendon Press, Oxford.
Rasiowa, Helena (1947). 'Axiomatisation d'un système partiel de la théorie de la déduction', *Comptes rendus des Séances de la Société des Sciences et des Lettres de Varsovie, Classe III* 40, 22–37.
Rasiowa, Helena (1955). 'O pewnym fragmencie implikacyjnego rachunku zdań (On a Fragment of the Implicative Propositional Calculus)', *Studia Logica* III, 208–226.
Rose, Alan (1951). 'The Degree of Completeness of a Partial System of the 2-Valued Propositional Calculus', *Mathematische Zeitschrift* 54, 181–183.
Russell, Bertrand (1906). 'The Theory of Implication', *American Journal of Mathematics* XXVIII, 158–202.
Schönfinkel, Moses (1924). 'Über die Bausteine der mathematischen Logik', *Mathematische Annalen* 92(3/4), 305–316.
Słupecki, Jerzy (1953). 'St. Leśniewski's Protothetics', *Studia Logica* I, 44–112. [Cf. paper V in this volume – Ed.]
Sobociński, Bolesław (1934). 'O kolejnych uproszczeniach aksjomatyki "ontologji" Prof. St. Leśniewskiego (On Successive Simplifications of the Axiom-System of Prof. St. Leśniewski's "Ontology")', in *Fragmenty Filozoficzne* (Księga pamiątkowa ku uczczeniu 15-lecia pracy nauczycielskiej w Uniwersytecie Warszawskim Prof. Tadeusza Kotarbińskiego), Vol. 1, pp. 143–160.

Sobociński, Bolesław (1939). 'Z badań nad prototetyką (An Investigation of Protothetic)', *Collectanea Logica* I, 171–177.

Sobociński, Bolesław (1949a). 'An Investigation of Protothetic', *Cahiers de l'Institut d'Études Polonaises en Belgique* 5, Bruxelles. [Cf. paper IV in this volume – Ed.]

Sobociński, Bolesław (1949b). 'L'analyse de l'antinomie russellienne par Leśniewski', *Methodos* I, 94–107, 220–228; 308–316; II (1950), 237–257.

Sobociński, Bolesław (1953). 'On a Universal Decision Element', *The Journal of Computing Systems* I, 71–80.

Sobociński, Bolesław (1954). 'Z badań nad aksjomatyką prototetyki Stanisława Leśniewskigo (Investigations on the Axiom-System of S. Leśniewski's Protothetic)', *Rocznik Polskiego Towarzystwa Naukowego na Obczyźnie, t. IV za rok 1953–54*, London, 18–20.

Sobociński, Bolesław (1955). 'Note on a Problem of Paul Bernays', *The Journal of Symbolic Logic* 20, 109–114.

Sobociński, Bolesław (1956a). 'In Memoriam, Jan Łukasiewicz (1878–1956)', *Philosophical Studies* (Ireland) VI, 3–49.

Sobociński, Bolesław (1956b). 'On Well Constructed Axiom-Systems', in *Rocznik Polskiego Towarzystwa Naukowego na Obczyźnie, t. VI, za rok 1955–56*, London, 54–65.

Stone, M.H. (1937). 'Note on Formal Logic', *American Journal of Mathematics* LIX, 506–514.

Tarski, Alfred (1923a). 'Sur le terme primitif de la Logistique', *Fundamenta Mathematicae* IV, 196–200.

Tarski, Alfred (1923b). 'O wyrazie pierwotnym logistyki (On the Primitive Term of Logistic)', *Przegląd Filozoficzny* XXVI, 68–89. [Cf. paper III in this volume – Ed.]

Tarski, Alfred (1924). 'Sur les truth-functions au sens de MM. Russell et Whitehead', *Fundamenta Mathematicae* V, 59–74.

Tarski, Alfred (1956). 'On the Primitive Term of Logistic', in *Logic, Semantics, Metamathematics*, translated by J.H. Woodger, Clarendon Press, Oxford, pp. 1–23.

Tarski, Alfred and Łukasiewicz, Jan (1956). 'Investigations into the Sentential Calculus', in *Logic, Semantics, Metamathematics*, translated by J.H. Woodger, Clarendon Press, Oxford, pp. 38–59. Cf. Łukasiewicz–Tarski (1930).

Wajsberg, Mordchaj (1931). 'Aksjomatyzacja trójwartościowego rachunku zdań (Axiomatization of the Three-Valued Propositional Calculus)', *Comptes rendus des Séances de la Société des Sciences et des Lettres de Varsovie, Classe III* 24, 126–145.

Wajsberg, Mordchaj (1937). 'Metalogische Beiträge. I', *Wiadomości Matematyczne* XLIII, 131–168.

Whitehead, Alfred N. and Russell, Bertrand (1925). *Principia Mathematicae*, Vol. 1, Second Edition, Cambridge.

Remarks: Tarski–Łukasiewicz (1956) is an English translation of Łukasiewicz–Tarski (1930); an English translation of Sobociński (1939) is a part of Sobociński (1949a); Tarski (1923a) and Tarski (1924) are French versions and Tarski (1956) is an English translation of Tarski (1923b).

V. FREDERICK RICKEY

AXIOMATIC INSCRIPTIONAL SYNTAX
PART II: THE SYNTAX OF PROTOTHETIC *

Part I** of this paper presented an axiomatized theory of general syntax, defined the usual notions of general syntax, and proved some typical theorems of general syntax. We claimed that this system is strong enough to precisely state the formative and deductive rules of formal languages. In Part II we support this claim by formulating, in a very precise way, the rule of procedure of Leśniewski's Protothetic. Section 1 is an informal introduction to Protothetic and its rule of procedure. Our metalogical system is extended there to a theory **MP** which is concerned explicitly with the syntax of Protothetic. This section is intended to motivate the Terminological Explanations of Section 2 which culminate in the statement of the rule of Protothetic. In Section 3 we sketch a proof that our formulation of the rule is equivalent to that given by Leśniewski (1929). This proof shows that our system is strong enough to conduct certain metalogical investigations.

There are several reasons why we chose Protothetic as a test case for our syntactical system: (1) Since the rule of Protothetic is as complicated as the deductive rule for any theory it provides a good test of the usefulness of our system for formalizing theories. (2) Most papers on Protothetic only discuss the deductive rule from an informal point of view. It is hoped that our formulation of the rule, which is simpler than that given by Leśniewski (1929), will serve to make this aspect of Leśniewski's work more accessible. (3) It was necessary, in order to make the comparisons that we do in Section 3, to use a theory whose deductive rule had previously been precisely stated. Protothetic is one of the few such theories whose rule had been formulated without making excessive use of ordinary language, set theory, or inductive techniques. And, as we stated in Part I, it is our conviction that the formalization of a system should be done using the weakest possible metalogic.

---

* Bibliographical Note. This paper originally appeared in the *Notre Dame Journal of Formal Logic* XIV (1973), pp. 1–52, by permission of the *Notre Dame Journal of Formal Logic*.

** Part I of this paper appeared in the *Notre Dame Journal of Formal Logic* XIII (1972), pp. 1–33. The author would like to thank Professor Bolesław Sobociński for considerable advice.

1

1.1. History and Informal Description of Protothetic

In 1923 Stanisław Leśniewski developed a bivalued system of logic which he called Protothetic. Besides including the full propositional calculus, Protothetic contains variables of all semantical categories which can be generated from propositions. To explain, let us say that Protothetic is a type theory whose lowest type is that of propositions:

$$p, q, r, \ldots .$$

These can be combined, using unary, binary, ..., functors, to form new propositions:

$$f(p), g(pg), h(pqr), \ldots .$$

Then we can form functors which take previously introduced types as variables, for example:

$$\varphi[fp], \psi\langle fgh\rangle, \ldots .$$

This process is continued indefinitely. After a constant functor with a given number of arguments is introduced we can use variables of that type. This will be explained more fully in Section 1.3.

Protothetic can be based on the law of simplification

$$(1)\ [pq] : p \supset .q \supset p$$

together with the principal of bivalence for each semantical category. These are theses of the form

$$(2)\ [fp] : .f(0). \supset : f(1). \supset .f(p)$$

$$(3)\ [Ff] :\cdot : F\langle \mathsf{N}\rangle. \supset :: F\langle \mathsf{Fl}\rangle. \supset : .F\langle \mathsf{Vr}\rangle. \supset : F\langle \mathsf{As}\rangle. \supset .F\langle f\rangle$$

There are, however, very grave difficulties with stating the rule of bivalence, for we must specify all the constants of a given semantical category before we can state the law of bivalence for that semantical category. This is not just a problem of specifying symbols for each of the constants, rather we must also state how they are defined or at least know that once the proper number have been defined that they are distinct. As this problem is still unsolved we use a different approach.

Another way to formulate Protothetic is to accept simplification (1), bivalence for propositions (2) and extensionality for every other semantical category.

The simplest form of extensionality is that for propositions

(4) $[pq] : .p \equiv q. \equiv: [f] : f(p). \equiv .f(q)$,

but this follows from (1) and (2). The theses of extensionality which are essential are those for all of the other semantical categories. These theses are of the form:

(5) $[fg] : .[p] : f(p). \equiv .g(p) :\equiv: [F] : F\langle f\rangle. \equiv .F\langle g\rangle$

(6) $[fg] : .[pq] : f(pq). \equiv .g(pq) :\equiv: [F] : F\,\langle f\rangle\,. \equiv .F\,\langle g\rangle$

Since we have a method of formulating the rule of extensionality[1] the difficulty is overcome.

(5) and (6) are stated using only the constants of equivalence. (1) and (2), however, use implication. It is highly desirable to base Protothetic on a single primitive term. Tarski (1923) showed that conjunction could be defined in terms of equivalence and quantification:[2]

(7) $[pq] :: p.q. \equiv: .[f] : .p. \equiv: f(p). \equiv .f(q)$

Since negation is easily defined:

(8) $[p] : . \sim p. \equiv: p. \equiv .[u].u$

we have all the connectives needed for the propositional calculus. This allows Protothetic to be based on equivalence (as sole primitive). This has the added advantage that definitions can be stated in the most natural form,[3] i.e., as equivalences:

$$\alpha \equiv \beta$$

The successive simplification of the axiomatization of Protothetic was a long and complicated process. The details of this work and their history can be found in Leśniewski (1929, 1938), and Sobociński (1960). The system of Protothetic which we will be concerned with is based on Sobociński's 1945 axiom:

(9) $\llcorner pq \lrcorner \ulcorner \phi(\phi(pq) \llcorner f \lrcorner \ulcorner \phi(f(pf(p \llcorner u \lrcorner \ulcorner u \urcorner)) \llcorner r \lrcorner \ulcorner \phi(f(qr)\phi(qp)) \urcorner) \urcorner) \urcorner$

or, in the more familiar Peano–Russellian notation:

(10) $[pq] :: p \equiv q. \equiv: .[f] : .f(pf(p0)). \equiv: [r] : f(qr). \equiv .q \equiv p$

This axiom is a dead object without some rules for deducing its consequences. We hinted above that we shall have rules of extensionality and definition. In addition, we shall need rules of detachment, substitution, and a rule for distributing quantifiers.

The precise statement of these rules occupies the whole of Section 2. In order to make the Terminological Explanations of that section understandable we wish first to give an informal statement of the rule. But first we must discuss semantical categories.

1.2. Semantical Categories

In the axiom of Protothetic (1, (9)) the basic unit is the biconditional[4]

$$\phi(pq)$$

or in the Peano–Russellian notation

$$p \equiv q.$$

Intuitively, equivalence is a functor which forms a proposition out of the two propositions which are its arguments. If we denote the semantical category of propositions by S then, using the notation of Ajdukiewicz (1935), we denote the semantical category of ϕ by

$$S/SS.$$

In general, the symbol

$$C/C_1C_2 \ldots C_n$$

represents the semantical category of a functor with n arguments of categories C_1, C_2, ..., C_n, respectively, which forms something of category C. For example, in the category S/S the only constant functors are the four unary predicates Vr, Fl, As, N; in the category S/SS we have the 16 binary functors. Semantically we want two functors to be in the same semantical category if they can be interchanged in all contexts without loss of meaning.[5] So for example, the functors in

$$\psi[fp], \quad \varphi[fp]$$

should belong to the same category, but we cannot decide to which until the categories of f and p are known. Similarly, in

$$[f] : f(p). \equiv . * \langle f \rangle$$

we want the two f's to be in the same category, but we cannot say which one until we know the category of p or $*$. All that we can say is that f belongs to the category S/C and $*$ to the category $S/(S/C)$, where C is the category of p.

Up to this point intersubstitutability has been the criteria for placing two functors in the same semantical category. This method will not suffice for our *syntactical* investigations for it uses the semantical notion of meaning.

Leśniewski was able to define semantical categories syntactically by letting the shape of the parentheses in which arguments are enclosed reveal the semantical categories of the arguments and their functors. He used the parentheses (– –) to enclose two sentential variables. So when we see

$$\phi(pq)$$

we know ϕ must be of the type S/SS since the parentheses tell us p and q are both in the category S and that the functor is sentential.

For the purpose of formalizing Protothetic we cannot use any semantical notions. Thus it is absolutely necessary to replace the semantical notion of a semantical category by some syntactical notion.

Following the methods of Leśniewski we shall let the parentheses reveal the semantical categories of the arguments and functors involved. We admit that the motivation is completely semantical, but still, our presentation must be completely syntactical. Accordingly we shall speak in the future only of syntactical categories. We are using the term " syntactical category" instead of "semantical category" at the suggestion of Chomsky (1953) and Bar-Hillel (1950). This notion is defined in Section 2 by purely syntactical means. But since our motivation is semantical it agrees with the semantical categories presented above.

Because it is necessary, if one wishes to understand what follows, to have a tentative notion of syntactical category we present the following "definition":

We shall say that two functors are in the same *syntactical category* if they have the same number of arguments, corresponding arguments are of the same syntactical categories, and the functors together with their arguments form objects of the same syntactical category.

1.3. Informal Statement of the Rule of Protothetic

As mentioned previously, the rule of procedure of Protothetic consists of five parts, viz., detachment, distribution of quantifiers, extensionality, substitution and definition. Let us consider them individually.

Detachment

If, at a given stage of development, Protothetic contains the thesis

(1) $\phi(AB)$

and also the thesis

(2) A

then we can add to Protothetic a thesis equiform to B. This rule is well enough known from other systems but we should remark that in (1) there is no main general quantifier preceding the biconditional. If we had

$$(3)\ \llcorner pq\ldots\lrcorner\ \ulcorner\phi(A(pq\ldots)B(pq\ldots))\urcorner$$

as a thesis and also

$$(4)\ \llcorner pq\ldots\lrcorner\ \ulcorner A(pq\ldots)\urcorner$$

then we would have to distribute the quantifier in (3) before detaching.

Distribution of Quantifiers

If Protothetic contains a thesis (3) then we can distribute the quantifiers and get

$$(5)\ \phi(\llcorner pq\ldots\lrcorner\ \ulcorner A(pq\ldots)\urcorner\ \llcorner pq\ldots\lrcorner\ \ulcorner B(pq\ldots)\urcorner)$$

The rule is even stronger than this. It is stated so that we can distribute not only all, but any of the variables in the main general quantifier (MGQ) of (3). Thus we can also get

$$(6)\ \llcorner q\ldots\lrcorner\ \ulcorner\phi(\llcorner p\lrcorner\ \ulcorner A(pq\ldots)\urcorner\ \llcorner p\lrcorner\ \ulcorner B(pq\ldots)\urcorner)\urcorner$$

$$(7)\ \llcorner p\ldots\lrcorner\ \ulcorner\phi(\llcorner q\lrcorner\ \ulcorner A(pq\ldots)\urcorner\ \llcorner q\lrcorner\ \ulcorner B(pq\ldots)\urcorner)\urcorner$$

$$(8)\ \llcorner\ldots\lrcorner\ \ulcorner\phi(\llcorner pq\lrcorner\ \ulcorner A(pq\ldots)\urcorner\ \llcorner pq\lrcorner\ \ulcorner B(pq\ldots)\urcorner)\urcorner$$

Protothetic also makes no use of vacuous quantifiers so if A or B contains no free occurrence of p, then p is omitted from the quantifier immediately preceding A or B after distribution of p. For example, from

$$(9)\ \llcorner pqr\lrcorner\ \ulcorner\phi(\llcorner p\lrcorner\ \ulcorner \mathsf{Vr}\,(p)\urcorner\ \llcorner s\lrcorner\ \ulcorner\psi(pqrs)\urcorner)\urcorner$$

we can get, among others,

$$(10)\ \llcorner qr\lrcorner\ \ulcorner\phi(\llcorner p\lrcorner\ \ulcorner \mathsf{Vr}\,(p)\urcorner\ \llcorner ps\lrcorner\ \ulcorner\psi(pqrs)\urcorner)\urcorner$$

$$(11)\ \phi(\llcorner p\lrcorner\ \ulcorner \mathsf{Vr}\,(p)\urcorner\ \llcorner pqrs\lrcorner\ \ulcorner\psi(pqrs)\urcorner)$$

$$(12)\ \llcorner p\lrcorner\ \ulcorner\phi(\llcorner p\lrcorner\ \ulcorner \mathsf{Vr}\,(p)\urcorner\ \llcorner qrs\lrcorner\ \ulcorner\psi(pqrs)\urcorner)\urcorner$$

Notice that the stronger statement of the rule of distribution permits the simpler statement of the rule of detachment.

The rule of quantification, as stated above, is strong enough so that we can derive the usual rules of quantification such as distribution of quantifiers through implication.

Extensionality
If the semantical categories

(13) $S/C_1\ C_2\ \ldots\ C_n, \quad S/(S/C_1\ C_2\ \ldots\ C_n)$

have already been introduced into Protothetic by way of definition (see below) then the rule of extensionality allows us to add the following as a new thesis of Protothetic:

(14) $\llcorner fq \lrcorner \ulcorner \phi(\llcorner x_1 \ldots x_n \lrcorner \ulcorner \phi(f \ddagger x_1 \ldots x_n \ddagger g \ddagger x_1 \ldots x_n \ddagger) \urcorner$
$\llcorner \psi \lrcorner \ulcorner \phi(\psi \prec f \succ \psi \prec g \succ) \urcorner) \urcorner$

where the variables $x_1, \ldots, x_n$ are of categories $C_1, \ldots, C_n$. To give a concrete example we state the law of extensionality for binary functors:

(15) $\llcorner fq \lrcorner \ulcorner \phi(\llcorner pq \lrcorner \ulcorner \phi(f(pq)g(pq)) \urcorner \llcorner F \lrcorner \ulcorner \phi(F \langle f \rangle\ F \langle g \rangle) \urcorner) \urcorner$

The rule does not allow us to add extensionality for the semantical category of propositions, since it is derivable from the axiom of Protothetic (cf. Sobociński, 1960).

The rule of extensionality allows us to add a thesis of extensionality for each semantical category (except for propositions), and not just those indicated above. We can also add extensionality when functors involving parameters (the so-called many link functors) are involved, e.g.,

$$\llcorner fg \lrcorner \ulcorner \phi(\llcorner pq \lrcorner \ulcorner \phi(f \prec p \succ (q) g \prec p \succ (q)) \urcorner \llcorner F \lrcorner \ulcorner \phi(F \ll f \gg\ F \ll g \gg) \urcorner) \urcorner$$

Substitution
The notion of substitution is well enough known from other systems of logic to need little comment here. The most important thing about substitution in Protothetic is that we never use incomplete symbols.[6] In other systems of logic if one had

(16) $\llcorner fp \lrcorner \ulcorner \Phi(f(p)) \urcorner$

and wanted to get

(17) $\llcorner p \lrcorner \ulcorner \Phi(\phi(pp)) \urcorner$

he would just say "substitute $\phi(p-)$ for f." On semantical grounds this is absurd, for f and ϕ clearly belong to different semantical categories; so we certainly cannot substitute ϕ for f. In Protothetic this substitution is accomplished via a definition:

(18) $\llcorner p \lrcorner \ulcorner \phi(\phi(pp)\mathsf{Vr}(p)) \urcorner$

First substitute Vr for f in (16) to get

(19) $\llcorner p \lrcorner \ulcorner \Phi(\mathsf{Vr}(p)) \urcorner$

This is clearly possible since f and Vr both belong to the category S/S. Then, using the logical basis so far developed, derive

(20) $\llcorner p \lrcorner \ulcorner \phi(\Phi(\mathsf{Vr}(p))\Phi(\phi(pp))) \urcorner$

Now distribute the quantifiers and detach (19) from (20) to get the desired (17).

Definition

In Protothetic definitions are biconditionals or generalizations of biconditionals with the definiendum on the right. Moreover, the definiendum is either a term (when we are defining a constant) or a function (when we are defining a constant depending on several parameters). The variables in the MGQ are precisely those occurring free in the definiens and also in the definiendum. Hence definitions have the form

(21) $\llcorner pq \ldots \lrcorner \ulcorner \phi(\psi(pq \ldots) * (pq \ldots)) \urcorner$.

The requirements imposed on definitions are similar to those in other systems of logic: $\psi(pq \ldots)$ must be well formed, $*$ is a *new* constant, etc. We concentrate now on the choice of parentheses in the definiendum.

The axiom of Protothetic contains only two semantical categories, the so-called basic semantic categories

$$S, \quad S/SS.$$

All other semantical categories must be introduced into Protothetic by defining a constant of that category. Once a semantical category has been introduced – and only then – does the rule of procedure allow us to use variables of that category. To introduce the categories

$$S/S, \quad S/SSS, \quad S/(S/S)$$

we could use, for example, the following definitions:

(22) $\llcorner p \lrcorner \ulcorner \phi(\phi(p \llcorner u \lrcorner \ulcorner u \urcorner) \sim(p)) \urcorner$

(23) $\llcorner pqr \lrcorner \ulcorner \phi(\phi(p\phi(qr))\phi(pqr)) \urcorner$

(24) $\llcorner f \lrcorner \ulcorner \phi(\llcorner p \lrcorner \ulcorner \sim(f(p)) \urcorner \sim\langle f \rangle) \urcorner$

These examples serve to indicate the freedom of the rule of definition. Notice that we have used equiform parentheses in

$$\phi(pq), \quad \sim(p), \quad \phi(pqr)$$

We did this – but we would not have had to – to show that these three functors act on sentential variables. Also we chose the symbol ϕ in (23) since the functor defined there has several properties in common with equivalence. There is no chance to confuse the two for they have different numbers of arguments.

In (24) we chose the symbolism $\sim\langle f \rangle$ to show analogy with $\sim(p)$. But here we could not have chosen the same kind of parentheses, for then we would have f and p in the same category. We were forced to choose a new kind of parentheses. Now if we wish to define Verum for one propositional argument we are forced to use parentheses equiform to those in the definiendum of (22):

(25) $\llcorner p \lrcorner \ulcorner \phi(\phi(pp)\mathsf{Vr}(p)) \urcorner$

If we took different parentheses in the definiendum – say $\nleftarrow$–$\nrightarrow$ – then Vr and $\sim$ would be in different categories; something we know is not true semantically, so we preclude it syntactically via the rule.

At first glance one might think our analogous constants and parentheses would cause the semantical categories to coalesce. For example, if we define

(26) $\llcorner fq \lrcorner \ulcorner \phi(\phi(\llcorner pq \lrcorner \ulcorner f(pq) \urcorner \llcorner pqr \lrcorner \ulcorner g(pqr) \urcorner) * \langle fg \rangle) \urcorner$

and then substitute ϕ from the Axiom for f, and ϕ from (23) for g, we would get:

(27) $\phi(\phi(\llcorner pq \lrcorner \ulcorner \phi(pq) \urcorner \llcorner pqr \lrcorner \ulcorner \phi(pqr) \urcorner) * \langle \phi\phi \rangle)$

It might be supposed that $*\langle \phi\phi \rangle$ contains two occurrences of the same constant, but that is false. To see this we look at the categories of the symbols in (27). By looking at the definiens of (26) and then back to our Axiom and (23) we see that f is of type S/SS, g of type S/SSS, and hence $*$ is of type $S/(S/SS, S/SSS)$. Thus the two ϕ's, although equiform, are not equisignificant.

Once a semantical category is introduced by an axiom or definition a type of parentheses is forever assigned to it. The axiom assigns the parentheses

(– –) to the category S/SS. The definitions (22), (23), (24), (26) dictate that a functor in the category

> S/S is always followed by parentheses of the type (–)
> S/SSS is always followed by parentheses of the type (– – –)
> $S/(S/S)$ is always followed by parentheses of the type ⟨–⟩
> $S/(S/SS, S/SSS)$ is always followed by parentheses of the type ⟨– –⟩.

In the future if we are defining a functor of one of these categories we *must* use parentheses of the indicated type. If a functor of a different type is being defined we must use different parentheses if the number of arguments are the same as one of those above.

For the purposes of substitution it does not always suffice to choose just one pair of parentheses in the definiendum. If we wanted to substitute in

(28) $f(p)$

to get

(29) ϙ(pq)

where ϙ represents conjunction, then a definition is necessary:

(30) ⌞pq⌟ ⌜ϕ(ϙ(pq)ϙ⸦q⸧(p))⌝

Then we could substitute ϙ⸦q⸧ for f (they are both of type S/S) in (28) to get

(31) ϙ⸦q⸧(p)

Then distribute the quantifier in (30), commute, and detach (31) to get the desired (29). This definition introduces a new category:

> $(S/S)/S$

i.e., ϙ is a functor of one sentential variable that forms a sentence forming functor of one sentential variable. Functors like ϙ in (31) are called *many-link-functors*. Many of the complications in stating the rule of definition arise from many-link-functors. We only remark here that a many-link-functor never has two sets of parentheses which are equiform, i.e., the links are enclosed in different shaped parentheses.

These informal remarks have brought out how interconnected the five parts of the rule are. Before detaching we usually distribute quantifiers; before stating a thesis of extensionality we have to give several definitions; before making a substitution we sometimes have to give a definition, and then use distribution, detachment and sometimes extensionality before obtaining the desired result. This indicates why we speak of "the *rule* of Protothetic" rather than "the *rules* of Protothetic."

The rule of Protothetic is formulated so that it automatically adjusts to the last thesis of the system. In any theory with definitions it is necessary to examine all previous definitions before adding a new one to make sure the constant has not been used before in that syntactical category.

1.4. The System MP

Chapters I and II [of Rickey (1972)] dealt with general syntax, whereas the next section is concerned with the syntax of Protothetic. Consequently, we must augment system M with terms applicable specifically to Protothetic.*

Since the rule of Protothetic allows us to add (under certain conditions) a new thesis A immediately after thesis B, which is the last thesis of Protothetic, and because the rule refers to some of the theses before B we need a syntactical term to express this. The term

$$A\ \varepsilon\ \mathsf{thp}(B)$$

which is read "A is a thesis of Protothetic with respect to thesis B," shall be used for this purpose. The general name $\mathsf{thp}(B)$ denotes those theses which have been derived before B. We use "before" here in the weak sense of "before or equal." This convention will save cumbersome phraseology later. Using this notion it is easy to define "thesis of Protothetic."

D3.4.1 $[A] : .A\ \varepsilon\ \mathsf{thp}. \equiv: [\exists B].A\ \varepsilon\ \mathsf{thp}(B) : A\ \varepsilon\ A$

It is important to understand that thp is a *growing name*.[7] At the beginning the Axiom is the only thesis. The rule then allows us to add another expression (actually it is one of many possible) as a new thesis immediately after the axiom. Then thp denotes the axiom and this new thesis. At this stage the rule allows us to add a new thesis. The number of inscriptions denoted by thp is always finite, but unbounded. In effect then a thesis of Protothetic is either the axiom or something added in accordance with the rule. But the rule cannot be stated without the term $\mathsf{thp}(B)$ and so we take it as primitive in our system MP. The axioms which this term satisfies will be stated later in this section.

As seen in the previous sections Protothetic makes heavy use of parentheses to indicate syntactical categories. Since the number of categories is unlimited we will need more than a fixed number of parentheses. Consequently we need a term to denote parentheses. One way would be to accept the general name

prnt

* [Ed. Note: For the definition of system M as well as of all the terms used but undefined in this paper the reader is referred to Rickey (1972).]

as a denoter of parentheses. We shall say each of

(, [, ⟨, ⋲,]

is a parenthesis. We are not too concerned about the shape of parentheses and can use most any symbol as a parenthesis. For example, we could write

pξq

instead of the more customary

(ξ)

but then ‘p’ and ‘q’ could never be used as variables. We are more concerned with availability, for (some) new syntactical categories require new kinds of parentheses.

The term prnt does not suffice however, for we must be able to mate parentheses. Since the smallest things our syntax can refer to are words it is not capable of distinguishing left from right parentheses,[8] nor is it possible to say that two parentheses are symmetric. We say that the parenthesis (is symmetric to the parenthesis), but not to any of the parentheses

⟨, [, (, ⊄,)

Each of the following pairs of parentheses are symmetric:

(,);], [; ⟨, ⟩; [,],

so symmetry does not mean that the first is a left parenthesis and the second a right. Hence we shall take as primitive the term

$A \varepsilon$ prntsym(B)

which we shall read “A is a parenthesis symmetric to parenthesis B.” Intuitively this means that A is a parenthesis and B is its mirror image. One easily sees that prnt can be defined in terms of prntsym:

D3.4.2 $[A] : .A \ \varepsilon$ prnt. $\equiv: [\exists B].A \ \varepsilon$ prntsym$(B) : A \ \varepsilon \ A$

Hence we can just take prntsym as primitive; and we do, even though it is less intuitive.

Besides using these two primitive terms we will need to make explicit reference to the axiom of Protothetic. We give that axiom here and call it by the name $\mathfrak{A}\mathfrak{x}$:

$$\llcorner pq \lrcorner \ulcorner \phi(\phi(pq) \llcorner f \lrcorner \ulcorner \phi(f(pf(p \llcorner u \lrcorner \ulcorner u \urcorner)) \llcorner r \lrcorner \ulcorner \phi(f(qr)\phi(qp)) \urcorner) \urcorner) \urcorner$$

This axiom is an expression consisting of 54 words. In the next section we shall refer to the 1st, 4th, 5th, 9th, 10th, 11th, 12th and last of them. Moreover, we need to know that the 9th and 12th words are symmetrical parentheses and that certain words in the axiom are not equiform. Statements Q7–Q 11 use vocabulary which is not defined until the next section, but we prefer to state them here. Our assumptions about the axiom – which could either be called axioms or empirical statements – can be summarized as follows:

Q1 $\mathfrak{Ax}\ \varepsilon\ \textbf{expr}$

Q2 $\textbf{9vrb}(\mathfrak{Ax})\ \varepsilon\ \mathsf{prntsym}(\textbf{12vrb}(\mathfrak{Ax}))$

Q3 $[\exists B].B\ \varepsilon\ \textbf{Uvrb}(\mathfrak{Ax}).B\ \varepsilon\ \textbf{54vrb}(\mathfrak{Ax})$

Q4 $\textbf{1vrb}(\mathfrak{Ax})\ \varepsilon\ \mathcal{N}(\textbf{cnf}(\textbf{4vrb}(\mathfrak{Ax})))$

Q5 $\textbf{5vrb}(\mathfrak{Ax})\ \varepsilon\ \mathcal{N}(\textbf{cnf}(\textbf{1vrb}(\mathfrak{Ax})))$

Q6 $\textbf{5vrb}(\mathfrak{Ax})\ \varepsilon\ \mathcal{N}(\textbf{cnf}(\textbf{4vrb}(\mathfrak{Ax})))$

Q7 $\textbf{6vrb}(\mathfrak{Ax})\ \varepsilon\ \mathsf{trm}$

Q8 $\textbf{10vrb}(\mathfrak{Ax})\ \varepsilon\ \mathsf{trm}$

Q9 $\textbf{11vrb}(\mathfrak{Ax})\ \varepsilon\ \mathsf{trm}$

Q10 $\textbf{Cmpl}(\textbf{9vrb}(\mathfrak{Ax}) \cup \textbf{10vrb}(\mathfrak{Ax}) \cup \textbf{11vrb}(\mathfrak{Ax}) \cup \textbf{12vrb}(\mathfrak{Ax}))\ \varepsilon\ \mathsf{prntm}$

Q11 $\textbf{10vrb}(\mathfrak{Ax})\ \varepsilon\ \mathsf{arg}(\textbf{Cmpl}(\textbf{9vrb}(\mathfrak{Ax})\cup\textbf{10vrb}(\mathfrak{Ax})\cup\textbf{11vrb}(\mathfrak{Ax})\cup\textbf{12vrb}(\mathfrak{Ax})))$

We now state the axioms which our primitive terms thp and prntsym satisfy:

P1 $\mathfrak{Ax}\ \varepsilon\ \mathsf{thp}(\mathfrak{Ax})$

P2 $[B] : B\ \varepsilon\ \mathsf{thp}(\mathfrak{Ax}). \supset .B = \mathfrak{Ax}$

P3 $[A] : A\ \varepsilon\ \mathsf{thp}(A). \supset .\mathfrak{Ax}\ \varepsilon\ \mathsf{thp}(A)$

P4 $[AB] : A\ \varepsilon\ \mathsf{thp}(B). \supset .A\ \varepsilon\ \mathsf{thp}(A)$

P5 $[ABC] : A\ \varepsilon\ \mathsf{thp}(B).B\ \varepsilon\ \mathsf{thp}(C). \supset .A\ \varepsilon\ \mathsf{thp}(C)$

P6 $[AB] : A\ \varepsilon\ \mathsf{thp}(A).B\ \varepsilon\ \mathsf{thp}(B).A\ \varepsilon\ \textbf{prcd}(B). \supset .A\ \varepsilon\ \mathsf{thp}(B)$

P7 $[A] : A\ \varepsilon\ \mathsf{thp}(A). \supset .A\ \varepsilon\ \textbf{expr}$

P8 $[A] : A\,\varepsilon\,\mathsf{thp}(A). \supset .\mathsf{Fin}\{\mathsf{thp}(A)\}$

P9 $[AB] : A\,\varepsilon\,\mathsf{thp}(B). \supset .B\,\varepsilon\,\mathsf{thp}(B)$

P10 $[AB] : .A\,\varepsilon\,\mathsf{thp}(B). \supset: A\,\varepsilon\,\mathbf{prcd}(B). \vee .A = B$

P11 $[AB] : .A\,\varepsilon\,\mathsf{prntsym}(B). \equiv: A\,\varepsilon\,\mathbf{vrb}.B\,\varepsilon\,\mathsf{prntsym}(A). \sim (A\,\varepsilon\,\mathbf{cnf}(B)) :$
$[C] : B\,\varepsilon\,\mathsf{prntsym}(C). \supset .A\,\varepsilon\,\mathbf{cnf}(C)$

The system **MP** consists of system **M** augmented by the terms

$\mathsf{thp}(B) \qquad \mathsf{prntsym}(B)$

and the axioms P1–P11, Q1–Q11[9] stated above. It is within this system that we formulate the rule of procedure of Protothetic in the next section.

We now make several remarks about the axioms. The denotation of $\mathsf{thp}(B)$ cannot be axiomatically prescribed as this term has its precise meaning only after statement of the rule. The axioms stated here are all of the properties of $\mathsf{thp}(B)$ which will be used in Section 3. From P11 we can derive the following theorems

T3.4.1 $[A] : A\,\varepsilon\,\mathsf{prnt}. \supset .A\,\varepsilon\,\mathbf{vrb}$ [D3.4.2, P11]

T3.4.2 $[AB] : A\,\varepsilon\,\mathsf{prntsym}(B). \supset .A\,\varepsilon\,\mathsf{prnt}.B\,\varepsilon\,\mathsf{prnt}$ [D3.4.2, P11]

T3.4.3 $[AB] : A\,\varepsilon\,\mathsf{prntsym}(B). \supset .A \neq B$ [P11, T2.2.18]

T3.4.4 $[ABC] : A\,\varepsilon\,\mathsf{prntsym}(B).B\,\varepsilon\,\mathsf{prntsym}(C). \supset .A\,\varepsilon\,\mathbf{cnf}(C)$ [P11]

T3.4.5 $[AB] : A\,\varepsilon\,\mathsf{prntsym}(B). \supset .B\,\varepsilon\,\mathsf{prntsym}(A)$ [P11]

T3.4.6 $[AB] : A\,\varepsilon\,\mathsf{prntsym}(B). \supset . \sim (A\,\varepsilon\,\mathbf{cnf}(B))$ [P11]

T3.4.6 is the only one of these requirements on parentheses which is really useful. It prohibits symbols like

$=, \quad +, \quad \ominus, \quad \phi$

from being used as parentheses. The other theses are all satisfied by any non-self-symmetric symbol, where for a symmetric parenthesis we take the mirror image. It should be remarked that if we take P11 too seriously then variables and constants could only be self-symmetric symbols like

A, V, T, +, =, U.

Since we do not wish to observe this requirement in practice, we shall use as parentheses only those symbols which one normally calls parentheses.

2

To formalize a deductive theory is to stipulate directives for derivation of its theorems in purely syntactical terms. To do this it is necessary to give precise explanations (definitions) of the metalinguistic terms used in these directives.

In this section we give, in the most precise way that we know, a formulation of the rule of procedure of Protothetic. The rule itself can be stated quite succinctly (cf. Section 2.6 below), but the explanation of the terms used there will require a long sequence of *Terminological Explanations* (*T.E.*'s).

Our Terminological Explanations are nothing more than definitions of metalinguistic terms which apply expressly to Protothetic, and as such shall use, besides the terms of system M, only those primitive terms which are peculiar to Protothetic, viz., $\mathsf{thp}(B)$ and $\mathsf{prntsym}(B)$, which were explained in Section 1.4.

In order to make the *T.E.*'s more perspicuous we shall include brief (and approximate!) renderings into English and shall give some of the semantical motivation behind these definitions. Finally, we shall prove several consequences of the *T.E.*'s which are either interesting in themselves or useful later. This material is intended to explain and supplement the *T.E.*'s, but only what is in the *T.E.*'s is necessary for the statement of the rule.

The general methodology of the *T.E.*'s is to delimit the possible senseful expressions by excluding certain undesirable expressions from consideration, rather than to use inductive techniques similar to those commonly used in propositional and functional calculi to build up senseful expressions (wffs). Since we wish to construct a language as free and rich as possible, all non-essential choices are left open to the individual. Accordingly, we shall not use specific variables or constants which are prescribed ahead of time.

2.1. Quantifiers and Variables (TE1–12)

Most theses of Protothetic are generalizations, i.e., expressions of the form

(1) $\llcorner pq \ldots \lrcorner \ulcorner \psi(pq \ldots) \urcorner$

which we customarily read "for all $p, q, \ldots, \psi(pq \ldots)$." The first part of (1), viz., the part equiform to

(2) $\llcorner pg \ldots \lrcorner$

is called a (general) quantifier (more precisely, the quantifier of generalization (1)), and the remainder, viz., the part equiform to

(3) $\ulcorner \psi(pq \ldots) \urcorner$

is called a subquantifier (more precisely, the subquantifier of generalization (1)). The words in the quantifier are called binders of the words equiform to them in the subquantifier which are called variables. Without further ado we give precise definitions of these, and some related, syntactical notions.

TE1[10] $[A] : A \,\varepsilon\, \mathsf{trm}. \equiv .$

$A \,\varepsilon\, \textbf{vrb}.$

$A \,\varepsilon\, \mathcal{N}(\mathsf{prnt}).$

$A \,\varepsilon\, \mathcal{N}(\textbf{cnf}(\textbf{1vrb}(\mathfrak{A}\mathfrak{x}))).$

$A \,\varepsilon\, \mathcal{N}(\textbf{cnf}(\textbf{4vrb}(\mathfrak{A}\mathfrak{x}))).$

$A \,\varepsilon\, \mathcal{N}(\textbf{cnf}(\textbf{5vrb}(\mathfrak{A}\mathfrak{x}))).$

$A \,\varepsilon\, \mathcal{N}(\textbf{cnf}(\textbf{Uvrb}(\mathfrak{A}\mathfrak{x})))$

A *term* is any word which is not a parenthesis or corner, where by corner we mean any word equiform to one of the following words:

$\llcorner, \lrcorner, \ulcorner, \urcorner.$

It should be clear that corners are not parentheses, for otherwise the last four conjuncts[11] of TE1 would not be necessary.

TE2 $[A] : A \,\varepsilon\, \mathsf{qntf}. \equiv .$

$\textbf{1vrb}(A) \,\varepsilon\, \textbf{cnf}(\textbf{1vrb}(\mathfrak{A}\mathfrak{x})).$

$\textbf{Uvrb}(A) \,\varepsilon\, \textbf{cnf}(\textbf{4vrb}(\mathfrak{A}\mathfrak{x})).$

$\textbf{Int}(A) \,\varepsilon\, \textbf{Cmpl}(\mathsf{trm} \cap \textbf{int}(A)).$

$A \,\varepsilon\, \textbf{non-rep}$

A *quantifier* is a string of non-equiform terms enclosed in lower corners. Thus

$\llcorner p \lrcorner, \quad \llcorner p f \psi \lrcorner, \quad \llcorner q \circ f \phi \lrcorner$

are examples of quantifiers, but the following are not:

$\lrcorner p \lrcorner, \quad \llcorner \ \lrcorner, \quad \llcorner p (q \lrcorner, \quad \llcorner q q \lrcorner.$

By TE2.3 $\textbf{Int}(A)$ is an individual and hence any quantifier contains at least three words. Note that the terms in a quantifier are neither repeated or separated by commas, as there is no need to do so. Variables of different syntactical categories may be in the same quantifier, thus eliminating the need of adjacent quantifiers. Later *T.E.*'s will prevent constants like $\circ$ and ϕ from occurring in quantifiers.

When we say quantifier we always mean general quantifier. The particular quantifier is not definable in Protothetic.

T4.1.1 $[A] : A \varepsilon$ qntf. $\supset .[\exists B].B \varepsilon$ **int**(A)

Hyp(1) . $\supset$.

2. **Int**$(A) \varepsilon$ **Kl**(**int**(A)). [TE2, D14]

   $[\exists B].B \varepsilon$ **int**(A) [2, T2.3.4]

T4.1.2 $[AB] : A \varepsilon$ qntf.$B \varepsilon$ **int**(A). $\supset .B \varepsilon$ trm

Hyp(2) . $\supset$.

3. **Int**$(A) \varepsilon$ **Cmpl**(trm $\cap$ **int**(A)). [1, TE2]

4. $B \varepsilon$ **vrb**(**Int**(A)). [T2.3.9, 2, D13, T2.2.9, D14]

5. trm $\cap$ **int**$(A) \subset$ **vrb**. [TE1]

   $B \varepsilon$ trm [T2.5.10, 5, 4, 3, T2.5.7]

T4.1.3 $[A] : A \varepsilon$ qntf. $\supset .A \varepsilon$ **expr** [TE2, T2.5.15]

The next few lemmas are used to prove that overlapping quantifiers are identical.

T4.1.4 $[ABCD] : .A \varepsilon$ qntf.$B \varepsilon$ qntf.$D \varepsilon$ **pr**(**1vrb**(B)).$D \varepsilon$ **vrb**(A).
$C \varepsilon$ **vrb**$(A) \cap$ **vrb**(B) : **1vrb**$(B) \varepsilon$ **pr**(C). $\vee$.**1vrb**$(B) =$
$C :\supset .D \varepsilon$ **vrb**(B)

Hyp(6) . $\supset$.

7. $A \varepsilon$ **expr** [1, T4.1.3]

8. **1vrb**$(B) \varepsilon$ **vrb**(A). [7, D4, 4, 5, 3, 6]

9. **1vrb**$(B) \varepsilon \mathcal{N}$(**1vrb**$(A)$). [T2.1.33, 3, 4]

10. **1vrb**$(B) \varepsilon$ **int**$(A) \cup$ **Uvrb**(A). [D13, 8, 7, 9]

11. $\mathbf{1vrb}(B)\ \varepsilon\ \mathbf{cnf}(\mathbf{1vrb}(\mathfrak{Ar}))$. [2, TE2]

12. $\mathbf{1vrb}(B)\ \varepsilon\ \mathcal{N}\langle\mathsf{trm}\rangle$. [11, TE1]

13. $\mathbf{1vrb}(B)\ \varepsilon\ \mathbf{Uvrb}(A)$. [T4.1.2, 1, 10, 12]

14. $\mathbf{Uvrb}(A)\ \varepsilon\ \mathbf{cnf}(\mathbf{4vrb}(\mathfrak{Ar}))$. [1, TE2]

15. $\mathbf{1vrb}(\mathfrak{Ar})\ \varepsilon\ \mathbf{cnf}(\mathbf{4vrb}(\mathfrak{Ar}))$. [11, 13, 14, T2.2.20, T2.2.21]

$D\ \varepsilon\ \mathbf{vrb}(B)$ [15, Q4]

In the proof of this theorem the use of Q4 is essential. It states that the words $\llcorner$ and $\lrcorner$ are not equiform.

T4.1.5 $[ABCD]$: $A\ \varepsilon$ qntf.$B\ \varepsilon$ qntf.$D\ \varepsilon\ \mathbf{pr}(C).D\ \varepsilon\ \mathbf{vrb}(A).C\ \varepsilon\ \mathbf{vrb}(A) \cap \mathbf{vrb}(B). \supset .D\ \varepsilon\ \mathbf{vrb}(B)$

Hyp(5) . ⊃:

6. $B\ \varepsilon$ **expr.** [1, T4.1.3]

7. $\mathbf{1vrb}(B)\ \varepsilon\ \mathbf{1vrb}(B)$. [2, T2.2.27, T2.2.30]

8. $\sim (C\ \varepsilon\ \mathbf{pr}(\mathbf{1vrb}(B)))$: [D3, 5, 7]

9. $\mathbf{1vrb}(B)\ \varepsilon\ \mathbf{pr}(C). \vee .\mathbf{1vrb}(B) = C$: [T2.2.16, 8, 5, 7, T2.2.9]

10. $\mathbf{1vrb}(B)\ \varepsilon\ \mathbf{pr}(D). \vee .D\ \varepsilon\ \mathbf{pr}(\mathbf{1vrb}(B)). \vee .D = \mathbf{1vrb}(B)$: [T2.2.16, 7, 4, D3, T2.2.9]

$D\ \varepsilon\ \mathbf{vrb}(B)$ [10, 6, D7, 3, 5, T4.1.4, 1, 2, 4, 5, 9, D3]

T4.1.6 $[ABCD]$: $A\ \varepsilon$ qntf.$B\ \varepsilon$ qntf.$C\ \varepsilon\ \mathbf{pr}(D).D\ \varepsilon\ \mathbf{vrb}(A)$.
$C\ \varepsilon\ \mathbf{vrb}(A) \cap \mathbf{vrb}(B). \supset .D\ \varepsilon\ \mathbf{vrb}(B)$ [*Sim.*, T4.1.5]

T4.1.7 $[ABCD]$: $A\ \varepsilon$ qntf.$B\ \varepsilon$ qntf.$C\ \varepsilon\ \mathbf{vrb}(A) \cap \mathbf{vrb}(B)$.
$D\ \varepsilon\ \mathbf{vrb}(A). \supset .D\ \varepsilon\ \mathbf{vrb}(B)$ [T2.2.16, T4.1.5, T4.1.6, T2.2.9]

T4.1.8 $[ABC]$: $A\ \varepsilon$ qntf.$B\ \varepsilon$ qntf.$C\ \varepsilon\ \mathbf{vrb}(A) \cap \mathbf{vrb}(B). \supset .A = B$ [T4.1.7, T2.2.2]

TE3 $[A] : .A\ \varepsilon$ sbqntf. $\equiv$:

$A\ \varepsilon$ **expr-w-int** :

$[C] : C\ \varepsilon$ **hd**(A). $\supset .C\ \varepsilon$ **Mtch(Uvrb**$(\mathfrak{A}\mathfrak{x})$**5vrb**$(\mathfrak{A}\mathfrak{x})$) :

$[C] : C\ \varepsilon$ **tl**(A). $\supset .C\ \varepsilon$ **Mtch(5vrb**$(\mathfrak{A}\mathfrak{x})$**Uvrb**$(\mathfrak{A}\mathfrak{x})$)

A *subquantifier* is an expression enclosed by upper corners such that upper corners occurring in it are properly paired. TE3.2 stipulates that any initial segment of a subquantifier contains more left upper corners than right upper corners. Thus

$\ulcorner p \urcorner, \quad \ulcorner \phi(pp) \urcorner, \quad \ulcorner \phi(((\urcorner$

are subquantifiers, while

$\ulcorner\ \urcorner, \quad \ulcorner\ulcorner p \urcorner, \quad \ulcorner p \urcorner \ulcorner q \urcorner$

are not. Notice that there is no requirement about parentheses being matched.

T4.1.9 $[A] : A\ \varepsilon$ sbqntf. $\supset .A\ \varepsilon$ **expr** [TE3, T2.5.13]

Now we shall prove the first (last) word of a subquantifier is a left (right) upper corner and that a subquantifier contains the same number of left and right upper corners.

T4.1.10 $[A] : A\ \varepsilon$ sbqntf. $\supset$.**1vrb**$(A)\ \varepsilon$ **cnf(5vrb**$(\mathfrak{A}\mathfrak{x})$)

Hyp(1) . $\supset$.

2. $A\ \varepsilon$ **expr**. [T4.1.9, 1]

3. $A\ \varepsilon\ \mathcal{N}$(**vrb**). [TE3, 1, D15, T2.5.18]

4. **1vrb**$(A)\ \varepsilon$ **hd**(A). [T2.5.50, 2, 3]

5. (**vrb**(**1vrb**(A)) $\cap$ **cnf(Uvrb**$(\mathfrak{A}\mathfrak{x})$))) $<$ (**vrb**(**1vrb**(A)) $\cap$ **cnf(5vrb**$(\mathfrak{A}\mathfrak{x})$))). [TE3, 1, 4, D20]

6. (**1vrb**(A) $\cap$ **cnf(Uvrb**$(\mathfrak{A}\mathfrak{x})$))) $<$ (**1vrb**(A) $\cap$ **cnf(5vrb**$(\mathfrak{A}\mathfrak{x})$))). [5, T2.2.10, T2.2.30, D3]

7. $\rightharpoondown$ (**1vrb**(A) $\cap$ **cnf(5vrb**$(\mathfrak{A}\mathfrak{x})$))). [*Ont.* 9, 4]

8. **1vrb**(A) $\cap$ **cnf(5vrb**$(\mathfrak{A}\mathfrak{x})$) ε V [*Ont.* 10, 7, 6]

$\mathbf{1vrb}(A)\ \varepsilon\ \mathbf{cnf}(\mathbf{5vrb}(\mathfrak{A}\mathfrak{x}))$. [*Ont.* 11, 4, 8]

T4.1.11 $[A] : A\ \varepsilon$ sbqntf. $\supset .\mathbf{Uvrb}(A)\ \varepsilon\ \mathbf{cnf}(\mathbf{Uvrb}(\mathfrak{A}\mathfrak{x}))$ [*Sim.*, T4.1.10]

T4.1.12 $[A] : A\ \varepsilon$ sbqntf. $\supset .(\mathbf{vrb}(A) \cap \mathbf{cnf}(\mathbf{5vrb}(\mathfrak{A}\mathfrak{x}))) \infty (\mathbf{vrb}(A) \cap \mathbf{cnf}(\mathbf{Uvrb}(\mathfrak{A}\mathfrak{x})))$

Hyp(1) . $\supset$.

2. $A\ \varepsilon\ \mathbf{expr}$. [T4.1.9, 1]

3. $\mathbf{Uvrb}(A)\ \varepsilon\ \mathbf{cnf}(\mathbf{Uvrb}(\mathfrak{A}\mathfrak{x}))$. [T4.1.11, 1]

$[\exists B]$.

4. $B\ \varepsilon\ \mathbf{int}(A)$. [TE3, 1, T2.5.12]

5. $B\ \varepsilon\ \mathbf{vrb}(A)$. [4, D13]

6. $B\ \varepsilon\ \mathbf{pr}(\mathbf{Uvrb}(A))$. [T2.2.36, 5, 4, D13]

$[\exists C]$.

7. $C\ \varepsilon\ \mathbf{Kl}(\mathbf{vrb}(A) \cap \mathbf{pr}(\mathbf{Uvrb}(A)))$. [T2.1.1, 5, 6]

8. $\mathbf{Uvrb}(A)\ \varepsilon\ \mathcal{N}(\mathbf{vrb}(C))$. [7, T2.3.10, T2.2.13, T2.2.9, 3]

9. $C\ \varepsilon\ \mathbf{expr}$. [T2.5.47, 2, 7]

10. $C\ \varepsilon\ \mathbf{Cmpl}(\mathbf{vrb}(A) \cap \mathbf{pr}(\mathbf{Uvrb}(A)))$. [D12, 7, 9, T2.2.9, T2.4.2, T2.5.2]

11. $C\ \varepsilon\ \mathbf{hd}(A)$. [T2.5.49, 10, 2, 3]

12. $\mathbf{vrb}(C) \cap \mathbf{cnf}(\mathbf{Uvrb}(\mathfrak{A}\mathfrak{x})) < \mathbf{vrb}(C) \cap \mathbf{cnf}(\mathbf{5vrb}(\mathfrak{A}\mathfrak{x}))$. [TE3, 1, 11, D20]

13. $(\mathbf{vrb}(C) \cup \mathbf{Uvrb}(A)) \circ \mathbf{vrb}(A)$. [7, D1, D5, T2.2.7, T2.2.36]

14. $\mathbf{Uvrb}(A) \cup (\mathbf{vrb}(C) \cap \mathbf{cfn}(\mathbf{Uvrb}(\mathfrak{A}\mathfrak{x}))) < \mathbf{Uvrb}(\mathfrak{A}\mathfrak{x}) \cup (\mathbf{vrb}(C) \cap \mathbf{cnf}(\mathbf{5vrb}(\mathfrak{A}\mathfrak{x})))$. [12, 8, *Ont.* 12]

15. $(\mathbf{Uvrb}(A) \cup \mathbf{vrb}(C)) \cap (\mathbf{Uvrb}(A) \cup \mathbf{cnf}(\mathbf{Uvrb}(\mathfrak{A}\mathfrak{x}))) <$
$(\mathbf{Uvrb}(A) \cup \mathbf{vrb}(C)) \cap (\mathbf{Uvrb}(A) \cup \mathbf{cnf}(\mathbf{5vrb}(\mathfrak{A}\mathfrak{x})))$. [14]

16. $\mathbf{vrb}(A) \cap \mathbf{cnf}(\mathbf{Uvrb}(\mathfrak{A}\mathfrak{x})) < \mathbf{vrb}(A) \cap (\mathbf{Uvrb}(A) \cup \mathbf{cnf}(\mathbf{5vrb}(\mathfrak{A}\mathfrak{x})))$.
[15, 13, 3]

17. $\mathbf{vrb}(A) \cap \mathbf{cnf}(\mathbf{Uvrb}(\mathfrak{A}\mathfrak{x})) < (\mathbf{vrb}(A) \cap \mathbf{Uvrb}(A)) \cup (\mathbf{vrb}(A) \cap$
$\mathbf{cnf}(\mathbf{5vrb}(\mathfrak{A}\mathfrak{x})))$. [16]

18. $\mathbf{vrb}(A) \cap \mathbf{cnf}(\mathbf{Uvrb}(\mathfrak{A}\mathfrak{x})) \leq \mathbf{vrb}(A) \cap \mathbf{cnf}(\mathbf{5vrb}(\mathfrak{A}\mathfrak{x}))$.
[17, *Ont.* 13, 8, D5]

19. $\mathbf{vrb}(A) \cap \mathbf{cnf}(\mathbf{5vrb}(\mathfrak{A}\mathfrak{x})) \leq \mathbf{vrb}(A) \cap \mathbf{cnf}(\mathbf{Uvrb}(\mathfrak{A}\mathfrak{x}))$. [*Sim.*, 18]

$\mathbf{vrb}(A) \cap \mathbf{cnf}(\mathbf{5vrb}(\mathfrak{A}\mathfrak{x})) \leq \mathbf{vrb}(A) \cap \mathbf{cnf}(\mathbf{Uvrb}(\mathfrak{A}\mathfrak{x}))$
[18, 19, *Ont.* 20]

The following theorems, which are not used later, are stated without proof.

T4.1.13 $[AB] : A \,\varepsilon$ sbqntf.$B \,\varepsilon$ sbqntf.$\mathbf{1vrb}(A) = \mathbf{1vrb}(B). \supset .A = B$

T4.1.14 $[ABC] : .A \,\varepsilon$ sbqntf.$B \,\varepsilon$ sbqntf.$C \,\varepsilon\, \mathbf{vrb}(A) \cap \mathbf{vrb}(B). \supset:$
$A \,\varepsilon\, \mathbf{ingr}(B). \vee .B \,\varepsilon\, \mathbf{ingr}(A)$

This last theorem cannot be strengthened to identify for consider the subquantifier

$\ulcorner \phi(p\phi(\llcorner u \lrcorner \ulcorner u \urcorner p)) \urcorner$

which contains properly a subquantifier equiform to

$\ulcorner u \urcorner.$

T4.1.14 rules out situations like

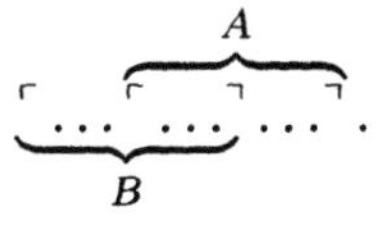

TE4 $[ABC] : A \,\varepsilon$ Gnrl$(B). \equiv .$
$B \,\varepsilon$ qntf.

$C \; \varepsilon$ sbqntf.

$A \; \varepsilon$ **Concat**(BC)

A is a *generalization* with quantifier B and subquantifier C.

TE5 $[AB] : A \; \varepsilon \; \text{Qntf}(B). \equiv .A \; \varepsilon \; A.[\exists C].B \; \varepsilon \; \text{Gnrl}(AC)$

A is the *quantifier of* generalization B.

TE6 $[AB] : A \; \varepsilon \; \text{Sbqntf}. \equiv .A \; \varepsilon \; A.[\exists C].B \; \varepsilon \; \text{Gnrl}(CA)$

A is the *subquantifier of* generalization B.

TE7 $[A] : A \; \varepsilon \; \text{gnrl}. \equiv .A \; \varepsilon \; A.[\exists BC].A \; \varepsilon \; \text{Gnrl}(BC)$

A is a *generalization.*

A generalization is an expression of the form

$$\llcorner \cdots \lrcorner \; \ulcorner --- \urcorner,$$

i.e., it consists of a quantifer followed by a subquantifier. We will prove that every generalization has a unique quantifier and subquantifier. However, a generalization may contain many quantifiers and subquantifiers. Thus in the generalization

$$\llcorner p \lrcorner \ulcorner \phi(\phi(p \llcorner r \lrcorner \ulcorner r \urcorner) \llcorner r \lrcorner \ulcorner r \urcorner) \urcorner$$

the two expressions equiform to

$$\llcorner r \lrcorner$$

are quantifiers *in* the generalization, but not *of* the generalization. Hence "having" must be distinguished from "containing." The following are examples of expressions which are not generalization.

$$\llcorner p \lrcorner * \ulcorner \varphi(p) \urcorner, \quad \llcorner p \lrcorner \llcorner q \lrcorner \ulcorner \varphi(pq) \urcorner.$$

Instead of the second of these, Protothetic uses

$$\llcorner pq \lrcorner \ulcorner \varphi(pq) \urcorner.$$

Hence adjacent quantifiers will never occur in Protothetic.

T4.1.15 $[ABCD] : A \; \varepsilon \; \text{Gnrl}(BC).D \; \varepsilon \; \text{Gnrl}(BC). \supset .A = D$

[TE4, D16, T2.5.7]

T4.1.16 $[ABC] : A \varepsilon$ Gnrl$(BC). \supset .A \varepsilon$ **expr** [TE4, D16, D12]

T4.1.17 $[ABC] : A \varepsilon$ Gnrl$(BC). \supset .B \varepsilon$ **ingr**(A) [TE4, T2.5.30]

T4.1.18 $[ABC] : A \varepsilon$ Gnrl$(BC). \supset .C \varepsilon$ **ingr**(A) [*Sim.*, T4.1.17]

T4.1.19 $[ABC] : A \varepsilon$ Gnrl$(BC). \supset .$**1vrb**$(A) =$ **1vrb**(B) [TE4, T2.5.33]

T4.1.20 $[ABC] : A \varepsilon$ Gnrl$(BC). \supset .$**Uvrb**$(A) =$ **Uvrb**(C) [*Sim.*, T4.1.19]

T4.1.21 $[A] : A \varepsilon$ gnrl. $\equiv .[\exists B] : B \varepsilon$ Qntf(A) [TE7, TE5, TE4]

T4.1.22 $[A] : A \varepsilon$ gnrl. $\equiv .[\exists C].C \varepsilon$ Sbqntf(A) [TE7, TE6, TE4]

T4.1.23 $[AB] : A \varepsilon$ Qntf$(B). \supset .A \varepsilon$ qntf [TE5, TE4]

T4.1.24 $[AB] : A \varepsilon$ Sbqntf$(B). \supset .A \varepsilon$ sbqntf [TE6, TE4]

After the rule of Protothetic is stated we will be able to prove partial converses of the above two theorems: If (sub)quantifier A is an ingredient of a thesis of Protothetic, then A is the (sub)quantifier of some generalization.

TE8 $[AB] : .A \varepsilon$ Essnt$(B). \equiv: A \varepsilon A : A \varepsilon$ **Int**(Sbqntf$(B)).\vee.A = B.A \varepsilon$ **expr**.
$A \varepsilon \mathcal{N}$(gnrl)

A is the *essence* of B.

The essence of B is the interior of the subquantifier of B if B is a generalization, and B otherwise.

T4.1.25 $[A] : A \varepsilon$ gnrl. $\supset$.**vrb**(Essnt$(A)) \circ$ **int**(Sbqntf(A))

Hyp(1) . $\supset$.

2. $[B] : B \varepsilon$ Essnt$(A). \equiv .B \varepsilon$ **Int**(Sbqntf(A)) : [1, TE8]

3. Essnt$(A) \circ$ **Int**(Sbqntf(A)). [2]

4. **vrb**(Essnt$(A)) \circ$ **vrb**(**Int**(Sbqntf(A))). [3]

vrb(Essnt$(A)) \circ$ **int**(Sbqntf(A)) [4, T2.5.22, T2.3.9, D13, T2.2.9, D14]

TE9 $[AB] : A \varepsilon$ bd$(B). \equiv .A \varepsilon$ **int**(Qntf(B))

A is a *binder* in B.

A binder is a word in the quantifier of a generalization. This T.E. is purely abbreviatory yet frequently used.

T4.1.26 $[AB] : A\,\varepsilon\,\mathsf{bd}(B).\supset.A\,\varepsilon\,\mathbf{vrb}(B)$

Hyp(1) . ⊃ .

2. $A\,\varepsilon\,\mathbf{int}(\mathsf{Qntf}(B))$. [TE9, 1]

$[\exists C]$.

3. $C\,\varepsilon\,\mathsf{Qntf}(B)$. [2, D13]

4. $A\,\varepsilon\,\mathbf{vrb}(C)$. [2, 3, D13]

$[\exists D]$.

5. $B\,\varepsilon\,\mathsf{Gnrl}(CD)$. [3, TE5]

6. $C\,\varepsilon\,\mathbf{ingr}(B)$. [5, TE4, T2.5.30]

$A\,\varepsilon\,\mathbf{vrb}(B)$ [T2.4.10, 4, 6]

T4.1.27 $[AB] : A\,\varepsilon\,\mathsf{bd}(B).\supset.[\exists C].C\,\varepsilon\,\mathsf{qntf}.A\,\varepsilon\,\mathbf{int}(C)$

Hyp(1) . ⊃ .

2. $A\,\varepsilon\,\mathbf{int}(\mathsf{Qntf}(B))$ [TE9, 1]

$[\exists C]$.

3. $A\,\varepsilon\,\mathbf{int}(C)$. [2, D13]

4. $C\,\varepsilon\,\mathsf{Qntf}(B)$. [2, D13]

5. $C\,\varepsilon\,\mathsf{qntf}$. [4, T4.1.23]

$[\exists C].A\,\varepsilon\,\mathbf{int}(C).C\,\varepsilon\,\mathsf{qntf}$ [3, 5]

T4.1.28 $[AB] : A\,\varepsilon\,\mathsf{bd}(B).\supset.B\,\varepsilon\,\mathsf{gnrl}$ [TE9, D13, T4.1.21]

It is important to remember that binders are not just words in quantifiers, but words in quantifiers or generalizations.

T4.1.29 $[ABC] : A \,\varepsilon\, \mathsf{bd}(B) . C \,\varepsilon\, \mathsf{bd}(B) . A \,\varepsilon\, \textbf{cnf}(C) . \supset . A = C$

Hyp(3) . $\supset$.

4. $A \,\varepsilon\, \textbf{int}(\mathsf{Qntf}(B))$. [1, TE9]

5. $C \,\varepsilon\, \textbf{int}(\mathsf{Qntf}(B))$. [2, TE9]

$[\exists D]$.

6. $D \,\varepsilon\, \mathsf{Qntf}(B)$. [4, D13]

7. $A \,\varepsilon\, \textbf{int}(D)$. [4, 6]

8. $C \,\varepsilon\, \textbf{int}(D)$. [5, 6]

9. $D \,\varepsilon\, \mathsf{qntf}$. [6, T4.1.23]

$A = C$ [9, TE2, D19, 3, 7, 8, D13]

TE10 $[ABC] : . A \,\varepsilon\, \mathsf{var}(BC) . \equiv :$

$B \,\varepsilon\, \mathsf{bd}(C)$.

$A \,\varepsilon\, \textbf{int}(\mathsf{Sbqntf}(C))$.

$A \,\varepsilon\, \textbf{cnf}(B) :$

$[DE] : D \,\varepsilon\, \textbf{ingr}(C) . E \,\varepsilon\, \mathsf{bd}(D) . A \,\varepsilon\, \textbf{cnf}(E) . A \,\varepsilon\, \textbf{vrb}(D) . \supset . D = C$

A is a *variable bound* by B in generalization C. This very important notion will be used constantly, so we give several examples. Consider the following generalization

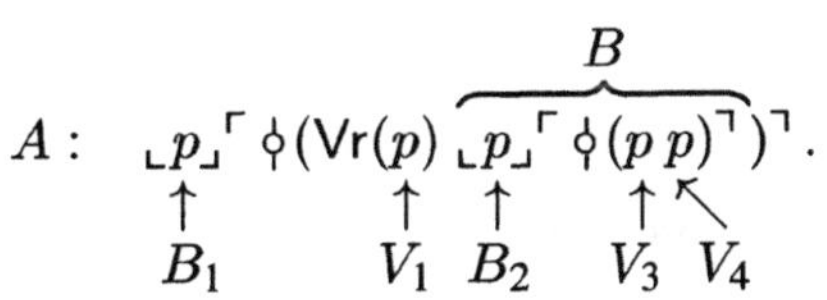

In this example we have pointed to certain words to name them. This will avoid cumbersome descriptions like "B_1 is the name of the second word of A," "B is the name of the Complex of the 11th through 20th words of A." Referring now to this diagram we can say $V_1 \,\varepsilon\, \mathsf{var}(B_1\ A)$, and, moreover, V_1

is the *only* variable of A. $V_3 \,\varepsilon\, \mathsf{var}(B_2\ B)$ and $V_4 \,\varepsilon\, \mathsf{var}(B_2\ B)$, so it should be clear that one binder can bind several variables. But neither V_3 nor V_4 is a variable in A. Since quantifiers contain no equiform words no variable can be bound by more than one binder. This prompts us to prove

T4.1.30 $[ABCDE] : A \,\varepsilon\, \mathsf{var}(DC).B \,\varepsilon\, \mathsf{var}(EC).A \,\varepsilon\, \mathbf{cnf}(B). \supset .D = E$

Hyp(3) . $\supset$.

4. $D \,\varepsilon\, \mathsf{bd}(C)$. [1, TE10]

5. $A \,\varepsilon\, \mathbf{cnf}(D)$. [1, TE10]

6. $E \,\varepsilon\, \mathsf{bd}(C)$. [2, TE10]

7. $B \,\varepsilon\, \mathbf{cnf}(E)$. [2, TE10]

8. $D \,\varepsilon\, \mathbf{cnf}(E)$. [3, 5, 7, T2.2.20, T2.2.21]

$D = E$ [T4.1.29, 4, 6, 8]

T4.1.31 $[ABC] : A \,\varepsilon\, \mathsf{var}(BC). \supset .B \,\varepsilon\, \mathbf{vrb}(C)$ [TE10, T4.1.26]

T4.1.32 $[ABC] : A \,\varepsilon\, \mathsf{var}(BC). \supset .C \,\varepsilon\, \mathsf{gnrl}$ [TE10, T4.1.28]

TE11 $[ABC] : .A \,\varepsilon\, \mathsf{cnvar}(BC). \equiv: A \,\varepsilon\, A.[\exists D].A \,\varepsilon\, \mathsf{var}(DC).B \,\varepsilon\, \mathsf{var}(DC)$

A is an *equiform variable* with B in C.

TE12 $[AB] : A \,\varepsilon\, \mathsf{var}(B). \equiv .A \,\varepsilon\, \mathsf{cnvar}(AB)$

A is a *variable in* the generalization B. As an example consider the generalization A:

$$\begin{array}{l} \llcorner pq \lrcorner \ulcorner \phi(\phi(pq)\ \phi(qp))\urcorner . \\ \uparrow\ \nwarrow \qquad\quad \uparrow\ \nwarrow\ \ \uparrow\ \nwarrow \\ B_1\ \ B_2 \qquad\ \ V_1\ \ V_2\ V_3\ \ V_4 \end{array}$$

We have

$V_1 \,\varepsilon\, \mathsf{cnvar}(V_4 A), \quad V_2 \,\varepsilon\, \mathsf{cnvar}(V_3 A)$

$V_1 \,\varepsilon\, \mathsf{var}(B_1 A), \quad V_3 \,\varepsilon\, \mathsf{var}(B_2 A)$

$V_i \,\varepsilon\, \mathsf{var}(A), \quad (i = 1, 2, 3, 4)$

The following theorems are obvious:

T4.1.33 $[ABC] : A \,\varepsilon\, \text{cnvar}(BC). \supset .A \,\varepsilon\, \text{cnvar}(AC)$ [TE11]

T4.1.34 $[ABC] : A \,\varepsilon\, \text{cnvar}(BC). \supset .A \,\varepsilon\, \text{var}(C)$ [T4.1.33, TE12]

T4.1.35 $[AB] : A \,\varepsilon\, \text{var}(B). \supset .B \,\varepsilon\, \text{gnrl}$ [TE12, TE11, T4.1.32]

T4.1.36 $[AB] : A \,\varepsilon\, \text{var}(B). \equiv .[\exists C].A \,\varepsilon\, \text{var}(CB)$ [TE12, TE11]

T4.1.37 $[ABC] : A \,\varepsilon\, \text{cnvar}(BC). \supset .B \,\varepsilon\, \text{cnvar}(AC)$ [TE11]

T4.1.38 $[ABC] : A \,\varepsilon\, \text{cnvar}(BC). \supset .A \,\varepsilon\, \mathbf{cnf}(B)$
[TE11, TE10, T2.2.20, T2.2.21]

T4.1.39 $[AB] : A \,\varepsilon\, \text{var}(B). \supset .B \,\varepsilon\, \text{trm}$ [T4.1.36, TE10, TE9, T4.1.2, TE1]

2.2. *Functions and Arguments (TE13–29)*

We previously remarked that almost all theses of Protothetic are generalizations of the form

(1) $\llcorner pq\ldots\lrcorner \ulcorner\psi(pq\ldots)\urcorner$.

More precisely, the essence of (1) is usually a many-link-function, i.e., an expression of the form

(2) $\Phi \nleftarrow pq\ldots \nrightarrow \leftarrow fg\ldots \rightarrow \ldots (fp\ldots)$

In (2) the links, i.e., the expressions like

(3) $\nleftarrow pq\ldots \nrightarrow, \leftarrow fg\ldots \rightarrow, (pf\ldots)$

are called parenthemes and consist of an expression enclosed between symmetric parentheses. Each parentheme can be uniquely decomposed into an expression of the form

(4) $\leftarrow a_1\, a_2\, \ldots\, a_n \rightarrow$

where each a_i is a term, generalization, or another function like (2). These a's are called the arguments of the parentheme (4). Besides these notions we shall develop vocabulary for comparing parenthemes and the arguments of different parenthemes.

TE13 $[A] : .A \,\varepsilon\, \text{prntm.} \equiv :$

$A \mathrel{\varepsilon} \textbf{expr-w-int}$:

$[C] : C \mathrel{\varepsilon} \textbf{hd}(A). \supset .C \mathrel{\varepsilon} \textbf{Mtch}(\textbf{Uvrb}(A)\textbf{1vrb}(A))$:

$[C] : C \mathrel{\varepsilon} \textbf{tl}(A). \supset .C \mathrel{\varepsilon} \textbf{Mtch}(\textbf{1vrb}(A)\textbf{Uvrb}(A))$:

$\textbf{1vrb}(A) \mathrel{\varepsilon} \text{prntsym}(\textbf{Uvrb}(A))$

A is a *parentheme*. In a parentheme the first and last word are symmetric parentheses, and those parentheses within A, which are equiform to either the first or last parentheses of A, are properly mated. As examples of parenthemes consider

$$(p), \quad \not(\varphi * \not), \quad \langle\llcorner u \lrcorner \ulcorner f(u) \urcorner)()\rangle.$$

One should also see the remarks concerning subquantifiers (TE3) as the *T.E.*'s are quite similar. The following theorems are analogous to those following TE3 and so no proofs are given.

T4.2.1 $[A] : A \mathrel{\varepsilon} \text{prntm}. \supset .\textbf{1vrb}(A) \mathrel{\varepsilon} \text{prnt}$

T4.2.2 $[AB] : A \mathrel{\varepsilon} \text{prntm}.B \mathrel{\varepsilon} \text{prntm}.\textbf{1vrb}(A) \mathrel{\varepsilon} \textbf{vrb}(B). \supset .A \mathrel{\varepsilon} \textbf{ingr}(B)$

T4.2.3 $[AB] : A \mathrel{\varepsilon} \text{prntm}.B \mathrel{\varepsilon} \text{prntm}.\textbf{1vrb}(A) = \textbf{1vrb}(B). \supset .A = B$

T4.2.4 $[AB] : A \mathrel{\varepsilon} \text{prntm}. \supset .A \mathrel{\varepsilon} \textbf{expr}$

We cannot prove yet that if two parenthemes have a word in common, then one is an ingredient of the other, for consider

$$B$$
$$\not(p \not(q \not) s \not)$$
$$A$$

Here parenthemes A and B are not disjoint, yet neither is part of the other.

TE14 $[Aa] : .A \mathrel{\varepsilon} \text{Fnct}(a). \equiv: \textbf{1vrb}(A) \mathrel{\varepsilon} \text{trm}$:

$[C] : C \mathrel{\varepsilon} a. \supset .C \mathrel{\varepsilon} \text{prntm}$:

$A \mathrel{\varepsilon} \textbf{Cmpl}(\textbf{1vrb}(A) \cup a)$

A is a *function with parenthemes* *a*.

TE15 $[A] : A \mathrel{\varepsilon} \text{fnct}. \equiv .A \mathrel{\varepsilon} A.[\exists a].A \mathrel{\varepsilon} \text{Fnct}(a).!\{a\}$

A is a *function*. A function is of the form

$$\Phi \not(\ldots \not) \not(\ldots \not) \ldots (\ldots).$$

The conjunct $!\{a\}$ in TE15 prevents terms from being functions. Hence we know a function has at least one parentheme. The many-link-functions we spoke of earlier are functions with more than one parentheme.

T4.2.5 $[A] : A \,\varepsilon\, \mathsf{trm}. \supset .A \,\varepsilon\, \mathsf{Fnct}(\Lambda)$

Hyp(1) . ⊃ .

2. $A \,\varepsilon\, \mathbf{vrb}.$ [TE1, 1]

3. $\mathbf{1vrb}(A) \,\varepsilon\, \mathbf{vrb}(A).$ [1, T2.2.27, T2.2.30, D3]

4. $\mathbf{1vrb}(A) = A.$ [3, 2, T2.2.10]

5. $\mathbf{1vrb}(A) \,\varepsilon\, \mathsf{trm}.$ [1, 4]

6. $\mathbf{1vrb}(A) \,\varepsilon\, \mathbf{expr}.$ [3, T2.4.2, T2.2.9]

7. $A \,\varepsilon\, \mathbf{Cmpl}(\mathbf{vrb}(A)).$ [T2.5.8, 6, 4]

8. $\mathbf{vrb}(A) \circ \mathbf{1vrb}(A) \cup \Lambda.$ [4, 3, T2.2.10]

9. $A \,\varepsilon\, \mathbf{Cmpl}(\mathbf{1vrb}(A) \cup \Lambda).$ [7, 8]

$A \,\varepsilon\, \mathsf{Fnct}(\Lambda)$ [TE14, 5, 9]

T4.2.6 $[Aab] : A \,\varepsilon\, \mathsf{Fnct}(a).A \,\varepsilon\, \mathsf{Fnct}(b). \supset .a \circ b$ [TE14]

T4.2.7 $[A] : A \,\varepsilon\, \mathsf{fnct}. \supset .\mathbf{2vrb}(A) \,\varepsilon\, \mathsf{prnt}$

Hyp(1) . ⊃ .

$[\exists a].$

2. $A \,\varepsilon\, \mathsf{Fnct}(a).$ [1, TE15]

3. $!(a).$ [1, TE15]

$[\exists C].$

4. $C \,\varepsilon\, a.$ [3]

5. $C \,\varepsilon\, \mathsf{prntm}.$ [2, 4, TE14]

$[\exists D]$.

6. $D \varepsilon \textbf{int}(C)$. [5, TE13, T2.5.12]

7. $C \varepsilon \textbf{ingr}(A)$. [4, 2, TE14, T2.5.9]

8. $D \varepsilon \textbf{int}(A)$. [6, 7, D10, D13]

9. $A \varepsilon \mathcal{N}(\textbf{vrb})$. [8, T2.5.18]

10. $[\exists B].B \varepsilon \textbf{2vrb}(A)$. [9, T2.2.29]

11. $\textbf{2vrb}(A) \varepsilon \textbf{vrb}(A)$. [10, T2.2.40, D4]

$[\exists B]$.

12. $B \varepsilon (\textbf{1vrb}(A) \cup a)$. [11, 2, TE14, D12, D1]

13. $\textbf{2vrb}(A) \varepsilon \textbf{vrb}(B)$. [11, 2, TE14, D12, D1]

14. $B \varepsilon a$. [12, 13, T2.2.39]

15. $\textbf{1vrb}(A) \varepsilon \mathcal{N}(\textbf{vrb}(B))$. [12, 13, T2.2.39]

16. $B \varepsilon \text{prntm}$. [TE14, 14, 2]

17. $B \varepsilon \textbf{ingr}(A)$. [14, 2, TE14, T2.5.9]

18. $\textbf{2vrb}(A) = \textbf{1vrb}(B)$. [13, 15, 17, T2.2.41]

19. $\textbf{1vrb}(B) \varepsilon \text{prnt}$. [16, T4.2.1]

$\textbf{2vrb}(A) \varepsilon \text{prnt}$ [18, 19]

T4.2.8 $[A] : A \varepsilon \text{fnct}. \supset .\textbf{Uvrb}(A) \varepsilon \text{prnt}$ [*Sim.*, T4.2.7]

TE16 $[AB] : A \varepsilon \text{prntm}(B). \equiv .A \varepsilon A.[\exists a].B \varepsilon \text{Fnct}(a).A \varepsilon a$

A is a *parentheme* of the function B.

T4.2.9 $[AB] : A \varepsilon \text{prntm}(B). \supset .B \varepsilon \text{fnct}$ [TE16, TE15]

T4.2.10 $[AB] : A \varepsilon \text{prntm}(B). \supset .A \varepsilon \text{prntm}$ [TE16, TE14]

Clearly a function can have several parenthemes, but it is not so clear that a parentheme can be a parentheme of several functions. Consider the function A:

$$\underbrace{\Phi\ \overbrace{\not{\not(}\,pq\,\not{\not)}}^{C}\ \not(\,t\,\not)}_{B}(ps)$$

C is a parentheme of A and also of the function B.

TE17 $[Aa] : .A \ \varepsilon$ P-arg$(a). \equiv:$

$A \ \varepsilon$ prntm :

$[C] : C \ \varepsilon \ a. \supset .C \ \varepsilon$ trm $\cup$ gnrl $\cup$ fnct :

Int$(A) \ \varepsilon$ **Cmpl**(a)

A is a *parentheme with arguments* a.

TE18 $[AB] : A \ \varepsilon$ arg$(B). \equiv .A \ \varepsilon \ A.[\exists a].B \ \varepsilon$ P-arg$(a).A \ \varepsilon \ a$

A is an *argument* of parentheme B. The most important thing to remember is that arguments can only be terms, generalizations or functions.

T4.2.11 $[AB] : A \ \varepsilon$ arg$(B). \supset .A \ \varepsilon$ trm $\cup$ gnrl $\cup$ fnct [TE18, TE17]

T4.2.12 $[AB] : A \ \varepsilon$ arg$(B). \supset .B \ \varepsilon$ prntm [TE18, TE17]

Only parenthemes have arguments. We will not speak of functions having arguments, but only of their parenthemes having arguments. We cannot prove that all parenthemes have arguments, for consider the parentheme:

$(\ulcorner)$.

$\ulcorner$ is the only thing which has a chance of being an argument, but it is neither a term, generalization nor function. So this parentheme has no arguments.

The requirement that arguments be terms, generalizations or functions is quite natural from a semantical point of view; we certainly do not want corners, parentheses, quantifiers or any meaningless expressions to occur as arguments. Syntactically it is easy to say to which of these three classes an argument belongs. Terms are always single letters, but not corners or parentheses. Generalizations always begin with $\llcorner$ and end with $\urcorner$. Because of the condition on matching corners we can always find the end of a generalization. Functions always consist of a term followed by a number of parenthemes. All of this is intended to convince one that we do not need commas between the arguments. Consider, for example, the following parentheme A:

$$\overbrace{(\ q}^{A_1\downarrow}\ \overbrace{\mathsf{Vr}(p)}^{A_2\downarrow}\ \overbrace{\llcorner p \lrcorner \ulcorner \phi(pp)\urcorner}^{A_3\downarrow}\ \overbrace{\Phi\{f\}}^{A_4\downarrow})$$

The first argument must begin with the first word in **Int**(A), i.e., q. Since the third word of A is a term, the first argument of A must be just the second word of A. As the fourth word of A is a parenthesis, the third word cannot be an argument by itself. So the fourth word is part of the second argument and, moreover, it begins a parentheme of the function which must be the second argument of A. This parentheme ends with the sixth word of A. As the seventh word is equiform to $\llcorner$, it cannot begin a new parentheme of the function which is the second argument and so begins the third argument. Since words equiform to $\llcorner$ can only start generalizations the third argument can only be a generalization. Going to the end of A_3 we find that the next word is a term and so is either an argument by itself or begins a function. Since the word after Φ is a parenthesis, we see that the last argument is A_4. This argument should convince one of the plausibility of the following theorem which states that parenthemes with arguments have those arguments uniquely.

T4.2.13 $[Aab] : A\ \varepsilon\ \mathsf{P\text{-}arg}(a) : A\ \varepsilon\ \mathsf{P\text{-}arg}(b). \supset .a \circ b$

We also have

T4.2.14 $[ABC] : A\ \varepsilon\ \mathsf{arg}(B).A\ \varepsilon\ \mathsf{arg}(C). \supset .B = C$

We now begin a sequence of *T.E.*'s which will enable us to discuss corresponding arguments of parenthemes and analogous functions. The semantical reason for these syntactic requirements is as follows: We want the number of arguments and shape of the parentheses of a parentheme to indicate the syntactical categories of its arguments.

The idea is to say – and this is the purpose of the next paragraph – that corresponding arguments enclosed in the same type of parentheses are of the same syntactical category. Thus, for example, when we write

$$\psi(pq), \quad \psi(\varphi\llcorner u\lrcorner \ulcorner u\urcorner)$$

we want p and φ, and also q and $\llcorner u\lrcorner \ulcorner u\urcorner$ to belong to the same category.

We avoid things like

$$f(p), \quad \Phi(f)$$

since we could conclude that p and f and also f and Φ are in the same category. Hence f and p are in the same category. But a functor and an argument of its parentheme being in the same category smacks of the Russell Antinomy.

TE19 $[AB] : A\ \varepsilon$ simprntm$(B). \equiv .$

$A\ \varepsilon$ prntm.

$B\ \varepsilon$ prntm.

1vrb$(A)\ \varepsilon$ **cnf**(**1vrb**(B)).

arg$(A) \propto$ arg(B)

A is a *parentheme similar* to B. Because of T4.2.12, i.e., the only things with arguments are parenthemes, it seems TE19.2 could be dropped. This is indeed the case if we knew parentheme A had any arguments. Consider $\dashv\!(\llcorner\,)\!\vdash$ and $\dashv\!(\dashv\!(\llcorner\,)\!\vdash$. The first of these is a parentheme with no arguments. The second also has no arguments, since it is not even a parentheme. This establishes the independence of TE19.2 (or TE19.1).

The following parenthemes are similar

(pq), $(\llcorner p \lrcorner\ \ulcorner ϙ(pp) \urcorner s)$, $(f(t)\ \Phi\langle f \rangle)$,

but none of these is similar to any of the following

(p), (pqr), $\dashv\!(\llcorner p)$

T4.2.15 $[AB] : A\ \varepsilon$ prntm. $\supset .A\ \varepsilon$ simprntm(A) [TE19, *Ont.* 14, T2.2.18]

T4.2.16 $[AB] : A\ \varepsilon$ simprntm$(B). \supset .B\ \varepsilon$ simprntm(A) [TE19, *Ont.* 15, T2.2.20]

T4.2.17 $[ABC] : A\ \varepsilon$ simprntm$(B).B\ \varepsilon$ simprntm$(C). \supset .A\ \varepsilon$ simprntm(C) [TE19, *Ont.* 16, T2.2.21]

TE20 $[ABCD] : A\ \varepsilon$ Anarg$(BCD). \equiv .$

$A\ \varepsilon$ arg(C).

$B\ \varepsilon$ arg(D).

(arg$(C) \cap$ **prcd**$(A)) \propto$ (arg$(D) \cap$ **prcd**(B)).

$C\ \varepsilon$ simprntm(D)

A is an *analogous argument* to B, which are arguments of C and D, respectively. Intuitively two arguments are analogous when they are both the n-th argument of their corresponding parenthemes. For example, in the similar parenthemes

$(pf(q)\llcorner u \lrcorner\ \ulcorner u \urcorner)$

and

$(t * \Phi\langle f \rangle)$

p is analogous to t, $f(q)$ to $*$ and $\llcorner u \lrcorner \ulcorner u \urcorner$ to $\Phi\langle f\rangle$.

T4.2.18 $[ABCD] : A \varepsilon \text{Anarg}(BCD). \supset .C \varepsilon \text{prntm}.D \varepsilon \text{prntm}$ [TE20, TE19]

T4.2.19 $[ABCD] : A \varepsilon \text{Anarg}(BCD). \supset .A \varepsilon \text{Anarg}(ACC)$

[TE20, *Ont.* 14, T4.2.18, T4.2.15]

T4.2.20 $[AC] : A \varepsilon \text{arg}(C). \supset .A \varepsilon \text{Anarg}(ACC)$

[TE20, *Ont.* 14, T4.2.15, T4.2.12]

TE21 $[ABCDEF] : A \varepsilon \text{Anfnct}(BCDEF). \equiv .$

$C \varepsilon \textbf{Concat}(AE).A \varepsilon A.$

$D \varepsilon \textbf{Concat}(BF).$

$E \varepsilon \text{simprntm}(F).$

$C \varepsilon \text{fnct}.$

$D \varepsilon \text{fnct}$

TE22 $[ABCD] : A \varepsilon \text{Anfnct}(BCD). \equiv .A \varepsilon A.[\exists EF].A \varepsilon \text{Anfnct}(BCDEF)$

A is an *analogous functor* to B in C and D, respectively. By a *functor* we mean either a function or term which is followed by a parentheme. Two functors are analogous when the parenthemes following them are similar. For example, in the functions

$$\text{Vr}\,(pq), \quad \Phi\langle f\rangle(t \llcorner u \lrcorner \ulcorner u \urcorner)$$

the functors Vr and $\Phi\langle f\rangle$ are analogous, but are not analogous to any functor of the functions

$$f(p), \quad \text{Vr}\,[fq], \quad \chi\langle pq\rangle(stu).$$

T4.2.21 $[ABCD] : A \varepsilon \text{Anfnct}(BCD). \supset .A \varepsilon \text{Anfnct}(ACC)$ [TE22, TE21]

TE23 $[ABCD] : .A \varepsilon \text{An}(BCD). \equiv: A \varepsilon A : A \varepsilon \text{Anarg}(BCD). \vee .$

$A \varepsilon \text{Anfnct}(BCD)$

A is *analogous* to B which are in C and D, respectively, iff A and B are either analogous arguments of parenthemes C and D, respectively, or analogous functors of C and D, respectively.

T4.2.22 $[ABCD] : A \varepsilon \mathsf{An}(BCD). \supset .A \varepsilon \mathsf{An}(ACC)$

[TE23, T4.2.19, T4.2.21]

T4.2.23 $[ABCD] : A \varepsilon \mathsf{An}(BCD). \supset .B \varepsilon \mathsf{An}(ADC)$

[TE23, TE22, TE21, TE20]

TE24 $[AB] : A \varepsilon \mathsf{1arg}(B). \equiv .A \varepsilon \mathsf{Anarg}(\mathbf{10vrb}(\mathfrak{Ax})\ B\ \mathbf{Cmpl}(\mathbf{9vrb}(\mathfrak{Ax}) \cup \mathbf{10vrb}(\mathfrak{Ax}) \cup \mathbf{11vrb}(\mathfrak{Ax}) \cup \mathbf{12vrb}(\mathfrak{Ax})))$

A is the *first argument* of parentheme B.

TE25 $[AB] : A \varepsilon \mathsf{2arg}(B). \equiv .A \varepsilon \mathsf{Anarg}(\mathbf{11vrb}(\mathfrak{Ax})\ B\ \mathbf{Cmpl}(\mathbf{9vrb}(\mathfrak{Ax}) \cup \mathbf{10vrb}(\mathfrak{Ax}) \cup \mathbf{11vrb}(\mathfrak{Ax}) \cup \mathbf{12vrb}(\mathfrak{Ax})))$

A is the *second argument* of parentheme B. It is important to remember that we are speaking of the arguments of very special parenthemes, viz., those similar to

(pq)

which is equiform to the Complex of the 9th through 12th words of our axiom $\mathfrak{Ax}$.

TE26 $[AB] : .A \varepsilon \mathsf{1eqvl}(B). \equiv: \mathbf{1vrb}(B) \varepsilon \mathbf{cnf}(\mathbf{6vrb}(\mathfrak{Ax})) : [\exists C].A \varepsilon \mathsf{1arg}(C).B \varepsilon \mathbf{Concat}(\mathbf{1vrb}(B)C) : A \varepsilon A$

A is the *first equivalence* of B.

TE27 $[AB] : .A \varepsilon \mathsf{2eqvl}(B). \equiv: \mathbf{1vrb}(B) \varepsilon \mathbf{cnf}(\mathbf{6vrb}(\mathfrak{Ax})) : [\exists C].A \varepsilon \mathsf{2arg}(C).B \varepsilon \mathbf{Concat}(\mathbf{1vrb}(B)C) : A \varepsilon A$

A is the *second equivalence* of B. B is a function of the form

$\phi(--)$,

where ϕ is to be interpreted as the biconditional.

T4.2.24 $[AB] : A \varepsilon \mathsf{1eqvl}(B). \supset .B \varepsilon \mathsf{fnct}$

Hyp(1) . $\supset$.

2. $\mathbf{1vrb}(B) \varepsilon \mathbf{cnf}(\mathbf{6vrb}(\mathfrak{Ax}))$. [1, TE26]

3. $\mathbf{1vrb}(B)$ ε trm. [2, TE1, Q7]

$[\exists C]$.

4. A ε $\mathsf{1arg}(C)$. [1, TE26]

5. B ε $\mathbf{Concat}(\mathbf{1vrb}(B)C)$. [1, TE26]

6. C ε prntm. [4, TE24, TE20, TE19]

7. B ε $\mathsf{Fnct}(C)$. [TE14, 3, 6, 5]

B ε fnct [6, 7, TE15]

TE28 $[AB] : A\ \varepsilon\ \mathsf{1trm}(B). \equiv .A\ \varepsilon\ \mathsf{1eqvl}(\mathsf{Essnt}(B))$

A is the *first term* of B.

TE29 $[AB] : A\ \varepsilon\ \mathsf{2trm}(B). \equiv .A\ \varepsilon\ \mathsf{2eqvl}(\mathsf{Essnt}(B))$

A is the *second term* of B. There are two cases to consider here. If B is a generalization, then it is of the form

$$\llcorner \cdots \lrcorner \ulcorner \phi(--)\urcorner$$

where the first dash indicates its first term and the second dash its second term. If B is not a generalization, then $\mathsf{Essnt}(B) = B$ and hence is of the form

$$\phi(--).$$

These forms are extremely important in Protothetic. All definitions have one of these two forms, although the second is used only in defining constants. All theses of extensionality have the first form.

2.3. Syntactical Categories (TE30–33)

No one today would consider the expression

$$f(f)$$

as meaningful, yet there is still considerable debate as to how to preclude such expressions from arising in formal languages. Leśniewski's solution was to introduce the concept of semantical category into his logical theories. The main semantical criteria for two expressions to be of the same semantical category is that they be intersubstitutable in all contexts. We cannot use this as a criteria for defining semantical categories, but it is the main motivational force behind what we now do.

TE30 $[AB] : A \varepsilon \text{ingrthp}(B). \equiv .A \varepsilon A.[\exists C].C \varepsilon \text{thp}(B).A \varepsilon \textbf{ingr}(C)$

A is an *ingredient of a thesis of Protothetic* with respect to B. This term is purely abbreviatory so we make no further comment.

TE31 $[AB] : .A \varepsilon \text{frp}(B). \equiv: A \varepsilon A : A \varepsilon \text{thp}(B). \vee .[\exists C].C \varepsilon \text{ingrthp}(B).$
$C \varepsilon \text{sbqntf}.A \varepsilon \textbf{Int}(C). \vee .[\exists C].C \varepsilon \text{ingrthp}(B).A \varepsilon \text{1arg}(C). \vee .[\exists C].$
$C \varepsilon \text{ingrthp}(B).A \varepsilon \text{2arg}(C)$

A is a (propositional) *phrase* with respect to B. This is a start toward building up the syntactical category of propositions. Clearly, we should include all theses and essences of generalizations. We also include the arguments of parenthemes similar to

(5) (pq)

i.e., both p and q are to be called propositions. One should realize that $\text{frp}(B)$ does not include everything which is in the syntactical category of propositions, for consider the definition of Verum:

(6) $\llcorner p \lrcorner \ulcorner \phi(\phi(pp)\mathsf{Vr}(p))\urcorner$

It is not the case that the p in $\mathsf{Vr}(p)$ is a phrase. Since we want it to be of the same category as the p's in $\phi(pp)$ we must enlarge our notions so that equiform variables are of the same category.

Immediately after the above definition we could make the following definition.

(7) $\llcorner p \lrcorner \ulcorner \phi(\llcorner f \lrcorner \ulcorner f(p)\urcorner * (p))\urcorner$

The question of what syntactical category f belongs to cannot be answered on the basis of equiform variables. It is the case that the f in $f(p)$ is analogous (in the sense of TE22) to Vr in $\mathsf{Vr}(p)$ and hence it is natural to say they are in the same syntactical category. These two remarks motivate

TE32 $[ABC] : .A \varepsilon \text{1homosemp}(BC). \equiv: A \varepsilon A : A \varepsilon \text{frp}(C).$
$B \varepsilon \text{frp}(C). \vee .[\exists D].D \varepsilon \text{ingrthp}(C).A \varepsilon \text{cnvar}(BD). \vee .$
$[\exists DE].D \varepsilon \text{ingrthp}(C).E \varepsilon \text{ingrthp}(C).A \varepsilon \text{An}(BDE)$

A is a *direct homoseme* of B with respect to thesis C. This means that A and B are in the same syntactical category.

The following theorems are easily proved.

T4.3.1 $[ABC] : A \,\varepsilon\, \mathsf{thp}(B).C \,\varepsilon\, \mathsf{thp}(B). \supset .A \,\varepsilon\, \mathsf{1homosemp}(CB)$

[TE32, TE31]

T4.3.2 $[AB] : A \,\varepsilon\, \mathsf{thp}(B). \supset .A \,\varepsilon\, \mathsf{1homosemp}(AB)$ [T4.3.1]

T4.3.3 $[ABC] : A \,\varepsilon\, \mathsf{1homosemp}(CB). \supset .A \,\varepsilon\, \mathsf{1homosemp}(AB)$

[TE32, T4.1.33, T4.2.22]

T4.3.4 $[ABC] : A \,\varepsilon\, \mathsf{1homosemp}(BC). \supset .B \,\varepsilon\, \mathsf{1homosemp}(AC)$

[TE32, T4.1.37, T4.2.23]

These last two theorems show reflexivity and symmetry, but transitivity is still lacking. The following example shows why

(8) $$\llcorner p \lrcorner \ulcorner \phi(\phi(pp)\ \mathsf{Vr}(p))\urcorner$$

(9) $$\llcorner f \lrcorner \ulcorner \phi(\llcorner p \lrcorner \ulcorner f(p)\urcorner\ \Phi\langle f\rangle)\urcorner$$

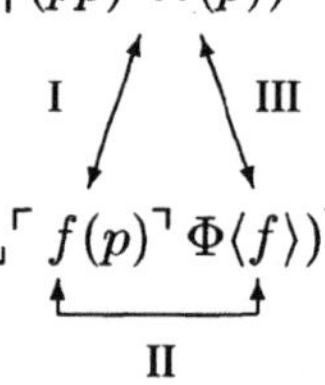

Vr and f are in the same category since they are analogous functors (arrow I). The two f's in (9) are in the same category (arrow II) since they are equiform variables. But 1homosemps are connected by analogy or by equiform variables, but not by both. Hence, Vr and the second f in (9) are not 1homosemps (arrow III).

In order to obtain transitivity, and hence arrow III, we must induct. As we have no recursive definitions we use the Frege–Dedekind technique of ancestral relations. Before giving TE33, let us see how it works. We want A and B to be in the same class iff there is a finite sequence

$$A, C_1, C_2, \ldots, C_n, B$$

such that adjacent terms are 1homosemps. We insure this in the following way. Let a be any class which includes B and which is closed under the relation of 1homosemp, i.e.,

$$[DE] : D \,\varepsilon\, a.E \,\varepsilon\, \mathsf{1homosemp}(DC). \supset .E \,\varepsilon\, a$$

(this is the condition which guarantees transitivity). Now take the smallest class which satisfies these properties. (For a simpler example of how this works, see Leśniewski (1931).)

Finally, we remark that the clause

$$[D] : D \varepsilon a. \supset .D \varepsilon \text{1homosemp}(DC)$$

is included in TE33 so that the intersection is finite. This is true since we consider only a's which are subsets of $\text{1homosemp}(DC)$. These sets are finite (for any given D and C) since, at any given time, Protothetic contains only a finite number of theses with respect to B and each of these contains only a finite number of expressions and hence 1homosemps. If we were willing to take an unrestricted quantification we could drop this part of the *T.E.*

TE33 $[ABC] :: A \varepsilon \text{homosemp}(BC). \equiv: .A \varepsilon A.$
$B \varepsilon \text{1homosemp}(BC) : .$
$[a] : .B \varepsilon a : [D] : D \varepsilon a. \supset .D \varepsilon \text{1homosemp}(DC) : [DE] : D \varepsilon a.$
$E \varepsilon \text{1homosemp}(DC). \supset .E \varepsilon a :\supset .A \varepsilon a$

A is of the *same syntactical category* as B with respect to thesis C.

T4.3.5 $[AC] : A \varepsilon \text{1homosemp}(AC). \supset .A \varepsilon \text{homosemp}(AC)$ [TE33]

We shall now prove transitivity.

T4.3.6 $[ABCDa] :: A \varepsilon \text{homosemp}(BC).B \varepsilon \text{homosemp}(DC).D \varepsilon a : [E] :$
$E \varepsilon a. \supset .E \varepsilon \text{1homosemp}(EC) : [EF] : E \varepsilon a.F \varepsilon \text{1homosemp}(EC). \supset .$
$F \varepsilon a : . \supset .A \varepsilon a.$

Hyp(5) . $\supset: \cdot :$

6. $[a] : .D \varepsilon a : [E] : E \varepsilon a. \supset .E \varepsilon \text{1homosemp}(EC) : [EF].E \varepsilon a.$
$F \varepsilon \text{1homosemp}(EC). \supset .F \varepsilon a : . \supset .B \varepsilon a : .$ [TE33, 2]

7. $B \varepsilon a.$ [6, 3, 4, 5]

$A \varepsilon a$ [1, TE33, 7, 4, 5]

T4.3.7 $[ABCD] : A \varepsilon \text{homosemp}(BC).B \varepsilon \text{homosemp}(DC). \supset .$
$A \varepsilon \text{homosemp}(DC)$ [TE33, T4.3. 6]

2.4. *Special Terms (TE34–45)*

In this section we give several very technical Terminological Explanations. They are designed for special uses in the next section and, consequently, we shall discuss them in the context that they will be used.

TE34 $[ABCD] : .C \,\varepsilon\, \mathsf{Antrm}(DAB). \equiv: C \,\varepsilon\, C : C \,\varepsilon\, \mathsf{1trm}(A).D \,\varepsilon\, \mathsf{1trm}(B). \vee .$
$C \,\varepsilon\, \mathsf{2trm}(A).D \,\varepsilon\, \mathsf{2trm}(A)$

This is used only in TE47, the rule for distribution of quantifiers. This gives us a method to relate analogous terms of generalizations where the essence is an equivalence. The following diagram may be useful.

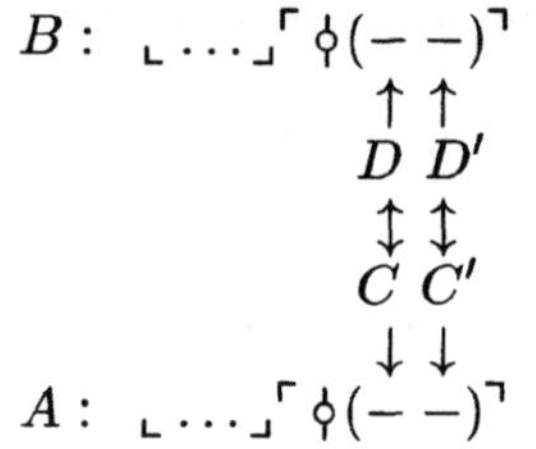

TE35 $[CDEa] : D \,\varepsilon\, \mathsf{Corr}(ECa). \equiv .$

$D \,\varepsilon\, D.$

$E \,\varepsilon\, a.$

$(a \cap \mathbf{prcd}(E)) \propto (\mathbf{int}(\mathsf{Sbqntf}(C)) \cap \mathbf{prcd}(D))$

This correspondence is used only in formulating the rule of substitution (cf. TE49 below). Taking advantage of some knowledge from TE49 we can say that Corr will be used only in the following situation

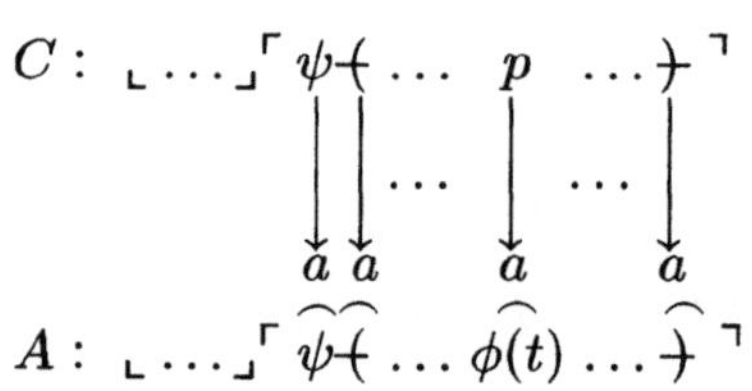

where D is one of the words in the essence of C and E is one of the a's in A; in particular, E is that a with the same number of a's preceding as D has words of $\mathsf{Essnt}(C)$ preceding.

TE36 $[AB] : .A \,\varepsilon\, \mathsf{fnctgen}(B). \equiv:$

$A \,\varepsilon\, \mathsf{fnct}.$

$\mathsf{prntm}(A) \leq \mathsf{prntm}(B) :$

$[CD] : C \,\varepsilon\, \mathsf{prntm}(A).D \,\varepsilon\, \mathsf{prntm}(B).(\mathsf{prntm}(A) \cap \mathbf{scd}(C)) \propto (\mathsf{prntm}(B) \cap \mathbf{scd}(D)). \supset .C \,\varepsilon\, \mathsf{simprntm}(D)$

A is a *function generated* by function B iff A has fewer links than B and each link of A is similar to the corresponding link in B. For example, if B is

$$\Phi\{p\}\{q\}(r)$$

then it generates the following functions

$$*\{p\}\{q\}(r), \quad \psi\{q\}(s), \quad \Phi(\llcorner u \lrcorner \ulcorner u \urcorner)$$

but none of

$$\phi\{p\}\{q\}\{r\}(s), \quad \phi(pq), \quad \oplus\{pq\}(r).$$

This *T.E.* will be of great use in formulating the directive concerning substitution: The only functions which can be substituted are those which are generated by functions already in the system. We state without proof

T4.4.1 $[A] : A \,\varepsilon\, \text{fnct.} \supset .A \,\varepsilon\, \text{fnctgen}(A)$

TE37 $[ABCDE] :: C \,\varepsilon\, \text{constp}(BADE). \equiv :.$

$D \,\varepsilon\, \text{homosemp}(EB).$

$C \,\varepsilon\, \mathbf{cnf}(D) :$

$[F] : F \,\varepsilon\, \text{ingrthp}(B). \supset .D \,\varepsilon\, \mathcal{N}\langle\text{var}(F)\rangle :$

$[\exists FG].F \,\varepsilon\, \mathbf{ingr}(A).G \,\varepsilon\, \text{ingrthp}(B).C \,\varepsilon\, \text{An}(EFG)$

TE38 $[ABC] : C \,\varepsilon\, \text{constp}(BA). \equiv .C \,\varepsilon\, C.[\exists DE].C \,\varepsilon\, \text{constp}(BADE)$

C is a *constant of Protothetic* with respect to thesis B and (occurring in) A. We now state without proof

T4.4.2 $[ABC] : C \,\varepsilon\, \text{constp}(BA). \supset .C \,\varepsilon\, \mathbf{vrb}(A)$

This term is used only three times, in TE46.3, 4, and TE49.8. TE37.3 states that no word equiform to a constant C has ever served as a variable. Thus, once we quantify a letter it may never be used as a constant. Condition TE37.4 guarantees that C is being used properly (if a functor, the right type of parentheme follows; if a variable, it occupies a spot which can hold something of that type) in expression A, which we intend to add as a thesis after B.

TE39 $[ABCDEF] : .A \,\varepsilon\, \text{q-homosemp}(BCDEF). \equiv :$

$E \,\varepsilon\, \text{homosemp}(FC).A \,\varepsilon\, A :$

$[\exists GI] : G \,\varepsilon\, \mathbf{ingr}(D).I \,\varepsilon\, \text{ingrthp}(C).A \,\varepsilon\, \text{An}(EGI) :$

$[\exists GI] : G \,\varepsilon\, \mathbf{ingr}(D).I \,\varepsilon\, \text{ingrthp}(C).B \,\varepsilon\, \text{An}(FGI)$

TE40 $[ABDF] : F \varepsilon \text{q-homosemp}(DBA). \equiv .F \varepsilon F.$
$[\exists GH].F \varepsilon \text{q-homosemp}(DBAGH)$

It is meaningful to say two expressions are of the same syntactical category only when they are ingredients of theses of Protothetic. Hence, when we are considering an inscription A as a possible thesis, it is not meaningful to say certain parts of it are in the same syntactical category. The term q-homosemp is used to circumvent this difficulty. We say F and D will be of the same syntactical category when A is added as a thesis after B if F and D are analogous to ingredients of theses which are already of the same syntactical category.

TE41 $[ABCDE] : .C \varepsilon \text{fnctp}(BADE). \equiv :$
$D \varepsilon \text{homosemp}(EB).$
$C \varepsilon \text{fnctgen}(D) :$
$[\exists FG].F \varepsilon \textbf{ingr}(A).G \varepsilon \textbf{ingrthp}(B).C \varepsilon \text{An}(EFG)$

TE42 $[ABFGHI] : .F \varepsilon \text{varp}(GBAHI). \equiv :$
$H \varepsilon \text{homosemp}(GB) :$
$[\exists JK] : K \varepsilon \textbf{ingr}(A).J \varepsilon \text{ingrthp}(B).H \varepsilon \text{An}(IJK) :$
$F \varepsilon \text{cnvar}(IA).$
$I \varepsilon \textbf{ingr}(\text{1trm}(A))$

TE43 $[ABCDE] : .C \varepsilon \text{prntmp}(BADE). \equiv :$
$D \varepsilon \text{homosemp}(BB).$
$E \varepsilon \text{prntm}(D).$
$C \varepsilon \text{prntm}(\text{2trm}(A)).$
$\text{arg}(C) \infty \text{arg}(E) :$
$[FG] : F \varepsilon \text{arg}(C).G \varepsilon \text{arg}(E).(\text{arg}(C) \cap \textbf{prcd}(F)) \infty (\text{arg}(E) \cap \textbf{prcd}(G)). \supset .[\exists HI].F \varepsilon \text{varp}(GBAHI)$

TE44 $[ABCDE] : C \varepsilon \text{1prntmp}(BADE). \equiv .$
$C \varepsilon \text{prntmp}(BADE).$
$\textbf{Uvrb}(D) \varepsilon \textbf{vrb}(E)$

TE45 $[ABCDEFG] : C \varepsilon \text{2prntmp}(BADEFG). \equiv .$

$C \varepsilon$ prntmp$(BADE)$.

$F \varepsilon$ prntm(D).

Uprcd$(F) \varepsilon$ **ingr**(E).

$G \varepsilon$ simprntm(F)

These technical terms will be useful for choosing parentheses in definitions. They will be explained later when used.

2.5. The T.E.'s for Definition, Quantification, Detachment, Substitution and Extensionality (TE46–51)

The previous forty-five Terminological Explanations were all in preparation for this section which is the key to Protothetic.

TE46 $[AB]$:·: $A \varepsilon$ defp(B). $\equiv$::

R1 **1vrb**(Essnt(A)) ε $\mathcal{N}$(var(A))$.A \varepsilon A$.

R2 **1vrb**(2trm(A)) ε $\mathcal{N}$(var(A)).

R3 **1vrb**(2trm(A)) ε $\mathcal{N}$(constp(BA)) : .

R4 $[C]$: $.C \varepsilon$ trm$.C \varepsilon$ **vrb**(1trm(A)). $\supset$: $[\exists G].C \varepsilon$ bd(G).$\vee$.$[\exists D].D \varepsilon$ **ingr**(A).
$C \varepsilon$ var(D). $\vee$ $.C \varepsilon$ constp(BA) : .

R5 $[CD]$: $D \varepsilon$ **ingr**$(A).C \varepsilon$ bd(D). $\supset$ $.[\exists E].E \varepsilon$ var(CD) :

R6 $[C]$: $C \varepsilon$ bd(A). $\supset$ $.[\exists D].D \varepsilon$ **vrb**(1trm(A))$.D \varepsilon$ var(CA) :

R7 $[C]$: $C \varepsilon$ bd(A). $\supset$ $.[\exists D].D \varepsilon$ **vrb**(2trm(A))$.D \varepsilon$ var(CA) : .

R8 $[DEF]$: $.F \varepsilon$ **vrb**(1trm(A))$.D \varepsilon$ **vrb**(1trm(A))$.F \varepsilon$ cnvar(DE). $\supset$:
$D = F$. $\vee$ $.F \varepsilon$ q-homosemp(DBA) : .

R9 $[C]$: $C \varepsilon$ gnrl$.C \varepsilon$ **ingr**(**Int**(A)). $\supset$ $.[\exists DEF].D \varepsilon$ homosemp(BB).
$E \varepsilon$ ingrthp$(B).F \varepsilon$ **ingr**$(A).D \varepsilon$ Anarg(CEF) : .

R10 $[CD]$: $C \varepsilon$ gnrl$.C \varepsilon$ **ingr**$(A).D \varepsilon$ Essnt(C). $\supset$: $D \varepsilon$ **vrb**.$\vee$.$[\exists E].E \varepsilon$ frp(B).
$D \varepsilon$ fnctgen(E) : .

R11 $[C]$: $C \varepsilon$ fnct$.C \varepsilon$ **ingr**(1trm(A)). $\supset$: $[\exists D].D \varepsilon$ gnrl$.C \varepsilon$ Essnt(D). $\vee$.

$[\exists DE].C \; \varepsilon \; \mathsf{fnctp}(BADE)$:

R12 $[CD] : D \; \varepsilon \; \textbf{int}(C).C \; \varepsilon \; \mathsf{prntm}(\mathsf{2trm}(A)). \supset .D \; \varepsilon \; \mathsf{var}(A)$:

R13 $\mathsf{2trm}(A) \; \varepsilon \; \textbf{non-rep}$:

R14 $[CDE] : C \; \varepsilon \; \mathsf{1prntmp}(BADE).\textbf{Uvrb}(\mathsf{2trm}(A)) \; \varepsilon \; \textbf{ingr}(C). \supset .$
$C \; \varepsilon \; \mathsf{simprntm}(E)$: .

R15 $[CDEFG] : C \; \varepsilon \; \mathsf{2prntmp}(BADEFG).G \; \varepsilon \; \textbf{ingr}(A).\textbf{Uprcd}(G) \; \varepsilon$
$\textbf{ingr}(C). \supset .C \; \varepsilon \; \mathsf{simprntm}(E)$: .

R16 $[CE] : C \; \varepsilon \; \mathsf{prntm}(\mathsf{2trm}(A)).\textbf{Uvrb}(\mathsf{2trm}(A)) \; \varepsilon \; \textbf{ingr}(C).E \; \varepsilon \; \mathsf{ingrthp}(B).$
$C \; \varepsilon \; \mathsf{simprntm}(E). \supset .[\exists D].C \; \varepsilon \; \mathsf{1prntmp}(BADE).$

R17 $[CEG] : C \; \varepsilon \; \mathsf{prntm}(\mathsf{2trm}(A)).G \; \varepsilon \; \mathsf{prntm}.G \; \varepsilon \; \textbf{ingr}(A).\textbf{Uprcd}(G) \; \varepsilon \; \textbf{ingr}(C).$
$E \; \varepsilon \; \mathsf{ingrthp}(B).C \; \varepsilon \; \mathsf{simprntm}(E). \supset .[\exists DF].C \; \varepsilon \; \mathsf{2prntmp}(BADEFG)$

A is a *definition* after thesis B.

Several conjuncts of TE46 apply to any prospective thesis of Protothetic. They are TE46.4, 5, 8, 9, 10, 11. Let us ignore them for the moment and concentrate on the conjuncts peculiar to definitions.

First remark that a definition need not be a generalization. But its essence, by TE46.2, must be an equivalence. The second term of A can contain no variables in this case (TE46.3) and is thus a term. In this case then, we are defining a constant without any parameters, or, as one usually says, a constant. The definition thus has the form

$$\phi(\psi \; *)$$

where $*$ is the newly defined constant and ψ is a well formed formula (the other conjuncts guarantee this) which contains no free variables.

As examples of this type of definition consider the following definitions of 0 and 1 (falsehood and truth)

$$\phi(\llcorner u \lrcorner \ulcorner u \urcorner 0)$$
$$\phi(\phi(\llcorner u \lrcorner \ulcorner u \urcorner \llcorner u \lrcorner \ulcorner u \urcorner)1)$$

These are not the only definitions of constants possible, but any constant defined will be equivalent to either 0 or 1.

Now we move to the more interesting case of where A is a generalization. From TE46.2, 5, 13 we can prove $\mathsf{2trm}(A) \; \varepsilon \; \mathsf{fnct}$ and so A has the form

$$\llcorner \ldots\ldots \lrcorner \ulcorner \phi(--\Phi\not\prec \ldots \not\succ \prec \ldots \succ(\ldots)) \urcorner$$

The first term of A, since it is an argument, is either a term (TE1), generalization, or function. All three cases can occur but if the definiens is a term then the definiendum has the shape

$$\mathsf{As}(x)$$

i.e., it has only one parameter. The definition is precisely (up to change of symbolism)

$$\llcorner x \lrcorner \ulcorner \phi(x\mathsf{As}(x))\urcorner,$$

which is usually called Assertium.

TE46.12 tells us that the words in the parenthemes of the function which is the definiendum are variables. Hence the parenthemes of the definiendum ($\mathsf{2trm}(A)$) look like

$$\dashv pq \ldots s \vdash,$$

and the definiendum looks like

$$\Phi \nprec pq \ldots s \nsucc \dashv fg \ldots \vdash \ldots (uv \ldots)$$

where all the symbols are different (TE46.13).

TE46.6, 7 tell us that every binder in A binds at least one term in the definiens and exactly one variable in the definiendum. These conditions are necessary to prevent contradictions (cf., e.g., Carnap, 1934).

Combining all of this information we see that definitions have the form

$$\llcorner pqfguv \ldots \lrcorner \ulcorner \phi(\psi(pqfguv \ldots)\ \Phi \nprec pq \ldots \nsucc \dashv fg \ldots \vdash \ldots (uv \ldots))\urcorner$$

It remains to show how TE46.14–17 govern the choice of parentheses in the definiendum. These conjuncts are very important for they insure that the syntactical categories are uniquely determined on syntactical grounds. In particular, if we know (1) the shape of the parentheses, and (2) the number of arguments of a parentheme, then we know the syntactical category of each argument and also the syntactical category of the functor which precedes this parentheme.

Conditions 14 and 15 force us to use certain parentheses; 16 and 17 prevent us from using others. Conditions 14 and 16 govern the choice of parentheses for the last (i.e., rightmost) parentheme in the definiendum. Conditions 15 and 17 govern the choice of the other parentheses in the definiendum in a manner akin to inductive definitions, but of course without using natural numbers.

The explanations of conditions 14 and 16 which follow refer to the following diagram:

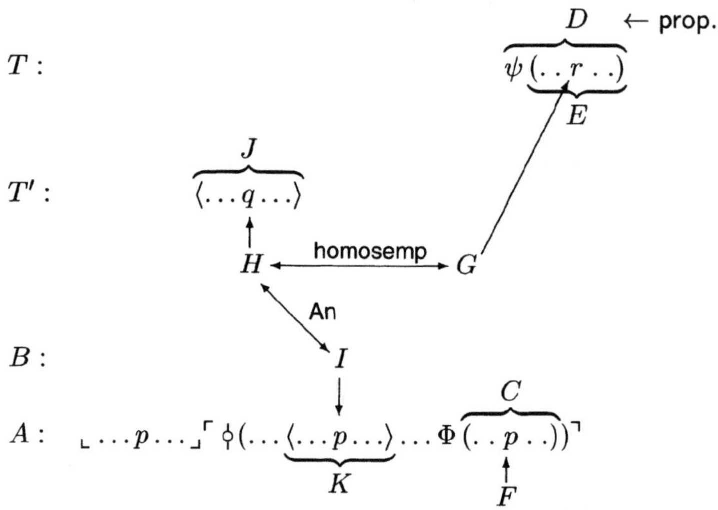

Condition 14 dictates that whenever there is a parentheme E, which is the last parentheme of a function D, which is in the syntactical category of propositions, and when corresponding arguments of E and C, which is the last parentheme of the definiendum (i.e., 2trm) of A, are destined to be in the same category, then one must use the same kind of parentheses for C as E has.

We speak of an argument F of C as being "destined to be" in a certain syntactical category since inscription A, of which C is a part, is not yet a thesis of Protothetic and accordingly its part F cannot be in any syntactical category. What we intend is that if and when A is added to Protothetic as a new thesis according to the rule of definition, then F will be in the same syntactical category as E.

The category to which F is destined to belong is established as follows: Relate F back to an equiform variable in the definiens, say I, which is in parentheme K (if I is a functor the argument is similar). Then I can be related to the variable H in parentheme J which is part of some thesis of Protothetic with respect to B. This is done in such a way that I and H are analogous arguments (or functors) in K and J, respectively. Now, since H and G are both ingredients of theses of Protothetic, it is meaningful to say that they are in the same syntactical category. Since H and I are analogous they should be in the same category, and since I and F are equiform variables they should be in the same category. Finally, by transitivity, F and G are destined to be in the same category.

Condition 16 states that when the last parentheme C of the definiendum is similar to the last parentheme E of a function D which is in the syntactical category of propositions then corresponding arguments of the parenthemes C and E are destined to belong to the same syntactical category. In particular this says that if the n-th argument of E is of a different category than the n-th argument of C, then the parenthemes E and C are not similar, i.e., either C and E have different numbers of arguments or they have non-equiform parentheses.

The following examples show how these two conditions govern the choice of parentheses in definitions.

The axiom contains only parentheses of the shape (– –) and these are preceded by a functor in the category S/SS. Condition 14 dictates that in any definition where we have a sentential function of two sentential variables, then the arguments must be enclosed in parentheses equiform to (– –). For example, in

(1) $\llcorner pq \lrcorner \ulcorner \phi(\phi(qp)\ \phi_1(pq))\urcorner$

the choice of parentheses in the definiendum is forced by condition 14 since the arguments of ϕ_1 are destined to be in the category S, and there are two of them. Any other choice of parentheses would violate 14.

(2) $\llcorner pqr \lrcorner \ulcorner \phi(\phi(p\ \phi(qr))\ *_1 +\!\!\!(p)\!\!\!+(qr))\urcorner$

Condition 14 forces the choice of the last parentheses of (2). But any type of parentheses, except (–), could be used for the second partheme. Condition 14 is satisfied for the following since its hypothesis is false:

(3) $\llcorner p \lrcorner \ulcorner \phi(\phi(pp)\ *_2 +\!\!\!(p)\!\!\!+)\urcorner$

However, this is unacceptable as a definition since the only parentheme in the definiendum is similar to the second parentheme in the definiendum of (2), which is not the last parentheme of (2) as is required by condition 16. The following is eligible as a definition immediately after (2):

(4) $\llcorner p \lrcorner \ulcorner \phi(\phi(pp)\mathsf{Vr}(p))\urcorner$

This satisfies both 14 and 16 vacuously, since there is no one argument parentheme in the system which is last in its definiendum. Actually, any choice of parentheses, except $+\!\!\!(-)\!\!\!+$, could be made here. However, in all further definitions where the last parentheme contains but one argument and that of the syntactical category of propositions, we must use parentheses equiform to those in (4). The freedom to choose the type of parenthesis in (4) enables one to display analogy between defined terms through the notation.

Both conditions 14 and 16 are necessary to keep the syntactical categories separate. If we dropped 14 then we could add

(5) $\llcorner p \lrcorner \ulcorner \phi(\phi(pp) *_3 \langle p \rangle)\urcorner$

and thus get the erroneous impression that Vr and $*_3$ belong to different categories. If we dropped 16 then we could add

(6) $\llcorner fp \lrcorner \ulcorner \phi(f(p) *_4 (fp))\urcorner$

which would give the impression that the functor $*_4$, which should be of the category $S/(S/S)S$, is of category S/SS. Thus if 16 were omitted the categories would coalese while if 14 were dropped the categories would multiply needlessly.

Conditions 15 and 17 govern the choice of parentheses for all but the last parentheme of the definiendum. The explanations which follow refer to the following diagram:

$$T: \quad \overbrace{\psi \underbrace{\Lsh \ldots\ldots \Rsh}_{E} \underbrace{\leftarrow \ldots\ldots \rightarrow}_{F} \ldots}^{D \leftarrow \text{prop.}}$$

$$B:$$

$$A: \llcorner \ldots \lrcorner \ulcorner \phi(-- \Phi \overbrace{\Lsh \ldots\ldots \Rsh}^{C} \overbrace{\leftarrow \ldots\ldots \rightarrow}^{G} \ldots)\urcorner$$

Condition 15 dictates that parenthemes C and E will have the same type of parentheses when they have the same number of arguments, when corresponding arguments are destined to be in the same syntactical category, and when the parenthemes following them (i.e., G and F, respectively) are similar.

Conversely, condition 17 stipulates that when the parenthemes C and E are similar and when C is followed by a parentheme G, then there is a parentheme F following E which is similar to G and, moreover, C and E have the same number of arguments and corresponding arguments are destined to be in the same category.

These conditions are very similar to 14 and 16 respectively and so we go right to examples. The following is a legitimate definition immediately after (1), (2), and (4) above:

(7) $\llcorner pqr \lrcorner \ulcorner \phi(\phi(\phi(pq)r) *_5 \leftarrow p \rightarrow (qr))\urcorner$

To use another shape of parentheses on the second parentheme would violate condition 15. This arrangement of parentheses was used in (2) and must always be used when a functor of type $(S/SS)/S$ is involved. We can also add the definition

$$(8)\ {}_{\llcorner}pqrs_{\lrcorner}\ulcorner \phi(\phi(\phi(pq)\phi(rs)) *_6 \mathrel{+\!\!\!\!-}pq\mathrel{-\!\!\!\!+}(rs))\urcorner$$

The use of the same shaped parentheses for the second parenthemes of (7) and (8) causes no confusion as these parenthemes contain different numbers of arguments. Both conditions 15 and 17 are satisfied as their hypotheses are false.

$$(9)\ {}_{\llcorner}pqr_{\lrcorner}\ulcorner \phi(\phi(p\,\phi(qr)) *_7 \mathrel{+\!\!\!\!-}pq\mathrel{-\!\!\!\!+}(r))\urcorner$$

This violates condition 17 since the second parenthemes of (8) and (9) are similar but the first partheme of (8) is not similar to the first of (9). So a different choice of parentheses must be made for the second parentheme of (9).

Condition 17 prevents us from taking

$$(10)\ {}_{\llcorner}pq_{\lrcorner}\ulcorner \phi({}_{\llcorner}f_{\lrcorner}\ulcorner f(pq)\urcorner *_8 (pq)\langle f\rangle)\urcorner$$

as a definition since the second parentheme of (10) is similar to the first parentheme of (2), whereas condition 17 requires that if the second parentheme of a proposed definition is similar to any other parentheme then that parentheme, among other things, must be followed by another parentheme. In particular 17 prevents us from having both $\mathrel{+\!\!\!\!-}-\mathrel{-\!\!\!\!+}(..)$ and $(..)\mathrel{+\!\!\!\!-}-\mathrel{-\!\!\!\!+}$ in the system.

Finally, let us consider the inscription

$$(11)\ {}_{\llcorner}pqr_{\lrcorner}\ulcorner \phi(\phi\ (p\,\phi(qr)) *_9 (p)(qr))\urcorner$$

This satisfies conditions 14–17, but does not satisfy 13 since the would-be definiendum contains two equiform symbols. Thus it is not eligible as a definition of Protothetic. Sobociński has remarked (orally) that Leśniewski would not have allowed this as a definition either. However, (11) does satisfy Leśniewski's formulation of the rule of definition as formulated in Leśniewski (1929). Thus it seems that Leśniewski's verbal statement of the rule does not agree with his printed one. This is the only point at which our rule differs from that given in Leśniewski (1929), and this point does not affect the deductive strength of the theory, but only the shape of parentheses in some theses.

TE47 $[AB] :\cdot: A\ \varepsilon\ \text{cnqntf}(B).\ \equiv::$

R1 $\text{Essnt}(1\text{trm}(A))\ \varepsilon\ \textbf{cnf}(\text{Essnt}(1\text{trm}(B))).A\ \varepsilon\ A.$

R2 $\text{Essnt}(2\text{trm}(A))\ \varepsilon\ \textbf{cnf}(\text{Essnt}(2\text{trm}(B)))\ :$

R3 $[C] : C\ \varepsilon\ \text{bd}(A).\ \supset .[\exists D].D\ \varepsilon\ \textbf{cnf}(C).D\ \varepsilon\ \text{bd}(B) :.$

R4 $[CDE] : .C \,\varepsilon\, \mathsf{Antrm}(DAB).E \,\varepsilon\, \mathsf{var}(B).E \,\varepsilon\, \mathbf{vrb}(D). \supset:$
$[\exists I] : I \,\varepsilon\, \mathbf{cnf}(E) : I \,\varepsilon\, \mathsf{bd}(A). \vee .I \,\varepsilon\, \mathsf{bd}(C) : .$

R5 $[CDE] : C \,\varepsilon\, \mathsf{Antrm}(DAB).E \,\varepsilon\, \mathsf{bd}(D). \supset .[\exists H].H \,\varepsilon\, \mathbf{cnf}(E).H \,\varepsilon\, \mathsf{bd}(C) : .$

R6 $[CDE] : .C \,\varepsilon\, \mathsf{Antrm}(DAB).E \,\varepsilon\, \mathsf{bd}(C). \supset .[\exists H].H \,\varepsilon\, \mathbf{cnf}(E).$
$H \,\varepsilon\, \mathbf{ingr}(D) : H \,\varepsilon\, \mathsf{var}(B). \vee .H \,\varepsilon\, \mathsf{bd}(D) : .$

R7 $[CDEF] : C \,\varepsilon\, \mathsf{Antrm}(DAB).F \,\varepsilon\, \mathsf{bd}(A).E \,\varepsilon\, \mathsf{bd}(C).F \,\varepsilon\, \mathbf{cnf}(E). \supset .[\exists G].$
$G \,\varepsilon\, \mathbf{cnf}(E).G \,\varepsilon\, \mathsf{bd}(D)$

A is a *consequence of B by quantification.* TE47.1, 2 tell us that both A and B are generalizations (admittedly both A and B could be equivalences but this case is both trivial and uninteresting since we are trying to distribute quantifiers), and, moreover, that they are generalizations of equivalences. Also, corresponding terms have equiform essences. Let us extend the schematic diagram following TE34:

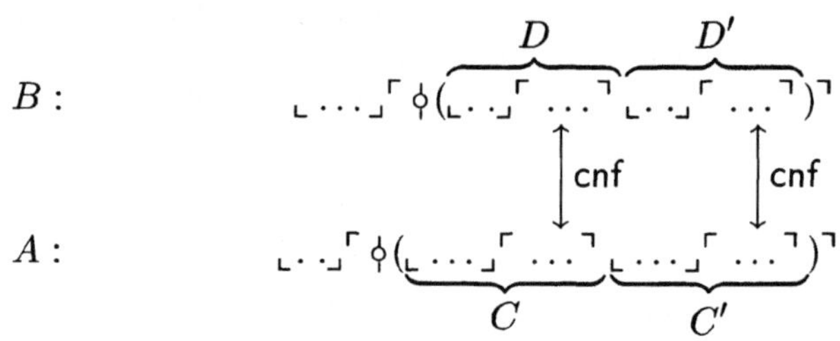

TE47.3 says any binder in A is equiform to a binder in B. Semantically, this means any binder which was not distributed is a binder in both A and B.

TE47.4 says that for every variable in B which occurs in D (D') there is a binder in A or C (C') which is equiform to that variable. This notation is quite cumbersome so we introduce the following.

We say the binders of A are an *equiformity subset* of the binders of B iff for every binder of A there is an equiform binder of B. Then TE47.3 merely says that the binders of A are an equiformity subset of the binders of B.

TE47.5 says that if C (C') is an analogous term to D (D') in A and B, respectively, then the binders of D are an equiformity subset of the binders of C.

TE47.6 says any binder in C has an equiform binder in D or there is some occurrence of that binder in D which is bound in B.

Finally, TE47.7 says that if the quantifiers of A and C contain equiform words, then there is a word in the quantifier of D equiform to it.

Let us give several examples. If B is

$$\llcorner pq \lrcorner \ulcorner \phi(\phi(pq)\,\phi(qp)) \urcorner$$

then A could be B or

$$\llcorner p \lrcorner \ulcorner \phi(\llcorner q \lrcorner \ulcorner \phi(pq) \urcorner \llcorner q \lrcorner \ulcorner \phi(qp) \urcorner) \urcorner$$

or a similar expression with p distributed or we could distribute both variables and get

$$\phi(\llcorner pq \lrcorner \ulcorner \phi(pq) \urcorner \llcorner pq \lrcorner \ulcorner \phi(qp) \urcorner)$$

In this case the order of the binders in either quantifier could be altered. As a more complicated example consider

$$\llcorner pq \lrcorner \ulcorner \phi(\llcorner p \lrcorner \ulcorner \phi(pq) \urcorner \phi(p \llcorner u \lrcorner \ulcorner u \urcorner)) \urcorner$$

Then A could be, if we distribute p,

$$\llcorner q \lrcorner \ulcorner \phi(\llcorner p \lrcorner \ulcorner \phi(pq) \urcorner \llcorner p \lrcorner \ulcorner \phi(p \llcorner u \lrcorner \ulcorner u \urcorner) \urcorner) \urcorner$$

or, if we distribute q,

$$\llcorner p \lrcorner \ulcorner \phi(\llcorner pq \lrcorner \ulcorner \phi(pq) \urcorner \phi(p \llcorner u \lrcorner \ulcorner u \urcorner)) \urcorner$$

TE48 $[ABC] : A\ \varepsilon\ \mathsf{cneqvl}(BC). \equiv .C\ \varepsilon\ \mathbf{cnf}(\mathsf{1eqvl}(B)).$
$A\ \varepsilon\ \mathbf{cnf}(\mathsf{2eqvl}(B))$

A is the result of *detaching* C from B.

Schematically, we have

$$B: \quad \begin{array}{c} \phi(-\ -) \\ \uparrow\ \uparrow \\ C\ \ A \end{array} \quad \longleftarrow \mathsf{cnf}$$

N.B. We are not considering detachment under quantifiers, i.e., we do not want the rule

$$\frac{\begin{array}{l} \llcorner \ldots \lrcorner \ulcorner C \urcorner \\ \llcorner \ldots \lrcorner \ulcorner \phi(CA) \urcorner \end{array}}{\llcorner \ldots \lrcorner \ulcorner A \urcorner}$$

but the rule

$$\frac{\begin{array}{c} C \\ \phi(CA) \end{array}}{A}$$

TE49 $[ABCa] :\cdot: A \,\varepsilon\, \mathsf{cnsbst}(BCa). \equiv::$

R1 $\mathsf{Essnt}(A) \,\varepsilon\, \mathbf{Cmpl}(a). A \,\varepsilon\, A.$

R2 $a \infty \mathbf{int}(\mathsf{Sbqntf}(C)) : .$

R3 $[DE] : .D \,\varepsilon\, \mathbf{int}(\mathsf{Sbqntf}(C)).D \,\varepsilon\, \mathsf{Corr}(ECa). \supset:$
$D \,\varepsilon\, \mathsf{var}(C). \vee .D \,\varepsilon\, \mathbf{cnf}(E) : .$

R4 $[DE] : D \,\varepsilon\, \mathbf{int}(\mathsf{Sbqntf}(C)).D \,\varepsilon\, \mathsf{Corr}(ECa). \supset .$
$E \,\varepsilon\, \mathsf{trm} \cup \mathsf{gnrl} \cup \mathsf{fnct} \cup \mathbf{cnf}(D) : .$

R5 $[DEFG] : D \,\varepsilon\, \mathsf{cnvar}(EC).D \,\varepsilon\, \mathsf{Corr}(FCa).E \,\varepsilon\, \mathsf{Corr}(GCa). \supset .$
$F \,\varepsilon\, \mathbf{cnf}(G) : .$

R6 $[DEFGHIKL] : D \,\varepsilon\, \mathbf{ingr}(\mathsf{Essnt}(C)).E \,\varepsilon\, \mathsf{bd}(D).F \,\varepsilon\, \mathsf{var}(KC).$
$F \,\varepsilon\, \mathbf{ingr}(D).E \,\varepsilon\, \mathsf{Corr}(GCa).F \,\varepsilon\, \mathsf{Corr}(HCa).L \,\varepsilon\, \mathbf{ingr}(A).$
$I \,\varepsilon\, \mathsf{var}(GL). \supset .I \,\varepsilon\, \mathsf{N}(\mathbf{ingr}(H)) : .$

R7 $[DE] : .D \,\varepsilon\, \mathsf{bd}(A).E \,\varepsilon\, \mathbf{cnf}(D).E \,\varepsilon\, \mathbf{ingr}(C). \supset: [\exists F].E \,\varepsilon\, \mathsf{bd}(F). \vee .$
$[\exists FG].F \,\varepsilon\, \mathbf{ingr}(C).E \,\varepsilon\, \mathsf{var}(GF) : .$

R8 $[D] : .D \,\varepsilon\, \mathsf{trm}.D \,\varepsilon\, \mathbf{vrb}(A). \supset: [\exists G].D \,\varepsilon\, \mathsf{bd}(G). \vee .[\exists G].D \,\varepsilon\, \mathsf{var}(G).$
$G \,\varepsilon\, \mathbf{ingr}(A). \vee .D \,\varepsilon\, \mathsf{constp}(BA) : .$

R9 $[DE] : E \,\varepsilon\, \mathbf{ingr}(A).D \,\varepsilon\, \mathsf{bd}(E). \supset .[\exists F].F \,\varepsilon\, \mathsf{var}(DE) : .$

R10 $[DEF] : .E \,\varepsilon\, \mathbf{ingr}(A).F \,\varepsilon\, \mathsf{cnvar}(DE). \supset:$
$F = D. \vee .F \,\varepsilon\, \mathsf{q\text{-}homosemp}(DBA) : .$

R11 $[D] : D \,\varepsilon\, \mathsf{gnrl}.D \,\varepsilon\, \mathbf{ingr}(A).D \neq A. \supset .[\exists EGH].E \,\varepsilon\, \mathsf{homosemp}(BB).$
$G \,\varepsilon\, \mathsf{ingrthp}(B).H \,\varepsilon\, \mathbf{ingr}(A).E \,\varepsilon\, \mathsf{Anarg}(DGH) : .$

R12 $[DE] : .D \,\varepsilon\, \mathsf{gnrl}.D \,\varepsilon\, \mathbf{ingr}(A).E \,\varepsilon\, \mathsf{Essnt}(D). \supset: E \,\varepsilon\, \mathbf{vrb}.$
$[\exists F].F \,\varepsilon\, \mathsf{frp}(B).E \,\varepsilon\, \mathsf{fnctgen}(F) : .$

R13 $[D] : .D \,\varepsilon\, \mathsf{fnct}.D \,\varepsilon\, \mathbf{ingr}(A). \supset: D = A. \vee .[\exists E].E \,\varepsilon\, \mathsf{gnrl}.$
$D \,\varepsilon\, \mathsf{Essnt}(E). \vee .[\exists EF].D \,\varepsilon\, \mathsf{fnctp}(BADE)$

TE50 $[ABC] : A \varepsilon \text{cnsub}(BC). \equiv .A \varepsilon A.[\exists a].A \varepsilon \text{cnsbst}(BCa)$

A is a consequence of substitution in C with respect to thesis B.

Contrary to the commonly held opinion that Hilbert and Bernays were the first (in 1934) to correctly state the rule of substitution for functional variables it should be noted that Leśniewski had already done it in 1929 (Leśniewski, 1929). Leśniewski cleverly solves the problem of substitution by not saying what is substituted for the variables. This paradoxical statement will be explained by means of an example. Suppose we have

(12) $\llcorner pqr \lrcorner \ulcorner \phi(\varphi(rqp)p)\urcorner$

and wish to substitute $\text{Vr}(p)$ for p. Everyone knows that we get

(13) $\llcorner pqr \lrcorner \ulcorner \phi(\varphi(rq\text{Vr}(p))\text{Vr}(p))\urcorner$

We do not care how one divines this but only need to give a method to verify that it is correct. Look at the essence of (13) and break it up into 10 parts (= the number of words in the subquantifier of (12)) in such a way that constants, corners, and parentheses in (12) correspond to equiform words in (13) and also so that variables in (12) correspond to *expressions* in (13). This nearly forces us to view (13) in the following way

(14)

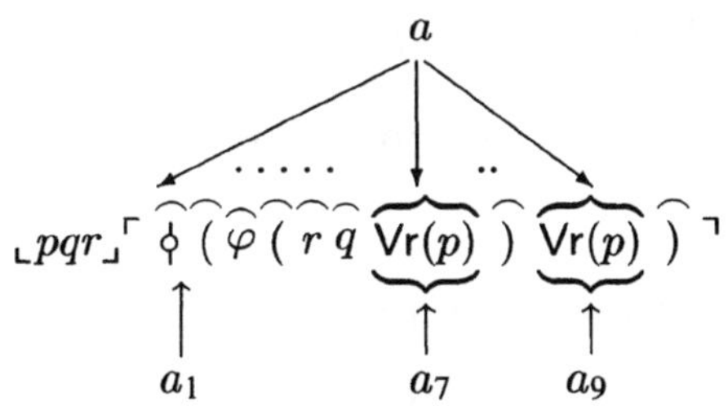

These ten expressions constitute a. The argument $D \varepsilon \text{Corr}(E\ C\ a)$ used in TE49.3, 4, 5, 6 corresponds words D in the subquantifier of C (i.e., (12) in our example) to expressions $E \varepsilon a$. TE49.3 says corresponding things are equiform except (possibly) in the case when D is a variable in C. The only places which are not equiform are those for which we substituted. Naturally enough the a's all must be terms, generalizations or functions, or equiform to that to which they correspond. Semantically speaking this says we can only substitute terms, generalizations or functions (cf. TE17).

TE49.5 guarantees that we substitute equiform expressions for equiform variables in C. This requirement is so natural as to defy comment.

TE49.6 is the famous and troublesome clause which states that interior quantifiers may never trap variables. If we have

$$\llcorner pq \lrcorner \ulcorner \psi(\llcorner p \lrcorner \ulcorner \phi(pq)\urcorner p)\urcorner$$

then we may not substitute p for q to get

$$\llcorner p \lrcorner \ulcorner \psi(\overbrace{\llcorner p \lrcorner \ulcorner \phi(p\ \ p)\urcorner}^{B} p)\urcorner$$

$$\uparrow v_1 \uparrow v_2 \uparrow v_3$$

for then we have both v_1 and v_2 bound in B whereas we wanted v_2 to be bound by the MGQ and hence to be an equiform variable to v_3.

The remaining conjuncts of TE49 concern arbitrary theses and will be discussed at the end of this section.

TE51 $[AB] :: A \varepsilon \text{extp}(B). \equiv: .$

R1 $\textbf{Int}(\text{Qntf}(A)) \varepsilon \mathcal{N}\langle\textbf{vrb}\rangle.A \varepsilon A :$

R2 $[CD] : D \varepsilon \textbf{ingr}(A).D \varepsilon \text{qntf}.C \varepsilon \textbf{int}(D). \supset .[\exists EF].E \varepsilon \text{var}(CF).$
$E \varepsilon \mathcal{N}\langle\textbf{cnf}(\textbf{1vrb}(\text{Essnt}(A)))\rangle :$

R3 $[\exists C].C \varepsilon \text{prntm}(\text{1trm}(\text{2trm}(A))).\textbf{1vrb}(\text{1eqvl}(\textbf{Int}(\text{Sbqntf}(\text{1trm}(A))))) \varepsilon$
$\text{cnvar}(\textbf{Int}(C)A).$

R4 $[\exists C].C \varepsilon \text{prntm}(\text{2trm}(\text{2trm}(A))).\textbf{1vrb}(\text{2trm}(\text{1trm}(A))) \varepsilon \text{cnvar}(\textbf{Int}(C)A) : .$

R5 $[C] : .C \varepsilon \text{fnct}.C \varepsilon \textbf{ingr}(A). \supset: [\exists D].C \varepsilon \textbf{Int}(\text{Sbqntf}(D)). \vee .[\exists DE].$
$C \varepsilon \text{fnctp}(BADE) : .$

R6 $[CDEF] : D \varepsilon \text{prntm}(\text{1trm}(\text{1trm}(A))).E \varepsilon \text{prntm}(\text{2trm}(\text{1trm}(A))).$
$F \varepsilon \text{Anarg}(CDE). \supset .F \varepsilon \text{cnvar}(C\ \text{1trm}(A)) :$

R7 $[CDF] : C \varepsilon \textbf{ingr}(A).F \varepsilon \text{cnvar}(DC). \supset .F \varepsilon \text{q-homosemp}(DBA) :$

R8 $[CD] : D \varepsilon \text{cnvar}(C\ \text{1trm}(A)). \supset .[\exists EF].D \varepsilon \text{Anarg}(CEF) :$

R9 $[CDE] : C \varepsilon \text{prntm}(\text{Essnt}(\text{2trm}(A))).D \varepsilon \text{arg}(C).E \varepsilon \text{Anfnct}(EDD). \supset .$
$E \varepsilon \text{var}(\textbf{Int}(\text{Qntf}(\text{2trm}(A)))\text{2trm}(A))$

A is a *thesis of extensionality* with respect to thesis B.

Let us begin with an example of a thesis of extensionality:

$$\llcorner fg \lrcorner \ulcorner \phi(\llcorner ab \lrcorner \ulcorner \phi(f\{a\}(b)\ g\{a\}(b)) \urcorner \llcorner \varphi \lrcorner \ulcorner \phi(\varphi \ll f \gg\ \varphi \ll g \gg) \urcorner) \urcorner$$

and examine what the rule tells us about it.

TE51.1 says that A is a generalization containing at least two binders. From TE51.3, especially from

$$C\ \varepsilon\ \mathsf{prntm}(\mathsf{1trm}(\mathsf{2trm}(A)))$$

we have that the essence of A is an equivalence and that the 2trm of this equivalence has an essence which is an equivalence. The rest of this conjunct tells us that **Int**(C) is a variable in A and hence a word. Thus, the 2trm(A) has the shape

$$\llcorner - \lrcorner \ulcorner \phi(- \ll f \gg\ - \ll g \gg) \urcorner$$

TE51.9 implies that the three dashes represent the same variable (φ).

TE51.6 and 8 tell us that the terms of the first term of A are equiform except for the first letter of each term.

The remaining conditions, TE51.5 and 7, apply to any thesis.

Finally, we mention several common conjuncts of the *T.E.*'s which are requirements we want any thesis to satisfy.

TE46.4, TE49.8: Terms are binders, variables or constants.

TE46.5, TE49.9: Binders actually bind variables. Hence every quantifier is part of a generalization.

TE46.8, TE49.10, TE51.7: This guarantees that equiform variables of inscriptions considered for addition to Protothetic as new theses will be in the same syntactical category.

TE46.9, TE49.11: Generalizations occupy argument places which are capable of having arguments from the category of propositions.

TE46.10, TE49.12: The essences of generalizations are either words or functions which have been generated by functions occurring in a previous thesis of Protothetic (this prevents substitution from introducing new syntactical categories).

2.6. *The Rule of Protothetic*

Under the assumption that B is the last thesis of this system of Protothetic one may add the expression A as a new thesis of Protothetic provided one of the following conditions is fulfilled:

(1) $A\ \varepsilon\ \mathsf{defp}(B)$.

(2) $[\exists C].C\ \varepsilon\ \mathsf{thp}(B).A\ \varepsilon\ \mathsf{cnqntf}(C)$.

(3) $[\exists CD].C\,\varepsilon\,\mathsf{thp}(B).D\,\varepsilon\,\mathsf{thp}(B).A\,\varepsilon\,\mathsf{cneqvl}(CD)$.

(4) $[\exists C].C\,\varepsilon\,\mathsf{thp}(B).A\,\varepsilon\,\mathsf{cnsub}(BC)$.

(5) $A\,\varepsilon\,\mathsf{extp}(B)$.

3

In this section we sketch the proof that our formulation of the rule of Protothetic has the same deductive strength as that which Leśniewski presented in Leśniewski (1929), pp. 63–76. The reader should have Leśniewski (1929) at hand.

3.1. Definability of Leśniewski's Primitive Terms in MP

The first step toward proving equivalence of our rule and Leśniewski's is to show that his primitive terms are definable in system **MP**. The definitions seem obvious; but there are subtle difficulties involved. To bring these out we describe again how the rule of Protothetic was formulated.

Leśniewski formulated the rule by giving a number of Terminological Explanations which prescribed exactly what he meant by detachment, substitution, etc. He realized that his *T.E.*'s did not belong to the system itself, and so it was his practice – one fortunately thwarted in Leśniewski (1929) by space limitations – to express the *T.E.*'s in ordinary language. Hence it is understandable that the undefined terms which he used to express the *T.E.*'s were explained by only a few words of commentary. It is here that our difficulty lies – for we have no way of being certain that our definitions of his primitive terms agree with the intuitive meanings that he attached to them.

What we do is to give definitions in MP of Leśniewski's undefined terms and then argue intuitively that we have captured the meanings which he intended. We shall use two devices for this:

(a) We refer to theorems in M and MP which agree with Leśniewski's commentary on his undefined terms.

(b) By examining Leśniewski's *T.E.*'s we deduce certain properties of his undefined terms which he must have intended if he was to be consistent with the remark (Leśniewski, 1929, p. 63) that the separate conjuncts of his *T.E.*'s are logically independent.

The following is a list of Leśniewski's undefined terms:

$A\,\varepsilon\,b$	$\mathsf{prntsym}(A)$
$\mathbf{Id}(A)$	$\mathbf{cnf}(A)$
$\mathcal{N}(a)$	$\mathfrak{A}\mathfrak{l}$
$a \cap b \cap \ldots \cap k$	thp

$a \cup b \cup \ldots \cup k$	**ingr**(A)
$a \infty b$	**prcd**(A)
$a < b$	**scd**(A)
vrb	**Uprcd**(A)
expr	**Uingr**(A)
prnt	**lingr**(A)
prntl	**2ingr**(A)

Of these ε, **Id** (i.e., $=$), $\wedge$, $\cap$, $\cup$, ∞, $<$ are terms of Ontology. Their definitions can be found in [Rickey (1972),] Chapter I.3.

Of the remaining terms $\mathfrak{Al}$ holds a special place, for it as an individual name, viz., the name of the axiom of Protothetic for which Leśniewski stated his rule (cf., Leśniewski, 1929, p. 59). This axiom is not the same one that we have used. For this reason it will be necessary to adjust Leśniewski's rule to the axiom $\mathfrak{Ar}$ which we use. These changes are minor and are necessitated by the different number of binders in the quantifiers of $\mathfrak{Al}$ and $\mathfrak{Ar}$.

We now proceed to define the rest of these terms in **MP**.

Df1.1 $[A] : A \,\varepsilon\, \mathbf{vrb_L} . \equiv . A \,\varepsilon\, \mathbf{vrb}$

This definition states that Leśniewski's term $\mathbf{vrb_L}$ corresponds exactly to our term **vrb**. We use the subscript L to indicate that the defined term is one of Leśniewski's undefined terms. In support of this definition we mention T2.3.2, T2.4.17, T2.4.18, and T2.4.26 which are the analogues in M (and hence in MP) of Leśniewski's commentary in Leśniewski (1929) on his term $\mathbf{vrb_L}$.

Df1.2 $[A] : A \,\varepsilon\, \mathbf{expr_L} . \equiv . A \,\varepsilon\, \mathbf{expr}$

The justification of defining Leśniewski's notion of expression as equivalent to ours is found in T2.4.2, T2.4.3, T2.4.4.

Df1.3 $[A] : A \,\varepsilon\, \mathrm{prnt_L} . \equiv . A \,\varepsilon\, \mathrm{prnt}$

Df1.4 $[A] : A \,\varepsilon\, \mathrm{prntl_L} . \equiv . A \,\varepsilon\, A . [\exists B] . B \,\varepsilon\, \mathrm{prntm} . A \,\varepsilon\, \mathbf{1vrb}(B)$

The first of these definitions identifies Leśniewski's notion of parenthesis with ours. The second defines left parenthesis, a term which we eliminated from our system, but can easily define as the first word of a parentheme. Actually, the explicit use of left parentheses has been eliminated in our presentation.

Df1.5 $[AB] : A \,\varepsilon\, \mathrm{prntsym_L}(B) . \equiv . A \,\varepsilon\, \mathrm{prntsym}(B)$

Hence our notion of symmetric parentheses agrees with Leśniewski's.

Up to now there has been little doubt about Leśniewski's intentions; his terms clearly agree with ours. For some of the remaining terms this is not the case and so care must be taken to observe the subscripted L's.

Df1.6 $[AB] : A \varepsilon \textbf{cnf}_{\textbf{L}}(B). \equiv .A \varepsilon \textbf{cnf}(B).$

$$A \varepsilon \textbf{expr}.$$

$$B \varepsilon \textbf{expr}$$

Leśniewski's notion of equiformity holds only between expressions, whereas ours is applicable to any inscriptions. He uses $\textbf{cnf}_{\textbf{L}}(A)$ as an abbreviation of "mit A gleichgestalteter *Ausdruck*" and later speaks of "Zwei miteinander gleichgestaltete *Ausdrücke*" (cf. Leśniewski, 1929, pp. 60, 62). Together these show that he intends

$$[AB] : A \varepsilon \textbf{cnf}_{\textbf{L}}(B). \supset .A \varepsilon \textbf{expr}_{\textbf{L}}$$

$$[AB] : A \varepsilon \textbf{cnf}_{\textbf{L}}(B). \supset .B \varepsilon \textbf{expr}_{\textbf{L}}$$

Together with Df1.2 these justify the above definition. For further evidence of agreement, see T2.4.14, T2.4.15.

Df1.7 $[A] : A \varepsilon \textsf{thp}_{\textsf{L}}. \equiv .A \varepsilon \textsf{thp}$

It should be remarked that whereas Leśniewski took $\textsf{thp}_{\textsf{L}}$ as primitive we have taken $\textsf{thp}(B)$. The reasons for this choice were explained in Section 1.4.

Df1.8 $[AB] : A \varepsilon \textbf{ingr}_{\textbf{L}}(B). \equiv: A \varepsilon A : [C] : C \varepsilon \textbf{vrb}(A). \supset .C \varepsilon \textbf{vrb}(B)$

Remembering that our notion of ingredient can hold only between expressions (cf., T2.4.7, T2.4.8) it is understandable that we can only prove

T5.1.1 $[AB] : A \varepsilon \textbf{ingr}_{\textbf{L}}(B).A \varepsilon \textbf{expr}.B \varepsilon \textbf{expr}. \equiv .A \varepsilon \textbf{ingr}(B)$

[D10, Df1.8]

Although Leśniewski did not explicitly state it we know he did not intend

$$[AB] : A \varepsilon \textbf{ingr}_{\textbf{L}}(B). \supset .B \varepsilon \textbf{expr}$$

for TEVI of Leśniewski (1929) contains the logically independent factors $B \varepsilon \textbf{expr}_{\textbf{L}}$ and $A \varepsilon \textbf{ingr}_{\textbf{L}}(B)$. That he does not intend

$$[AB] : A \varepsilon \textbf{ingr}_{\textbf{L}}(B). \supset .A \varepsilon \textbf{expr}$$

follows from TEXXII which contains the conjuncts $A \varepsilon \textbf{expr}_{\textbf{L}}$ and $A \varepsilon \textbf{ingr}_{\textbf{L}}(B)$. That he does not intend this is also indicated by the complexity of TEVII.4. Cf., also T2.4.6.

It is understandable that Leśniewski did not restrict $\mathbf{ingr_L}$ to expressions, for without this restriction we have just the usual notion of ingredient in a completely atomic Mereology. We used the more restrictive notion of ingredient to simplify our presentation of the *T.E.*'s.

Df1.9 $[AB] : A\ \varepsilon\ \mathbf{prcd_L}(B). \equiv .A\ \varepsilon\ \mathbf{prcd}(B)$

Leśniewski gives no commentary on his use of $\mathbf{prcd_L}$. It is clear that strict precedence is intended, i.e.,

$$[AB] : A\ \varepsilon\ \mathbf{prcd_L}(B). \supset .A \neq B$$

for TEXXXII contains the conjunct

$$A\ \varepsilon\ \mathbf{prcd_L}(B). \vee .A = B$$

which would be redundant with weak precedence. At all other times $\mathbf{prcd_L}$ is used in contexts like

$$A\ \varepsilon\ \mathbf{vrb_L} \cap \mathbf{prcd_L}(B)$$

or

$$A\ \varepsilon\ \mathbf{expr_L} \cap \mathbf{prcd_L}(B)$$

(cf., TEIX.3, XVI.3, XXV.4, XLVII.3, 4, 5, 6, IL.1) and so we cannot decide if he intends

$$[AB] : A\ \varepsilon\ \mathbf{prcd_L}. \supset .A\ \varepsilon\ \mathbf{expr}.B\ \varepsilon\ \mathbf{expr}$$

The interpretation of the rule does not depend on whether we consider precedence as restricted to expressions, or holding for arbitrary inscriptions. We have opted for the latter.

After similar argumentation we define

Df1.10 $[AB] : A\ \varepsilon\ \mathbf{scd_L}(B). \equiv .A\ \varepsilon\ \mathbf{scd}(B)$

We make no comments about the remaining definitions as there is no reason to question their correctness.

Df1.11 $[AB] : A\ \varepsilon\ \mathbf{Uprcd_L}(B). \equiv .A\ \varepsilon\ \mathbf{Uprcd}(B)$

Df1.12 $[AB] : A\ \varepsilon\ \mathbf{Uingr_L}(B). \equiv .A\ \varepsilon\ \mathbf{Uvrb}(B)$

Df1.13 $[AB] : A\ \varepsilon\ \mathbf{1ingr_L}(B). \equiv .A\ \varepsilon\ \mathbf{1vrb}(B)$

Df1.14 $[AB] : A\ \varepsilon\ \mathbf{2ingr_L}(B). \equiv .A\ \varepsilon\ \mathbf{2vrb}(B)$

We now make the claim that Leśniewski's and our primitive terms are interdefinable in system MP, and hence our statement of the rule and his rest on the same logical basis. This claim is partly justified by the argumentation above, but mainly by intimate knowledge of Leśniewski (1929). The methods used above have followed, in spirit, Goodman's comments on constructional definitions in Goodman (1951).

3.2. Methodological Scheme

Each of Leśniewski's Terminological Explanations is in the form of an ontological definition:

$$(1)\ [A\ldots] : A\,\varepsilon\,\Phi(\ldots). \equiv .\psi(A\ldots)$$

where Φ is the defined term. $\psi(A\ldots)$ is a well formed formula built up from Leśniewski's primitive and previously defined terms. As such, this is not a formula in MP. But since in Section 3.1 we defined Leśniewski's primitives in system MP, there is a natural translation of (1) into a formula (2) in MP:

$$(2)\ [A\ldots] : A\,\varepsilon\,\Phi_{\mathsf{L}}(\ldots). \equiv .\psi_{\mathsf{L}}(A\ldots)$$

where the subscripted L's indicate that we are using the analogues of Leśniewski's terms in MP. Thus whenever Leśniewski writes

$A\,\varepsilon\,\mathbf{ingr}(B)$

in Leśniewski (1929) we shall write

$A\,\varepsilon\,\mathbf{ingr}_{\mathbf{L}}(B)$

as its natural translation into MP.

In Section 2 we defined what we wish to prove are terms equivalent to Leśniewski's. To do this we prove, in each of the following paragraphs, theorems of the form

$$(3)\ [A\ldots] : A\,\varepsilon\,\Phi_{\mathsf{L}}(\ldots). \equiv .A\,\varepsilon\,\Phi_{\mathsf{R}}(\ldots)$$

where Φ_{R} is a term defined in one of the *T.E.*'s in Section 2. These theorems establish that our terminglogical explanations are equivalent to those given in Leśniewski (1929). Only when our terms are proved equivalent to Leśniewski's will it be clear that our statement of the rule of procedure for Protothetic is equivalent to that stated in Leśniewski (1929).

*3.3. Concerning TEI (*vrb1*)*

Df3 $[A] : A\,\varepsilon\,\mathsf{vrb}_{\mathsf{L}}1. \equiv .A\,\varepsilon\,\mathbf{cnf}_{\mathbf{L}}(\mathbf{lingr}_{\mathbf{L}}(\mathfrak{A}\mathfrak{x}))$

A is a *word of the first kind* (in the sense of Leśniewski). This definition is the exact analogue of Leśniewski's TEI. Since we have not defined "vrb1" in Section 2 we cannot prove a theorem of the desired form (3) without first defining "vrb1."

Df3.1 $[A] : A \, \varepsilon \, \mathsf{vrb1}. \equiv .A \, \varepsilon \, \mathbf{cnf}(\mathbf{1vrb}(\mathfrak{A}\mathfrak{r}))$

Since our notion of equiformity differs from Leśniewski's we need several lemmas.

T5.3.1 $[AB] : A \, \varepsilon \, \mathbf{1ingr_L}(B). \supset .A \, \varepsilon \, \mathbf{vrb}$ [Df1.13, D3, T2.2.9]

T5.3.2 $[A] : A \, \varepsilon \, \mathbf{vrb}. \supset .\mathbf{cnf}(A) \circ \mathbf{cnf_L}(A)$ [T2.4.2, T2.4.24, Df1.6]

T5.3.3 $[A] : A \, \varepsilon \, \mathsf{vrb_L1}. \equiv .A \, \varepsilon \, \mathsf{vrb1}$ [Df3, Df3.1, Df1.13, T5.3.1, T5.3.2]

3.4. *Concerning TEII, III, IV* (vrb2, vrb3, vrb4)

Df4.1 $[A] : A \, \varepsilon \, \mathsf{vrb_L2}. \equiv .A \, \varepsilon \, \mathbf{cnf_L}(\mathbf{4ingr_L}(\mathfrak{A}\mathfrak{r}))$

Df4.2 $[A] : A \, \varepsilon \, \mathsf{vrb_L3}. \equiv .A \, \varepsilon \, \mathbf{cnf_L}(\mathbf{5ingr_L}(\mathfrak{A}\mathfrak{r}))$

Df4.3 $[A] : A \, \varepsilon \, \mathsf{vrb_L4}. \equiv .A \, \varepsilon \, \mathbf{cnf_L}(\mathbf{Uingr_L}(\mathfrak{A}\mathfrak{r}))$

These are Leśniewski's TEII, III, IV, adjusted to Sobociński's Axiom. In complete analogy with Section 3.3 we need the definitions

Df4.4 $[A] : A \, \varepsilon \, \mathsf{vrb2}. \equiv .A \, \varepsilon \, \mathbf{cnf}(\mathbf{4vrb}(\mathfrak{A}\mathfrak{r}))$

Df4.5 $[A] : A \, \varepsilon \, \mathsf{vrb3}. \equiv .A \, \varepsilon \, \mathbf{cnf}(\mathbf{5vrb}(\mathfrak{A}\mathfrak{r}))$

Df4.6 $[A] : A \, \varepsilon \, \mathsf{vrb4}. \equiv .A \, \varepsilon \, \mathbf{cnf}(\mathbf{Uvrb}(\mathfrak{A}\mathfrak{r}))$

to prove the theorems

T5.4.1 $[A] : A \, \varepsilon \, \mathsf{vrb_L2}. \equiv .A \, \varepsilon \, \mathsf{vrb2}$

T5.4.2 $[A] : A \, \varepsilon \, \mathsf{vrb_L3}. \equiv .A \, \varepsilon \, \mathsf{vrb3}$

T5.4.3 $[A] : A \, \varepsilon \, \mathsf{vrb_L4}. \equiv .A \, \varepsilon \, \mathsf{vrb4}$

3.5. *Concerning TEV* (trm)

Df5 $[A] : A\,\varepsilon\, \mathsf{trm_L} . \equiv .$

$A\,\varepsilon\, \mathbf{vrb_L}.$

$A\,\varepsilon\, \mathcal{N}(\mathsf{prnt_L}).$

$A\,\varepsilon\, \mathcal{N}(\mathsf{vrb_L 1}).$

$A\,\varepsilon\, \mathcal{N}(\mathsf{vrb_L 2}).$

$A\,\varepsilon\, \mathcal{N}(\mathsf{vrb_L 3}).$

$A\,\varepsilon\, \mathcal{N}(\mathsf{vrb_L 4})$

T5.5.1 $[A] : A\,\varepsilon\, \mathsf{trm_L} . \equiv . A\,\varepsilon\, \mathsf{trm}$

[TE1, Df5, Df1.1, Df1.3, T5.3.3, T5.4.1, T5.4.2, T5.4.3]

Df5 is the exact analogue of Leśniewski's TEV. The theorem, which is of the desired form (3), proves that our TE1 has the same strength as Leśniewski's TEV.

3.6. *Concerning TEVI* (int)

Df6 $[AB] : A\,\varepsilon\, \mathbf{int_L}(B) . \equiv .$

$B\,\varepsilon\, \mathbf{expr_L}.$

$A\,\varepsilon\, \mathbf{vrb_L}.$

$A\,\varepsilon\, \mathbf{ingr_L}(B).$

$A\,\varepsilon\, \mathcal{N}(\mathbf{1ingr_L}(B)).$

$A\,\varepsilon\, \mathcal{N}(\mathbf{Uingr_L}(B))$

T5.6.1 $[AB] : A\,\varepsilon\, \mathbf{vrb}(B) . \supset . A\,\varepsilon\, \mathbf{vrb_L} . A\,\varepsilon\, \mathbf{ingr_L}(B)$

Hyp(1) . $\supset$.

2. $A\,\varepsilon\, \mathbf{vrb_L}.$ [1, T2.2.9, Df1.1]

3. $A\,\varepsilon\, \mathbf{ingr_L}(B).$ [1, T2.2.7, Df1.8]

$A\,\varepsilon\, \mathbf{vrb_L} . A\,\varepsilon\, \mathbf{ingr_L}(B)$ [2, 3]

T5.6.2 $[AB] : A\,\varepsilon\, \mathbf{vrb_L} . A\,\varepsilon\, \mathbf{ingr_L}(B) . \supset . A\,\varepsilon\, \mathbf{vrb}(B)$

Hyp(2) . $\supset$.

3. $A \varepsilon$ **vrb**(A). [1, Df1.1, D2]

$A \varepsilon$ **vrb**(B) [2, Df1.8, 3]

T5.6.3 $[AB] : A \varepsilon \textbf{vrb}(B). \equiv .A \varepsilon \textbf{vrb}_\textbf{L}.A \varepsilon \textbf{ingr}_\textbf{L}(B)$ [T5.6.1, T5.6.2]

T5.6.4 $[AB] : A \varepsilon \textbf{int}_\textbf{L}(B). \equiv .A \varepsilon \textbf{int}(B)$

[Df6, D13, Df1.2, Df1.13, Df1.14, T5.6.3]

3.7. *Concerning TEVII* (**Cmpl**)

Df7 $[Aa] :: A \varepsilon \textbf{Cmpl}_\textbf{L}(A). \equiv: .A \varepsilon \textbf{expr}_\textbf{L} : .$
$[B] : B \varepsilon \textbf{vrb}_\textbf{L}.B \varepsilon \textbf{ingr}_\textbf{L}(A). \supset .[\exists C].C \varepsilon a.B \varepsilon \textbf{ingr}_\textbf{L}(C) : .$
$[BCD] : B \varepsilon a.C \varepsilon a.D \varepsilon \textbf{vrb}_\textbf{L}.D \varepsilon \textbf{ingr}_\textbf{L}(B).D \varepsilon \textbf{ingr}_\textbf{L}(C). \supset .$
$B = C :$
$[B] : B \varepsilon a. \supset .B \varepsilon \textbf{expr}_\textbf{L} \cap \textbf{ingr}_\textbf{L}(A)$

T5.7.1 $[a] : .\textbf{disj}(a). \equiv: [BCD] : B \varepsilon a.C \varepsilon a.D \varepsilon \textbf{vrb}_\textbf{L}.D \varepsilon \textbf{ingr}_\textbf{L}(B).$
$D \varepsilon \textbf{ingr}_\textbf{L}(C). \supset .B = C$ [D11, T5.6.3]

T5.7.2 $[ABa] : A \varepsilon \textbf{Kl}(a).B \varepsilon a. \supset .B \varepsilon \textbf{ingr}_\textbf{L}(A)$ [D1, Df1.8]

T5.7.3 $[Aa] : .A \varepsilon \textbf{Kl}(a).a \subset \textbf{expr}. \supset: [B] : B \varepsilon a. \supset .B \varepsilon \textbf{expr}_\textbf{L} \cap \textbf{ingr}_\textbf{L}(A)$

[T5.7.2, Df1.2]

T5.7.4 $[ABa] : A \varepsilon \textbf{Kl}(a).B \varepsilon \textbf{vrb}_\textbf{L}.B \varepsilon \textbf{ingr}_\textbf{L}(A). \supset .$
$[\exists C].C \varepsilon a.B \varepsilon \textbf{ingr}_\textbf{L}(C)$

Hyp(3) . $\supset$.

$[\exists C]$.

4. $C \varepsilon a.$ [1, D1, T5.6.3, 2, 3]

5. $B \varepsilon$ **vrb**(C). [1, D1, T5.6.3, 2, 3]

$[\exists C].C \,\varepsilon\, a.B \,\varepsilon\, \mathbf{ingr_L}(C)$ [4, 5, T5.6.3]

T5.7.5 $[Aa] : A \,\varepsilon\, \mathbf{Cmpl}(a). \supset .A \,\varepsilon\, \mathbf{Cmpl_L}(a)$
[D12, Df7, Df1.2, T5.7.4, T5.7.1, T5.7.3]

T5.7.6 $[Aa] : .A \,\varepsilon\, A : [B] : B \,\varepsilon\, \mathbf{vrb_L}.B \,\varepsilon\, \mathbf{ingr_L}(A). \supset .[\exists C].C \,\varepsilon\, a.$
$B \,\varepsilon\, \mathbf{ingr_L}(C) : [B] : B \,\varepsilon\, a. \supset .B \,\varepsilon\, \mathbf{expr_L} \cap \mathbf{ingr_L}(A) :\supset .A \,\varepsilon\, \mathbf{Kl}(a)$

Hyp(3) . ⊃:

4. $[B] : B \,\varepsilon\, \mathbf{vrb}(A). \supset .[\exists C].C \,\varepsilon\, a.B \,\varepsilon\, \mathbf{vrb}(C) :$ [2, T5.6.3]

5. $[BC] : C \,\varepsilon\, a.B \,\varepsilon\, \mathbf{vrb}(C). \supset .B \,\varepsilon\, \mathbf{vrb}(A) :$ [3, Df1.8]

$A \,\varepsilon\, \mathbf{Kl}(a)$ [D1, 1, 4, 5]

T5.7.7 $[Aa] : A \,\varepsilon\, \mathbf{Cmpl_L}(a). \equiv .A \,\varepsilon\, \mathbf{Cmpl}(a)$
[D12, Df7, T5.7.6, Df1.2, T5.7.1, T5.7.5]

3.8. Concerning TEVIII (qntf)

Df8 $[A] : .A \,\varepsilon\, \mathsf{qntf_L}. \equiv:$
$\mathbf{1ingr_L}(A) \,\varepsilon\, \mathsf{vrb_L1}.$
$\mathbf{Uingr_L}(A) \,\varepsilon\, \mathsf{vrb_L2}.$
$[\exists B].B \,\varepsilon\, \mathbf{int_L}(A) :$
$[B] : B \,\varepsilon\, \mathbf{int_L}(A). \supset .B \,\varepsilon\, \mathsf{trm_L} :$
$[BC] : B \,\varepsilon\, \mathbf{int_L}(A).C \,\varepsilon\, \mathbf{int_L}(A).B \,\varepsilon\, \mathbf{cnf_L}(C). \supset .B = C$

T5.8.1 $[A] : \mathbf{1ingr_L}(A) \,\varepsilon\, \mathsf{vrb_L1}. \equiv .\mathbf{1vrb}(A) \,\varepsilon\, \mathbf{cnf}(\mathbf{1vrb}(\mathfrak{A}\mathfrak{x}))$
[Df1.13, T5.3.3, Df3.1]

T5.8.2 $[A] : \mathbf{Uingr_L}(A) \,\varepsilon\, \mathsf{vrb_L2}. \equiv .\mathbf{Uvrb}(A) \,\varepsilon\, \mathbf{cnf}(\mathbf{4vrb}(\mathfrak{A}\mathfrak{x}))$ [*Sim.*, T5.8.1]

T5.8.3 $[A] : \mathbf{Int}(A) \,\varepsilon\, \mathbf{Cmpl}(\mathsf{trm} \cap \mathbf{int}(A)). \supset .[\exists B].B \,\varepsilon\, \mathbf{int_L}(A)$
[D14, T2.3.4, T5.6.4]

T5.8.4 $[AB] : \mathbf{Int}(A) \,\varepsilon\, \mathbf{Cmpl}(\mathsf{trm} \cap \mathbf{int}(A)).B \,\varepsilon\, \mathbf{int_L}(A). \supset .B \,\varepsilon\, \mathsf{trm_L}$

Hyp(2) . ⊃ .

3. $B \varepsilon$ **int**(A). [2, T5.6.4]

4. $B \varepsilon$ **vrb**. [3, D13, T2.2.9]

5. $B \varepsilon$ **vrb**(**Int**(A)). [T2.3.9, 3, 4, D14]

$[\exists C]$.

6. $C \varepsilon$ trm. [1, 5, D12, D1]

7. $B \varepsilon$ **vrb**(C). [1, 5, D12, D1]

8. $C \varepsilon$ **vrb**. [6, TE1]

9. $B = C$. [7, 8, T2.2.10]

$B \varepsilon$ trm$_{\mathsf{L}}$ [6, 9, T5.5.1]

T5.8.5 $[ABC]$: $A \varepsilon$ **non-rep**.$B \varepsilon$ **int**$_{\mathbf{L}}(A)$.$C \varepsilon$ **int**$_{\mathbf{L}}(A)$.$B \varepsilon$ **cnf**$_{\mathbf{L}}(C)$. ⊃ . $B = C$

Hyp(4) . ⊃ .

5. $B \varepsilon$ **vrb**(A). [2, T5.6.4, D13]

6. $C \varepsilon$ **vrb**(A). [3, T5.6.4, D13]

7. $B \varepsilon$ **cnf**(C). [4, Df1.6]

$B = C$ [D19, 1, 5, 6, 7]

T5.8.6 $[A]$: $A \varepsilon$ qntf. ⊃ .$A \varepsilon$ qntf$_{\mathsf{L}}$ [TE2, Df8, T5.8.1–5]

T5.8.7 $[AB]$: .$B \varepsilon$ **int**$_{\mathbf{L}}(A)$: $[B]$: $B \varepsilon$ **int**$_{\mathbf{L}}(A)$. ⊃ .$B \varepsilon$ trm$_{\mathsf{L}}$:⊃ . **Int**$(A) \varepsilon$ **Cmpl**(trm ∩ **int**(A))

Hyp(2) . ⊃ .

3. $B \varepsilon$ **int**(A). [1, T5.6.4]

4. $\textsf{trm} \cap \textbf{int}(A) \circ \textbf{int}(A)$. [2, T5.5.1, T5.6.4]

$[\exists C]$.

5. $C \,\varepsilon\, \textbf{Kl}(\textbf{int}(A))$. [3, T2.1.1]

6. $C = \textbf{Int}(A)$. [5, T2.3.1, D14]

7. $C \,\varepsilon\, \textbf{Cmpl}(\textbf{int}(A))$. [6, T2.5.25]

8. $C \,\varepsilon\, \textbf{Cmpl}(\textsf{trm} \cap \textbf{int}(A))$. [4, 7]

$\textbf{Int}(A) \,\varepsilon\, \textbf{Cmpl}(\textsf{trm} \cap \textbf{int}(A))$ [6, 8]

In Leśniewski's presentation TEVIII (qntf) contains the clause

$$[BC] : B \,\varepsilon\, \textbf{int}_{\textbf{L}}(A).C \,\varepsilon\, \textbf{int}_{\textbf{L}}(A).B \,\varepsilon\, \textbf{cnf}_{\textbf{L}}(C). \supset .B = C$$

which is equivalent to

$$\textbf{Int}(A) \,\varepsilon\, \textbf{non-rep}$$

In TE2 we used

$$A \,\varepsilon\, \textbf{non-rep}$$

This is possible since a generalization begins and ends with (left and right respectively) corners, which are not terms and hence not equiform to any word in $\textbf{Int}(A)$.

From D13 we obtain at once

$$[AB] : .B \,\varepsilon\, \textbf{vrb}(A).A \,\varepsilon\, \textbf{expr}. \supset: B \,\varepsilon\, \textbf{int}(A). \vee .$$
$$B \,\varepsilon\, \textbf{1vrb}(A). \vee .B \,\varepsilon\, \textbf{Uvrb}(A)$$

Hence when we take two words in (but not necessarily interior to) A we obtain the following nine possibilities:

I. $B \,\varepsilon\, \textbf{int}(A).C \,\varepsilon\, \textbf{int}(A)$
II. $B \,\varepsilon\, \textbf{int}(A).C \,\varepsilon\, \textbf{1vrb}(A)$
III. $B \,\varepsilon\, \textbf{int}(A).C \,\varepsilon\, \textbf{Uvrb}(A)$
IV. $B \,\varepsilon\, \textbf{1vrb}(A).C \,\varepsilon\, \textbf{int}(A)$
V. $B \,\varepsilon\, \textbf{1vrb}(A).C \,\varepsilon\, \textbf{1vrb}(A)$
VI. $B \,\varepsilon\, \textbf{1vrb}(A).C \,\varepsilon\, \textbf{Uvrb}(A)$
VII. $B \,\varepsilon\, \textbf{Uvrb}(A).C \,\varepsilon\, \textbf{int}(A)$
VIII. $B \,\varepsilon\, \textbf{Uvrb}(A).C \,\varepsilon\, \textbf{1vrb}(A)$
IX. $B \,\varepsilon\, \textbf{Uvrb}(A).C \,\varepsilon\, \textbf{Uvrb}(A)$

We wish to establish theorems of the form

$[ABC] : A \,\varepsilon\, \mathsf{qntf_L}.B \,\varepsilon\, \mathbf{cnf}(C).Y. \supset .B = C$

where Y is one of I–IX. Taken together these yield

$[A] : A \,\varepsilon\, \mathsf{qntf_L}. \supset .A \,\varepsilon\, \mathbf{non\text{-}rcp}$

Of the nine cases IV, VII, VIII can be omitted by symmetry with II, III, VI, respectively. Case VI is impossible since ⌞ and ⌟ are not equiform. Cases V, IX are just T2.2.30, T2.2.31. So there remain three cases to consider.

T5.8.8 $[ABC] : .[BC] : B \,\varepsilon\, \mathbf{int_L}(A).C \,\varepsilon\, \mathbf{int_L}(A).B \,\varepsilon\, \mathbf{cnf_L}(C). \supset .$
$B = C : B \,\varepsilon\, \mathbf{vrb}(A).C \,\varepsilon\, \mathbf{vrb}(A).B \,\varepsilon\, \mathbf{cnf}(C).$
$B \,\varepsilon\, \mathbf{int}(A).C \,\varepsilon\, \mathbf{int}(A) :\supset .B = C$

Hyp(6) . ⊃ .

7. $B \,\varepsilon\, \mathbf{expr}.$ [2, T2.2.9, T2.4.2]

8. $C \,\varepsilon\, \mathbf{expr}.$ [3, T2.2.9, T2.4.2]

9. $B \,\varepsilon\, \mathbf{cnf_L}(C).$ [4, 7, 8, Df1.6]

10. $B \,\varepsilon\, \mathbf{int_L}(A).$ [5, T5.6.4]

11. $C \,\varepsilon\, \mathbf{int_L}(A).$ [6, T5.6.4]

$B = C$ [1, 10, 11, 9]

T5.8.9 $[ABC] : .\mathbf{1ingr_L}(A) \,\varepsilon\, \mathsf{vrb_L1} : [B] : B \,\varepsilon\, \mathbf{int_L}(A). \supset .$
$B \,\varepsilon\, \mathsf{trm_L} : B \,\varepsilon\, \mathbf{cnf}(C).B \,\varepsilon\, \mathbf{int}(A).C \,\varepsilon\, \mathbf{1vrb}(A) :\supset .B = C$

Hyp(5) . ⊃ .

6. $B \,\varepsilon\, \mathbf{int_L}(A).$ [4, T5.6.4]

7. $B \,\varepsilon\, \mathsf{trm_L}.$ [2, 6]

8. $B \,\varepsilon\, \mathcal{N}(\mathbf{vrb1}).$ [7, Df5]

9. $C \,\varepsilon\, \mathsf{vrb_L1}.$ [5, T2.2.30, Df1.13, 1]

10. $C \,\varepsilon\, \mathbf{cnf}(\mathbf{1vrb}(\mathfrak{A}\mathfrak{x})).$ [9, T5.3.3, Df3.1]

11. $B \varepsilon$ **cnf**(**1vrb**($\mathfrak{A}\mathfrak{x}$)). [3, 10, T2.2.1]

$B = C$ [11, 8, 15.3.3, Df3.1]

T5.8.10 $[ABC]$: .**Uingr**$_{\mathbf{L}}(A)$ ε vrb$_{\mathrm{L}}$2 : $[B]$: $B \varepsilon$ **int**$_{\mathbf{L}}(A)$. $\supset$.$B \varepsilon$ trm$_{\mathrm{L}}$:
$B \varepsilon$ **cnf**(C).$B \varepsilon$ **int**(A).$C \varepsilon$ **Uvrb**(A) :$\supset$.$B = C$ [*Sim.*, T5.8.9]

T5.8.11 $[ABC]$: $B \varepsilon$ **1vrb**(A).$C \varepsilon$ **Uvrb**(A).$B \varepsilon$ **cnf**(C).**1ingr**$_{\mathbf{L}}(A)$ ε vrb$_{\mathrm{L}}$1.
Uingr$_{\mathbf{L}}(A)$ ε vrb$_{\mathrm{L}}$2. $\supset$.$B = C$

Hyp(5) . $\supset$.

6. $B \varepsilon$ vrb$_{\mathrm{L}}$1. [1, T2.2.30, Df1.12,4]

7. $C \varepsilon$ vrb$_{\mathrm{L}}$2. [2, T2.2.31, Df1.13, 5]

8. **1vrb**($\mathfrak{A}\mathfrak{x}$) ε **cnf**(**4vrb**($\mathfrak{A}\mathfrak{x}$)).
[T5.3.3, T5.4.3, 6, 7, Df3.1, Df4.6, T2.2.20, T2.2.21]

$B = C$ [8, Q4]

Finally, we are able to state the desired theorem

T5.8.11.1 $[A]$: .$A \varepsilon$ qntf$_{\mathrm{L}}$.$B \varepsilon$ **vrb**(A).$C \varepsilon$ **vrb**(A).$B \varepsilon$ **cnf**(C). $\supset$.$B = C$

Hyp(4) . $\supset$:

5. $A \varepsilon$ **expr** : [1, Df8, T5.6.4, D13]

6. $B \varepsilon$ **int**(A). $\vee$.$B \varepsilon$ **1vrb**(A). $\vee$.$B \varepsilon$ **Uvrb**(A) : [2, 5, D13]

7. $C \varepsilon$ **int**(A). $\vee$.$C \varepsilon$ **1vrb**(A). $\vee$.$C \varepsilon$ **Uvrb**(A) : [3, 5, D13]

$B = C$ [6, 7, Df8, T5.8.8, T5.8.9, T5.8.10, T5.8.11, T2.2.30, T2.2.31]

T5.8.12 $[A]$: $A \varepsilon$ qntf$_{\mathrm{L}}$. $\supset$.$A \varepsilon$ **non-rep** [T5.8.11.1, D19]

T5.8.13 $[A]$: $A \varepsilon$ qntf$_{\mathrm{L}}$. $\supset$.$A \varepsilon$ qntf
[Df8, TE2, T5.8.1, T5.8.2, T5.8.7, T5.8.12]

T5.8.14 $[A]$: $A \varepsilon$ qntf$_{\mathrm{L}}$. $\equiv$.$A \varepsilon$ qntf [T5.8.6, T5.8.13]

We now continue in this fashion introducing definitions of Leśniewski's terms into our system and then proving that our terms are equivalent to his. In this way we obtain theorems which show that each of the terms used in the statement of the rule of Protothetic that we gave is equivalent to a term used in Leśniewski's statement. Hence the two statements of the rule are equivalent in the sense that they allow precisely the same deductions. These proofs, the details of which are too long to include here, can be found in Rickey (1968).

CONCLUSION

As we have stressed earlier the most important part of this work is the first, particularly Chapter II [of Rickey (1972)]. There we presented and developed a system M of metalogic. Our aim in constructing this system was twofold. We wished it to be adequate for the statement of procedural rules of formalized systems, and wished it also to be adequate for conducting some metalogical investigations. Our claim is that, with varying degrees of success, both of these aims have been accomplished.

Clearly a proof that we can formulate rules for *any* formalized system in M, or even in extensions of M like MP, is out of the question at this time. To do so would require a precise notion of "rule of procedure" which is, to our knowledge, not available at this time. Probably if such a definition did exist we could easily establish whether all such rules were expressible in M. Since such a proof was beyond our grasp we adopted the following technique.

Protothetic is a system with rules which are complicated to state precisely, and strong enough so that they encompass the rules of most logical theories. Thus there can be little doubt that the statement of the rule of Protothetic is a challenging test of the first part of our claim.

For one familiar with Leśniewski's systems it should be clear that the rule of Ontology can be presented analogously to the rule of Protothetic. Moreover, using the results of Section 2 we could simplify the statement of the rule of Ontology given in Leśniewski (1930).

It seems almost banal to remark that system M is sufficient to state the rules for propositional calculi. If the only rules are substitution and detachment, then the statement of the rule is trivial; the propositional constants all occur in the axioms and so we can just specify all symbols not equiform to one of those to be variables. If the calculus contains definitions, then more care is required (cf. Leśniewski, 1931).

By modifying the techniques of Sections 1, 2 we can present the rules for functional calculi of any order and for set theories. For Gentzen-like systems it is not so obvious that we can use similar techniques. The main difficulties are the linearization of the language and allowing for expressions of arbitrary (even infinite) length. This would require us to drop the axiom concerning the finiteness of inscriptions.

In regard to the other half of our claim we have shown in Section 2 that system **M**, or really **MP**, is adequate for some metalogical investigations, viz., proof of the equivalence of different statements of rules. A most interesting question is what type of metalogical investigations can actually be carried out in system **M**. Probably consistency and completeness of propositional systems could be carried out completely in **M**, but it seems highly doubtful that incompleteness results could be established within **M** without the use of variables of higher types and new axioms involving these types.

NOTES

1 Cf. TE51, Section 2.5.

2 The following definition, due to Sobociński (1949), is more perspicuous:

$$[pq] : .p.q. \equiv: [f] : f(pq). \equiv .f(11)$$

3 In systems of Protothetic based on implication definitions must be stated either as pairs of implications:

$$\alpha \supset \beta$$
$$\beta \supset \alpha$$

or in the complicated form

$$\alpha \supset \beta. \supset: \beta \supset \alpha. \supset p : . \supset p$$

where both the definiens α and definiendum β are free in p, cf. Lejewski (1958).

4 The symbolism used here is that of Leśniewski. Once the key to the "wheel and spoke" notation is explained it is easy to remember. The symbol ϕ corresponds to the truth table

	0	1
0	1	0
*1	0	1

i.e., to equivalence. In general a spoke at the top (right, bottom, left) means a 1 in the (0-0)-place (resp. (0-1), (1-1), (1-0)-place) of the truth table. Thus ϙ is conjunction for it corresponds to

	0	1
0	0	0
*1	0	1

It should be understood that this symbolism, while very convenient, is not mandatory. Protothetic is formulated in a way which leaves all possible choices of symbolism up to the individual.

5 Interchanging functors in the same category need not preserve truth, and generally it does not.

[6] One reason for rejecting incomplete symbols is that the rule of substitution has never been strictly formulated using them.
[7] Just like 'man' is a growing name. Every time a baby is born the name 'man' denotes one more object.
[8] Whether a parenthesis is called left or right is a matter of having a familiar looking notation. We could just as well write ')p(' as '(p)'. No confusion can arise as the introductory thesis for a new pair of parentheses determines which of the two is called left.
[9] We have used different enumerations on the axioms to indicate that they are of different natures. The P axioms concern the primitive terms of **MP**. The Q axioms are statements about the axiom 𝔄𝔵 of Protothetic. They are of empirical origin, but must be introduced as axioms into system **MP**.
[10] The Terminological Explanations stated in this section are different than those in Leśniewski (1929). They will be proved equivalent in Section 3.
[11] By the n-th conjunct of a *T.E.* we mean the n-th conjunct on the right hand side of the main equivalence of the *T.E.* More specifically, *TEM*n will denote the n-th conjunct of *TEM*. The conjuncts of the *T.E.*'s are independent. This claim implies that the 1st, 4th, 5th and **Uvrb** of 𝔄𝔵 are not equiform.

REFERENCES

Ajdukiewicz, Kazimierz (1935). 'Die Syntaktische Konnexität', *Studia Philosophica* 1, 1–27. Translated in McCall (1967), pp. 207-231.

Bar-Hillel, Yehoshua (1950). 'On Syntactical Categories', *The Journal of Symbolic Logic* 15, 1–16.

Carnap, Rudolf (1934). *Logische Syntax der Sprache*, Julius Springer, Wien. Translated as *The Logical Syntax of Language*, London (1937).

Chomsky, Noam (1953). 'Systems of Syntactic Analysis', *The Journal of Symbolic Logic* 18, 242–256.

Goodman, Nelson (1951). *The Structure of Appearance*, Harvard University Press, Cambridge, MA (second edition 1966).

Lejewski, Czesław (1958). 'On Implicational Definitions', *Studia Logica* VIII, 189–211.

Leśniewski, Stanisław (1929). 'Grundzüge eines neuen Systems der Grundlagen der Mathematik', *Fundamenta Mathematicae* XIV, 1–81.

Leśniewski, Stanisław (1930). 'Über die Grundlagen der Ontologie', *Comptes rendus des Séances de la Société des Sciences et des Lettres de Varsovie, Classe III* 23, 111–132.

Leśniewski, Stanisław (1931). 'Über Definitionen in der sogenannten Theorie der Deduktion', *Comptes rendus des Séances de la Société des Sciences et des Lettres de Varsovie, Classe III* 24, 289–309. Translated in McCall (1967), pp. 170–187.

Leśniewski, Stanisław (1938). 'Einleitende Bemerkungen zur Fortsetzung meiner Mitteilung u. d. T. 'Grundzüge eines neuen Systems der Grundlagen der Mathematik", *Collectanea Logica* I, 1–60. Translated in McCall (1967), pp. 116–169. [All the Leśniewski's papers are translated in Leśniewski (1992), see the Protothetic Bibliography – Ed.]

McCall, Storrs, ed. (1967). *Polish Logic 1920–1939*, Clarendon Press, Oxford.

Rickey, V. Frederick (1968). 'An Axiomatic Theory of Syntax', Ph.D. Thesis, University of Notre Dame, Notre Dame, IN.

Rickey, V. Frederick (1972). 'Axiomatic Inscriptional Syntax. Part I: General Syntax', *Notre Dame Journal of Formal Logic* XIII, 1–33.

Sobociński, Bolesław (1949). 'An Investigation of Protothetic', *Cahiers de l'Institut d'Études Polonaises en Belgique* No. 5, Bruxelles. Reprinted in McCall (1967), pp. 201–206. Unfortunately, the reprint omits the "Introduction (pp. 7–27) of unusual expository and bibliographical interest, giving an account of the ill-fated Polish periodical *Collectanea*

Logica." See the review by Church, *The Journal of Symbolic Logic* 15 (1950), 64. [See also paper IV in this volume – Ed.]

Sobociński, Bolesław (1960). 'On the Single Axioms of Protothetic, I, II, III', *Notre Dame Journal of Formal Logic* I, 52–73, II (1961), 111–126, 129–148. [Cf. paper VI in this volume – Ed.]

Tarski, Alfred (1923). 'O wyrazie pierwotnym logistyki', *Przegląd Filozoficzny* XXVI, 68–89. Translated in Tarski (1956), pp. 1–23. [Cf. paper III in this volume – Ed.]

Tarski, Alfred (1956). *Logic, Semantics, Metamathematics, Papers from 1923–1938*, translated by J.H. Woodger, Clarendon Press, Oxford.

AUDOËNUS LE BLANC

INVESTIGATIONS IN PROTOTHETIC *

In this article I present some results of five years' research into Leśniewski's protothetic.[1] I outline deductions from the axiom A_n considerably shorter than those previously known (see Sobociński, 1961a) and I derive the laws of implication from this axiom without using the rule of extensionality.[2] Since this paper can best be read in the light of articles by Professor Sobociński published in this Journal (see Sobociński, 1960, 1961a and 1961b),** I have largely adopted his conventions of symbolism, and the following symbols in particular:

α The rule permitting definitions of new constants (see Sobociński, 1960, pp. 58–59).‡
β The rule for distributing quantifiers (see Sobociński, 1960, p. 59).‡
0 An informal abbreviation for '$[u].u$'.
1 An informal abbreviation for '$[u].u. \equiv .[u].u$'.

From the axiom

$$A_n \ [pq] :: p \equiv q. \equiv .\therefore [f] .\therefore f(pf(p0)). \equiv : [r] : f(qr). \equiv .q \equiv p$$

we prove the following theorems:

$$\text{D1} \ [p].p \equiv \text{As}(p) \qquad [\alpha]$$

$$\text{L1} \ [fp] .\therefore f(pf(p0)). \equiv : [r] : f(\text{As}(p)r). \equiv .\text{As}(p) \equiv p \qquad [A_n \ q/\text{As}(p); \text{D1}]$$

$$\text{D2} \ [pr] .\therefore p \equiv r. \equiv .\text{As}(p \equiv r) :\equiv .\text{Vr}(pr) \qquad [\alpha]$$

* Bibliographical Note: This paper originally appeared in the *Notre Dame Journal of Formal Logic* 26 (1985), 483–489, by permission of the *Notre Dame Journal of Formal Logic*.
** [Ed. Note: Cf. paper VI in this volume.]
‡ [Ed. Note: Cf. paper VI in this volume, pp. 160–161.]

L2 $[pr].\mathrm{Vr}(pr)$ [D2; D1 $p/p \equiv r$]

L3 $[pr] : \mathrm{Vr}(\mathrm{As}(p)r). \equiv .\mathrm{As}(p) \equiv p$ [L1 f/Vr; L2 $r/\mathrm{Vr}(p0)$]

We can now establish the following four metarules of procedure:[3]

S I. *If we have in the system a thesis of the type*

$[x, y, \ldots] : A. \equiv .B$

(with or without an initial quantifier), then we can add the corresponding thesis

$[x, y, \ldots] : B. \equiv .A.$

(𝔞) $[x, y, \ldots]A. \equiv .B$ [The assumption]

(𝔟) $[x', y', \ldots] : A'. \equiv .B'$ [Substitution of 𝔞 to eliminate any free variables of shapes 'f' or 'r']

(𝔠) $[f, x', y', \ldots] :: f(A'f(A'0)). \equiv .\therefore[r].\therefore f(B'r). \equiv : B'. \equiv .A'$ [A_n $p/A', q/B'$; 𝔟]

(𝔡) $[r, x', y', \ldots].\therefore \mathrm{Vr}(B'r). \equiv .B'. \equiv .A'$ [𝔠 f/Vr; L2 $p/A', r/\mathrm{Vr}(A'0)$]

(𝔢) $[x', y', \ldots] : B'. \equiv .A'$ [𝔡; L2 p/B']

(𝔣) $[x, y, \ldots] : B. \equiv .A$ [Substitution of 𝔢 to restore the original variables]

S II. *If we have in the system two corresponding theses of types*

$[x, y, \ldots] : A. \equiv .B$

$[x, y, \ldots].B,$

then we can add the corresponding thesis

$[x, y, \ldots].A.$

This follows immediately from metarule S I.

S III. *If we have in the system two corresponding theses of types*

$[x, y, \ldots] : A. \equiv .B$

$[x, y, \ldots].F(A)$

(where '$F(A)$' represents any meaningful expression which contains at least one expression equiform with 'A' not falling within the scope of any internal quantifier), then we can add the corresponding thesis

$[x, y, \ldots].F(B)$.

(𝔞) $[x, y, \ldots] : A. \equiv .B$ [Assumption]

(𝔟) $[x, y, \ldots].F(A)$ [Assumption]

(𝔠) $[x', y', \ldots] : A'. \equiv .B'$ [These substitutions of 𝔞 and 𝔟

(𝔡) $[x', y', \ldots].F'(A')$ eliminate any free variables of shapes 'f', 'p', 'q', or 'r']

(𝔢) $[x', y', \ldots] : B'. \equiv .A'$ [S I: 𝔠]

(𝔣) $[x', y', \ldots].\mathrm{As}(F'(A'))$ [D1 $p/F'(A')$; 𝔡]

(𝔤) $[f, x', y', \ldots] :: f(A'f(A'0)). \equiv .\therefore [r].\therefore f(B'r). \equiv: B'. \equiv .A'$ [A_n p/A', q/B'; 𝔠]

(𝔥) $[p, q, r, x', y', \ldots] :: \mathrm{Vr}(\mathrm{As}(F'(p))r). \equiv: \mathrm{As}(F'(p)). \equiv .F'(q)$
$\therefore \equiv .G\{px'y' \ldots\}(qr)$ [α]

(𝔧) $[x', y', \ldots].G\{A'x'y' \ldots\}(A'G\{A'x'y' \ldots\}(A'0))$
[𝔥 p/A', q/A', $r/G\{A'x'y' \ldots\}(A'0)$;
L3 $p/F'(A')$, $r/G\{A'x'y' \ldots\}(A'0)$]

(𝔨) $[r, x', y', \ldots].\therefore G\{A'x'y' \ldots\}(B'r). \equiv: B'. \equiv .A'$
[𝔤 $f/G\{A'x'y' \ldots\}$; 𝔧]

(𝔩) $[r, x', y', \ldots].G\{A'x'y' \ldots\}(B'r)$ [S II: 𝔨; 𝔢]

(𝔪) $[r, x', y', \ldots].\therefore \mathrm{Vr}(\mathrm{As}(F'(A'))r). \equiv: \mathrm{As}(F'(A')). \equiv .F'(B')$
[S II: 𝔥 p/A', q/B'; 𝔩]

(𝔫) $[x', y', \ldots].F'(B')$ [𝔪; L2 $p/\mathrm{As}(F'(A'))$; 𝔣]

(𝔬) $[x, y, \ldots].F(B)$ [Substitution of 𝔫 to restore the original variables]

S IV. *If we have in the system corresponding theses of types*

$$[x, y, \ldots] : A. \equiv .B$$

$$[x, y, \ldots].F(B)$$

(with '$F(B)$' understood as in S III), then we can add the corresponding thesis

$$[x, y, \ldots].F(A).$$

This follows immediately from S I and S III.

Aided by these rules, we prove the following theses:[4]

L4 $[pr] : \mathrm{Vr}(pr). \equiv .p \equiv p$ [S IV: D1; L3]

L5 $[p].p \equiv p$ [L4; L2]

L6 $[fp]\therefore f(pf(p0)). \equiv: [r] : f(pr). \equiv .p \equiv p$ [A_n q/p; L5]

D3 $[pq]\therefore q. \equiv .q \equiv p :\equiv .Fb(pq)$ [α]

D4 $[pr] : Fb(\mathrm{Vr}(0r)p). \equiv .\mathrm{Vr}_1(pr)$ [α]

L7 $[r] : Fb(10). \equiv .\mathrm{Vr}_1(0r)$ [S III: L4 $p/0$; D4 $p/0$]

L8 $[r]\therefore Fb(10). \equiv .1 :\equiv: \mathrm{Vr}_1(0r). \equiv .1$

[S III: L7; L5 $p/Fb(10). \equiv .1$]

L9 $Fb(10). \equiv .1 :\equiv: [r] : \mathrm{Vr}_1(0r). \equiv .1$ [L8; β]

L10 $\mathrm{Vr}_1(0Vr_1(00)). \equiv: Fb(10). \equiv .1$ [S IV: L9; L6 $f/Vr_1, p/0$]

L11 $Fb(10). \equiv: Fb(10). \equiv .1$ [S IV: L7 $r/\mathrm{Vr}_1(00)$; L10]

L12 $Fb(1Fb(10))$ [D3 $q/Fb(10), p/1$; L11]

L13 $[r] : Fb(1r). \equiv .1 \equiv 1$ [L6 $f/Fb, p/1$; L12]

L14 $[p].Fb(1p)$ [S II: L13 r/p; L5 $p/1$]

L15 $[p] : p. \equiv .p \equiv 1$ [S II: D3 $p/1, q/p$; L14]

L16 $[pq]\therefore q \equiv 1. \equiv .p :\equiv: q \equiv p. \equiv .1$ [S III: L15 p/q; L15 $p/q \equiv p$]

At this point we can establish the following metatheorem:

S V. *If we have in our system two theses of types*

$F(0)$

$F(1)$

which are substitutions of some meaningful expression

$[p].F(p)$

with the expressions '0' and '1' replacing the variable 'p', then we can add this latter expression to the system as a thesis in the following way:

(𝔞) $F(0)$ [Assumption]

(𝔟) $F(1)$ [Assumption]

(𝔠) $[pq]\therefore F(p). \equiv .q \equiv q :\equiv .G(qp)$ [α]

(𝔡) $[p] : F(p). \equiv .G(0p)$ [S IV: L15 $p/F(p)$; 𝔠 $q/0$]

(𝔢) $G(00)$ [𝔡 $p/0$; 𝔞]

(𝔣) $G(01)$ [𝔡 $p/1$; 𝔟]

(𝔤) $G(00). \equiv .1$ [L15 $p/G(00)$; 𝔢]

(𝔥) $G(0G(00))$ [S IV: 𝔤; 𝔣]

(𝔧) $[r] : G(0r). \equiv .1$ [L6 $f/G, p/0$; 𝔥]

(𝔨) $[p].G(0p)$ [S II: 𝔧 r/p; L5 $p/0$]

(𝔩) $[p].F(p)$ [S II: 𝔡; 𝔨]

With the help of S V, we prove the following theorems:

L17 $[pq]\therefore q \equiv p. \equiv .0 :\equiv: q \equiv 0. \equiv .p$ [S V: L5 $p/q \equiv 0. \equiv .0$; L16 $p/0$]

L18 $[p]\therefore p. \equiv .p \equiv 0 :\equiv .0$ [S II: L17 $q/p, p/p \equiv 0$; L5 $p/p \equiv 0$]

L19 $0. \equiv .1 \equiv 0 :\equiv .0\therefore \equiv .0$ [S III: L18 $p/1$; L17 $q/0, p/1 \equiv 0$]

L20 $[p] :: p. \equiv .p \equiv 0 :\equiv \therefore 0. \equiv .1 \equiv 0 :\equiv .0$ [S IV: L19; L18]

D5 $[pq] :\cdot: p. \equiv .1 \equiv 0 :\equiv .q \equiv 0.\therefore \equiv .q \equiv 0 ::\equiv .Fc(pq)$ [α]

L21 $0. \equiv .1 \equiv 0 :\equiv .1.\therefore \equiv .Fc(00)$

[S IV: L15 $p/0. \equiv .1 \equiv 0$; D5 $p/0, q/0$]

L22 $0. \equiv .1 \equiv 0 :\equiv .Fc(00)$ [S IV: L15 $p/0. \equiv .1 \equiv 0$; L21]

L23 $Fc(00. \equiv .1 \equiv 0)$ [D5 $p/0, q/0. \equiv .1 \equiv 0$; L20 $p/0. \equiv .1 \equiv 0$]

L24 $Fc(0Fc(00))$ [S III: L22; L23]

L25 $[r] : Fc(0r). \equiv .1$ [L6 $f/Fc, p/0$; L24]

L26 $Fc(00)$ [S II: L25 $r/0$; L5 $p/0$]

L27 $0. \equiv .1 \equiv 0$ [S II: L22; L26]

L28 $0 \equiv 1. \equiv .1 \equiv 0$ [S III: L15 $p/0$; L27]

L29 $1 \equiv 0. \equiv .0 \equiv 1$ [S I: L28]

L30 $[p] : p \equiv 1. \equiv .1 \equiv p$ [S V: L28; L5 $p/1 \equiv 1$]

L31 $[p] : p \equiv 0. \equiv .0 \equiv p$ [S V: L5 $p/1$; L29]

L32 $[pq] : p \equiv q. \equiv .q \equiv p$ [S V: L31; L30]

L33 $[pq].\therefore q \equiv p. \equiv .1 :\equiv: q \equiv 1. \equiv .p$ [S I: L16]

L34 $[pqr].\therefore q \equiv p. \equiv .r :\equiv: q \equiv r. \equiv .p$ [S V: L17; L33]

L35 $[pqr].\therefore p \equiv q. \equiv .r :\equiv: q \equiv r. \equiv .p$ [S IV: L32; L34]

L36 $[pqr].\therefore p \equiv q. \equiv .r :\equiv: p. \equiv .q \equiv r$ [S IV: L32 $q/q \equiv r$; L35]

L37 $[pq].\therefore p. \equiv: q. \equiv .p \equiv q$ [L36 $r/p \equiv q$; L5 $p/p \equiv q$]

L38 $[pq].\therefore p. \equiv: q. \equiv .q \equiv p$ [L36 $r/q \equiv p$; L32]

L39 $[pq] : p. \equiv .Fb(pq)$ [S IV: L38; D3]

L40 $[pqr].\therefore p. \equiv: Fb(qr). \equiv .q \equiv p$ [S III: L39 $p/q, q/r$; L38]

L41 $[pqr] \therefore Fb(pFb(p0)). \equiv: Fb(qr). \equiv .q \equiv p$

[S III: L39 $q/Fb(p0)$; L40]

L42 $[pq] \therefore Fb(pFb(p0)). \equiv: [r] : Fb(qr). \equiv .q \equiv p$ [L41; β]

L43 $[fp] :\cdot: p. \equiv:: f(pf(p0)). \equiv: [r] : f(pr). \equiv .p \equiv p. \therefore \equiv .p$

[L37, q/p, $p/f(pf(p0)). \equiv: [r] : f(pr). \equiv .p \equiv p$; L6]

L44 $[p] :\cdot: p. \equiv:: [f] :: f(pf(p0)). \equiv: [r] : f(pr). \equiv .p \equiv p. \therefore \equiv .p$

[L43; β]

L45 $[p] :: p. \equiv \therefore [f] \therefore f(pf(p0)). \equiv: [r] : f(1r). \equiv .1 \equiv p$

[S IV: L15; A_n $q/1$]

L46 $[p] :\cdot: [f] \therefore f(pf(p0)). \equiv: [r] : f(1r). \equiv .1 \equiv p. \therefore \equiv:: [f] ::$
$f(pf(p0)). \equiv: [r].f(1r). \equiv .1 \equiv p. \therefore \equiv .1$

[L15 $p/f(pf(p0)). \equiv: [r] : f(1r). \equiv .1 \equiv p$; β]

L47 $[p] :\cdot: p. \equiv:: [f] :: f(pf(p0)). \equiv: [r] : f(1r). \equiv .1 \equiv p. \therefore \equiv .1$

[S IV: L45; L46]

D6[5] $[pq] :: p. \equiv:: [f] :: f(pf(p0)). \equiv: [r] : f(qr). \equiv .q \equiv p. \therefore \equiv .q$
$:\cdot:\equiv .p \supset q$ [α]

L48 $[p].\, p \supset p$ [D6 q/p; L44]

L49 $[p].\, p \supset 1$ [D6 $q/1$; L47]

L50 $[p].\, 0 \supset p$ [S V: L48 $p/0$; L49 $p/0$]

L51 $1 \supset 1. \equiv .1$ [L15 $p/1 \supset 1$; L48 $p/1$]

L52 $1. \supset .1 \supset 1$ [S IV: L51; L48 $p/1$]

L53 $0. \supset .1 \supset 0 :\equiv: 1. \supset .1 \supset 1$

[L37 $p/0. \supset .1 \supset 0$, $q/1. \supset .1 \supset 1$; L50 $p/1 \supset 0$; L52]

L54 $1. \supset .1 \supset 0 :\equiv \therefore 1. \supset: 1. \supset .0 \equiv 1$ [S III: L15 $p/0$; L5 $p/1. \supset .1 \supset 0$]

L55 $[p] :: p. \supset .1 \supset 0 :\equiv .\because 1. \supset: 1. \supset .0 \equiv p$ [S V: L53; L54]

L56 $1. \supset .\because 1. \supset: 0. \equiv .1 \supset 0$ [L55 $p/1 \supset 0$; L48 $p/1 \supset 0$]

At this point we prove the metatheorem:

S VI.[6] *If we have in the system corresponding theses*

$A \supset B$

$A,$

then we can add to the system the thesis

$B.$

($\mathfrak{a}$) $A \supset B$ [Assumption]

($\mathfrak{b}$) A [Assumption]

($\mathfrak{c}$) $A. \equiv:: [f] :: f(Af(A0)). \equiv: [r] : f(Br). \equiv .B \equiv A.\because \equiv .B$
[S II: D6 p/A, q/B; $\mathfrak{a}$]

($\mathfrak{d}$) $[f] :: f(Af(A0)). \equiv: [r] : f(Br). \equiv .B \equiv A.\because \equiv .B$ [$\mathfrak{c}$; $\mathfrak{b}$]

($\mathfrak{e}$) B [$\mathfrak{d}$ f/Fb; L42 p/A, q/B]

We then use this rule to prove

L57 $1. \supset: 0. \equiv .1 \supset 0$ [S VI: L56; L5 $p/0$]

L58 $0. \equiv .1 \supset 0$ [S VI: L57; L5 $p/0$]

This completes the proof that the laws of implication can be derived from A_n without using the rule of extensionality. Similar proofs may easily be constructed for A_m, A_o, A_p, and A_q (see Sobociński, 1960, p. 67 and Sobociński, 1961a, p. 122).* On the other hand, it seems that the deductions from axiom A_l and the proof of metatheorem A2[7] require the use of theses of higher extensionality. Observe the crucial use of the rule for distributing quantifiers at L9, L42, L44, and L46; it is largely this rule and the unusual definition of implication at D6 which allow us to prove the laws of implication L48, L49, and L58.

It is well known that theses L32 and L36 are an adequate axiom system for the E-calculus,[8] and so replace the thesis A78 proved by Professor

* [Ed. Note: Cf. paper VI in this volume, pp. 170–171 and 186.]

Sobociński. My research has concerned the axiom A_o, from which deductions are slightly shorter than those from A_n[9]; even a slight simplification of the known proof of A78 required 1,205 applications of the rule of inference of $\mathfrak{S}_5$, including definitions. The proof of L32 and L36 requires at most 583 applications of the rule, while the proof of the related system L34 and L59

L59 $[pqr] \therefore r. \equiv .p \equiv q :\equiv: r. \equiv .q \equiv p$

requires at most 568 applications of the rule. Hence the deductions outlined in this paper reduce to less than half its length the proof that A_o is adequate for the E-calculus.

NOTES

[1] The basic sources describing protothetic are Leśniewski (1929, 1939a) and (1939b). The articles Sobociński (1960, 1961a), and (1961b) are indispensable for serious study of the theory [cf. paper VI in this volume – Ed.].

[2] The proof given in L37 to L58 of this paper shows that derivations from this axiom need not rely on the theoretical considerations which led to its discovery. See Sobociński (1953), p. 19; Sobociński (1960), pp. 66–67 [see also paper VI in this volume, pp. 169–170 – Ed.]; Sobociński (1961b), p. 146, note 83 [see also note 81 on p. 214 in this volume – Ed.]; and especially Sobociński (1961a), p. 122: "On the other hand I am unable to prove that A_n ... is a single axiom of $\mathfrak{S}_5$ without the application of reasonings ... in which point ϵ plays an essential role." [See also p. 185 in this volume – Ed.]

[3] See Sobociński (1961a), pp. 114–117 and 119–120 [see also paper VI in this volume pp. 174–178 and 181–182 – Ed.]. My formulation of these rules is slightly broader, as I refer to meaningful expressions instead of functors.

[4] My deductions diverge significantly from those in Sobociński (1961a) from definition D4 onwards.

[5] This unusual definition differs from any of those previously known, and from any of the related definitions of conjunction. For previous definitions see Leśniewski (1929), p. 12, and Leśniewski (1939a), D3 on p. 93 and D4 on p. 107. For definitions of conjunction see Leśniewski (1929), p. 13; Leśniewski (1939b), pp. 138, 141 and 147; Sobociński (1939), pp. 202–203 [see also paper IV in this volume, p. 76 – Ed.]; and the discussion in Sobociński (1961b), pp. 134–144 [see also paper VI in this volume, pp. 194–208 – Ed.]. Compare the proof of L58 with the theorem A1 in Sobociński (1961b), pp. 134–139 [see also pp. 194–200 in this volume – Ed.], to which it bears a close resemblance. I do not know how to shorten the present definition of implication and still carry out the proof of L58.

[6] This rule is quite different from the SVI of Sobociński (1961b), p. 132 [see also paper VI in this volume, p. 192 – Ed.].

[7] This axiom is found in Sobociński (1960), p. 66, and discussed in Sobociński (1961a), pp. 123–124 [see also paper VI in this volume, pp. 170 and 187–189 – Ed.]. The metatheorem appears in Sobociński (1961b), pp. 139–144 [see also pp. 200–208 in this volume – Ed.].

[8] See Łukasiewicz (1939), p. 91. Initial deductions from a closely related axiom system are given in Wajsberg (1937), p. 314.

[9] For example, the thesis L5 can be proved from A_o in only 44 applications of the rule, but its proof from A_n appears to require at least (and certainly at most) 48 applications. Even the initial deductions proving L32 and L36 from A_o would require about 100 pages of print, so that I cannot see any possibility of publishing them.

REFERENCES

Leśniewski, Stanisław (1929). 'Grundzüge eines neuen Systems der Grundlagen der Mathematik', *Fundamenta Mathematicae* XIV, 1–81.

Leśniewski, Stanisław (1939a). 'Grundzüge eines neuen Systems der Grundlagen der Mathematik (Fortsetzung)', *Collectanea Logica* I, 61–144.

Leśniewski, Stanisław (1939b). 'Einleitende Bemerkungen zur Fortsetzung meiner Mitteilung u. d. T. "Grundzüge eines neuen Systems der Grundlagen der Mathematik"', *Collectanea Logica* I, 1–60. All page numbers cited are from the English translation in McCall (1967).

Łukasiewicz, Jan (1939). 'Der Äquivalenzenkalkül', *Collectanea Logica* I, 145–169. The citation refers to the pages of the English translation in McCall (1967).

McCall, Storrs, ed. (1967). *Polish Logic, 1920–1939*, Clarendon Press, Oxford.

Sobociński, Bolesław (1939). 'Z badań nad prototetyką', *Collectanea Logica* I, 171–177. The citation refers to the pages of the English translation in McCall (1967). An earlier translation appeared in *Cahiers de l'Institut d'Études Polonaises en Belgique*, no. 5 (1949), 29–39. [Cf. paper IV in this volume – Ed.]

Sobociński, Bolesław (1953). 'Z badań nad aksjomatyką prototetyki Stanisława Leśniewskiego', *Rocznik Polskiego Towarzystwa Naukowego na Obczyźnie* 4 (1953–1954), 18–20.

Sobociński, Bolesław (1960). 'On the Single Axioms of Protothetic', *Notre Dame Journal of Formal Logic* I, 52–73. [Cf., paper VI in this volume – Ed.]

Sobociński, Bolesław (1961a). 'On the Single Axioms of Protothetic', *Notre Dame Journal of Formal Logic* II, 111–126. [Cf. paper VI in this volume – Ed.]

Sobociński, Bolesław (1961b). 'On the Single Axioms of Protothetic', *Notre Dame Journal of Formal Logic* II, 129–148. The continuation of this article has not yet appeared. [Cf. paper VI in this volume – Ed.]

Wajsberg, Mordchaj (1937). 'Metalogische Beiträge', *Wiadomości Matematyczne* XLIII, 1–38. The citation refers to the page in the English translation in McCall (1967).

PROTOTHETIC BIBLIOGRAPHY

The abbreviations of names of serials which are used in the Protothetic Bibliography are the same as those used in *Mathematical Reviews* or *Philosophical Reviews*.

ANONYMOUS

(1976) *XXIInd Conference on the History of Logic*, Kraków 1976. Abstracts of papers (J. Surma (?) ed.), Jagiellonian University and Polish Academy of Sciences.

BOUDREAUX, J.C.

(1976) 'Set-Theoretical Models for Leśniewski's Logical Systems', in *Proc. of the XXIInd Conf. on the History of Logic*, Kraków 1976, pp. 2–5. Jagiellonian University and Polish Academy of Sciences.

CANTY, J.T.

(1963) 'Leśniewski's Terminological Explanations as Recursive Concepts', *Notre Dame J. Formal Logic* X, 337–369.

(1969) 'Ontology: Leśniewski's Logical Language', *Foundations of Language* 5, 455–469.

CHIKAWA, K.

(1967) 'On Equivalences of Laws in Elementary Protothetics, I, II', *Proc. Japan Acad.* XLIII (1967), 743–747; XLIV (1968), 56–59.

CHURCH, A.

(1956) *Introduction to Mathematical Logic*, Vol. I, Princeton University Press, Princeton, New Jersey.

CIRULIS, J.P.

(1979a) 'Protothetic without Typical Ambiguity of Expressions', *Latv. Mat. Ezhegodnik* 23, 166–178, 275 [in Russian].

(1979b) 'Traditional Means of Inference in Protothetic', *Latv. Mat. Ezhegodnik* 23, 179–193, 275 [in Russian].

(1982) 'On Axioms of Protothetic with an Extended Rule of Extensionality', *Latv. Mat. Ezhegodnik* 26, 264–270 [in Russian].

(1985) 'The Rule of Parametric Substitution in Protothetic', *Polish Acad. Sci. Inst. Philos. Sociol. Bull. Sect. Logic* 14(4), 134–139.

(1988) 'Definitions in Protothetic', *Latv. Mat. Ezhegodnik* 32, 90–101, 241–242 [in Russian].

CURRY, H.B.

(1950) *A Theory of Formal Deducibility*, Notre Dame Mathematical Lectures, No. 6, Notre Dame, IN.

CZEŻOWSKI, T.

(1949) *Logika. Podręcznik dla Studiujących Nauki Filozoficzne* (Logic. A Textbook for Philosophy Students), PZWS, Warszawa.

GILES-PETERS, A.R.

(1972) *Nominalistic Philosophy of Logic, with Particular Reference to the Systems of Stanisław Leśniewski*, Master of Arts Thesis, Philosophy Department, La Trobe University, Bundoora, Victoria.

GINISTI, J.-P.

(1991) 'La créativité des définitions dans les systèmes para-Euclidiens', *Math. Inform. Sci. Humaines* 116, 69–88.

GRZEGORCZYK, A.

(1955) 'The Systems of Leśniewski in Relation to Contemporary Logical Research', *Studia Logica* III, 77–97.

GRENIEWSKI, H.

(1950) 'Functors of the Propositional Calculus', *Ann. Soc. Polon. Math.* 22, 78–86.

HIŻ, H.

(1952) 'On Primitive Terms of Logic', *J. Symbolic Logic* 17, 156–157 (abstract).

(1960) 'The Intuitions of Grammatical Categories', *Methods* 12, 311–319.

(1965) 'Ontological Definitions in Augmented Protothetic', *J. Symbolic Logic* 31, 149–150 (abstract).

(1967) 'Grammar Logicism', *The Monist* 51, 110–127.

ISÉKI, K.

(1966) 'Algebraic Formulations of Propositional Calculi with Variable Forming Functors', *Proc. Japan Acad.* XLII, 1058–1059.

(1968) *Kigô ronrigaku (meidai ronri)* (Symbolic Logic (Propositional Logic)), Maki Shoten, Tokyo.

JORDAN, Z.A.

(1945) *The Development of Mathematical Logic and of Logical Positivism in Poland between the Two Wars*, Polish Science and Learning, no. 6, Clarendon Press, Oxford. Modified version in McCall (1967).

KEARNS, J.T.

(1967) 'The Contribution of Leśniewski', *Notre Dame J. Formal Logic* VIII, 61–93.

KOTARBIŃSKI, T.

(1957a) 'La philosophie dans la Pologne contemporaine', *Synthèse* 12(137), 29–38.

(1957b) *Wykłady z Dziejów Logiki* (Lectures on the History of Logic), Societas Scientiarum Łodziensis 28, Łódź.

(1964) *Leçons sur L'Historie de la Logique*, Press Universitaires de France, Paris.

(1967) 'Notes on the Development of Formal Logic in Poland in the Years 1900–39', pp. 1–14 in McCall (1967).

KÜNG, G.

(1977) 'Leśniewski's Systems', pp. 168–177 in Marciszewski (1981).

LE BLANC, A.V.

(1985) 'Investigations in Protothetic', *Notre Dame J. Formal Logic* 26, 483–489.

(1991) *Leśniewski's Computative Protothetic*, Ph.D. Dissertation at the University of Manchester.

LEJEWSKI, C.

(1956) Review of Słupecki (1953), *J. Symbolic Logic* 21, 188–191.

(1958) 'On Implicational Definitions', *Studia Logica* VIII, 189–211.

(1969) 'Consistency of Leśniewski's Mereology', *J. Symbolic Logic* 34, 321–328.

(1989) 'Formalization of Functionally Complete Propositional Calculus with the Functor of Implication as the Only Primitive Term', *Studia Logica* XLVIII, 479–494.

(1990) 'Stanisław Leśniewski (1886–1939)', *Wiadom. Mat.* 28(2), 153–182.

(1995) 'Remembering Stanisław Leśniewski', pp. 25–66 in Leśniewski (1995).

LEŚNIEWSKI, S.

(1929) 'Grundzüge eines neuen Systems der Grundlagen der Mathematik', *Fund. Math.* XIV, 1–81. English translation: Fundamentals of a New System of the Foundations of Mathematics, in Leśniewski (1992).

(1930) 'Über die Grundlagen der Ontologie', *Comptes rendus des Séances de la Société des Sciences et des Lettres de Varsovie, Classe III* 23, 111–132. English translation: On the Foundations of Ontology, in Leśniewski (1992).

(1938a) 'Einleitende Bemerkungen zur Fortsetzung meiner Mitteilung u.d.T. "Grundzüge eines neuen Systems der Grundlagen der Mathematik"', *Collectanea Logica* I (offprint 1938), 1–60. English translation: Introductory Remarks to the Continuation of my Article: "Grundzüge eines neuen Systems der Grundlagen der Mathematik", in McCall (1967) and Leśniewski (1992).

(1938b) 'Grundzüge eines neuen Systems der Grundlagen der Mathematik, §12', *Collectanea Logica* I (offprint 1938), 61–144.

(1985) *Stanisław Leśniewski's Ontology, Proc. of the Colloquium*, Neuenburg 1984, *Dialectica* 39(3).

(1988) *S. Leśniewski's Lecture Notes in Logic*, J.T.J. Srzednicki and Z. Stachniak, eds., Nijhoff International Philosophy Series, Vol. 24., Kluwer Academic Publishers, Dordrecht.

(1992) *Stanisław Leśniewski: Collected Works*, S.J. Surma, J.T.J. Srzednicki, and D.I. Barnett, eds.; with an Annotated Bibliography by V.F. Rickey, Nijhoff International Philosophy Series, Vol. 44, PWN – Polish Scientific Publishers, Warsaw, and Kluwer Academic Publishers, Dordrecht.

(1995) *Stanisław Leśniewski, Aujourd'hui*, D. Miéville and D. Vernant, eds.; Recherches sur la Philosophie et le langage, Université Pierre Mendes France, Grenoble, No. 16.

LIPPERT, B.M.

(1976) *Rekonstruktionen zur Leśniewski'schen Logik*, Zulassungsarbeit zum Staatsexamen, Konstanz.

ŁUKASIEWICZ, J.

(1951) 'On Variable Functors of Propositional Arguments', *Proc. Royal Irish Academy, Section A* 54, 25–35. Reprinted in Jan Łukasiewicz, *Selected Works*, L. Borkowski, ed., North-Holland Publishing Company, Amsterdam, 1970, pp. 311–324.

LUSCHEI, E.C.

(1962) *The Logical Systems of Leśniewski*, North-Holland Publishing Company, Amsterdam.

MARCISZEWSKI, W.

(1981) *Dictionary of Logic as Applied in the Study of Language. Concepts, Methods, Theories*, W. Marciszewski, ed., Nijhoff International Philosophy Series, Vol. 9, Martinus Nijhoff Publishers, Dordrecht.

(1988) 'The Place of Stanisław Leśniewski in Contemporary Logic-Philosophical Thought', *Wiadom. Mat.* 28(1), 65–68.

MCCALL, S.

(1967) *Polish Logic: 1920–1939*. S. McCall, ed., Clarendon Press, Oxford.

MIÉVILLE, D.

(1984) *Un développement des systèmes logiques de Stanisław Leśniewski. Protothétique–Ontologie–Méréologie*, Peter Lang.

(1985) 'Un aperçu des caractéristiques et de l'esprit des systèmes logiques de Stanisław Leśniewski', *Dialectica* 39(3), 165–179.

(1987) 'Axiomes et définitions chez Leśniewski: une manière génétique de développer les systèmes formels', *Theoria* (San Sebastian) 2(5–6), 285–307.

POLI, R.

(1992) *Ontologia Formale*, Marietti, Genova.

PRIOR, A.N.

(1953) 'On Propositions Neither Necessary Nor Impossible', *J. Symbolic Logic* 18, 105–108.

(1955) *Formal Logic*, Clarendon Press, Oxford.

(1965) Review of Sobociński (1960), *J. Symbolic Logic* 30, 245–246.

QUINE, W.V.

(1940) Reviews of Leśniewski (1938a) and Leśniewski (1938b), *J. Symbolic Logic* 5, 83–84.

RICKEY, V.F.

(1972) 'Axiomatic Inscriptional Syntax, Part I: General Syntax', *Notre Dame J. Formal Logic* XIII, 1–33.

(1973) 'Axiomatic Inscriptional Syntax, Part II: The Syntax of Protothetic', *Notre Dame J. Formal Logic* XIV, 1–52.

(1977) 'A Survey of Leśniewski's Logic', *Studia Logica* XXXVI, 407–426.

SCHARLE, T.W.

(1971) 'Completeness of Many-Valued Protothetic', *J. Symbolic Logic* 36, 363–364 (abstract).

SCHOLZ, H.

(1941) Review of Leśniewski (1938a), *Zentralblatt für Mathematik und Ihre Grenzgebiete* 23, 98–99.

SIMONS, P.

(1982) 'On Understanding Leśniewski', *Hist. Philos. Logic* 3, 165–191.

(1985a) 'A Semantics for Ontology', *Dialectica* 39(3), 193–216.

(1985b) 'Leśniewski's Logic and Its Relation to Classical and Free Logics', in *Foundations of Logic and Linguistics*, G. Dorn and P. Weingartner, eds., Plenum Press, New York, pp. 369–400.

(1989) 'Combinators and Categorial Grammar', *Notre Dame J. Formal Logic* 30, 241–261.

(1993) 'Nominalism in Poland', in *Polish Scientific Philosophy: The Lvov–Warsaw School*, F. Coniglione, R. Poli, and J. Woleński, eds., Poznań Studies in the Philosophy of the Sciences and the Humanities, Vol. 28, Rodopi, Amsterdam, pp. 207–232.

(1995) 'Leśniewski and Ontological Commitment', pp. 103–119 in Leśniewski (1995).

SŁUPECKI, J.

(1953) 'St. Leśniewski's Protothetics', *Studia Logica* I, 44–112.

SOBOCIŃSKI, B.

(1939) 'Z badań nad prototetyką (An Investigation of Protothetic)', *Collectanea Logica* I (offprint), 171–177.

(1949) 'An Investigation of Protothetic', *Cahiers de l'Institut d'Études Polonaises en Belgique* 5, Bruxelles.

(1954) 'Z badań nad aksjomatyką prototetyki Stanisława Leśniewskiego (An Investigation of the Axiomatics of Stanisław Leśniewski's Protothetic)', Polskie Towarzystwo Naukowe na Obczyźnie (Polish Society of Arts and Sciences Abroad), Yearbook for 1953–1954, Vol. IV, London, pp. 18–20.

(1960) 'On the Single Axioms of Protothetic. I, II, III', *Notre Dame J. Formal Logic* I (1960), 52–73; II (1961), 111–126 and 129–148.

(1967) 'An Investigation of Protothetic', pp. 203–206 in McCall (1967) (English translation of Sobociński, 1939).

SOLONIN, J.N.

(1969) 'The Main Features of St. Leśniewski's Logical-Mathematical Systems', *Vestnik Leningradskogo Universiteta, ser. Ekonomika, filosofia, pravo* 23, 93–103 [in Russian].

(1975) 'Propositional Calculus with Variable Functors', in *Contributed Papers to the Fifth International Congress of Logic Methodology and Philosophy of Science*, London, Canada, XII, pp. 53–54.

STACHNIAK, Z.

(1993) 'An Essay on Resolution Logics', *Studia Logica* LII, 309–322.

SURMA, S.J.

(1977a) 'On the Work and Influence of Stanisław Leśniewski', in *Logic Colloquium 76*, Oxford, Studies in Logic and Foundations of Math., Vol. 87, North-Holland Publishing Company, pp. 191–220.

(1977b) *On Leśniewski's Systems. Proceedings of XXII Conference on the History of Logic*, Kraków 1976, S.J. Surma, ed., *Studia Logica* XXXVI.

SUSZKO, R.

(1977) 'The Fregean Axiom and Polish Mathematical Logic in the 1920s', *Studia Logica* XXXVI, 377–380.

TANAKA, S.

(1969) 'Leśniewski's Protothetics $S1$, $S2$. I, II, III', *Proc. Japan Acad.* XLV, 97–101, 259–262, and 263–265.

TARSKI, A.

(1923a) 'Sur le terme primitif de la Logistique', *Fund. Math.* IV, 196–200.

(1923b) 'O wyrazie pierwotnym logistyki (On the Primitive Term of Logistic)', *Przegląd Filozoficzny* XXVI, 68–89. English translation based on this paper, on Tarski (1923a), and on Tarski (1924) in Tarski (1956) and in Tarski (1986).

(1924) 'Sur les truth-functions au sens de MM. Russell et Whitehead', *Fund. Math.* V, 59–74.

(1956) *Logic, Semantics, Metamathematics. Papers from 1923 to 1938*, Translated by J.H. Woodger, Clarendon Press, Oxford.

(1986) *Collected Papers*, Vol. 1–4, S. Givant and R. McKenzie, eds., Birkhäuser Verlag, Basel.

WATANABE, S.

(1974a) 'On Many-Valued Protothetic', Ph.D. Dissertation, University of Manchester.

(1974b) 'Many-Valued Protothetic', *J. Symbolic Logic* 39, 409–410 (abstract).

WOLEŃSKI, J.

(1985) *Filozoficzna Szkoła Lwowsko-Warszawska* (Lvov–Warsaw Philosophical School), PWN – Polish Scientific Publishers, Warsaw.

(1989) *Logic and Philosophy in the Lvov–Warsaw School*, Synthese Library, Vol. 198, Kluwer Academic Publishers, Dordrecht.

AUTHOR INDEX

MIX
Papier aus verantwortungsvollen Quellen
Paper from responsible sources
FSC® C105338

If you have any concerns about our products,
you can contact us on
ProductSafety@springernature.com

In case Publisher is established outside the EU,
the EU authorized representative is:
Springer Nature Customer Service Center GmbH
Europaplatz 3, 69115 Heidelberg, Germany

Printed by Libri Plureos GmbH
in Hamburg, Germany